METHODS IN MOLECULAR BIOLOGY

Series Editor
John M. Walker
School of Life and Medical Sciences
University of Hertfordshire
Hatfield, Hertfordshire, UK

For further volumes:
http://www.springer.com/series/7651

For over 35 years, biological scientists have come to rely on the research protocols and methodologies in the critically acclaimed *Methods in Molecular Biology* series. The series was the first to introduce the step-by-step protocols approach that has become the standard in all biomedical protocol publishing. Each protocol is provided in readily-reproducible step-by step fashion, opening with an introductory overview, a list of the materials and reagents needed to complete the experiment, and followed by a detailed procedure that is supported with a helpful notes section offering tips and tricks of the trade as well as troubleshooting advice. These hallmark features were introduced by series editor Dr. John Walker and constitute the key ingredient in each and every volume of the *Methods in Molecular Biology* series. Tested and trusted, comprehensive and reliable, all protocols from the series are indexed in PubMed.

DNA and RNA Origami

Methods and Protocols

Edited by

Julián Valero

Interdisciplinary Nanoscience Center (iNANO) and Department of Molecular Biology and Genetics (MBG), Aarhus University, Aarhus, Denmark

Humana Press

Editor
Julián Valero
Interdisciplinary Nanoscience Center (iNANO)
and Department of Molecular Biology
and Genetics (MBG)
Aarhus University
Aarhus, Denmark

ISSN 1064-3745 ISSN 1940-6029 (electronic)
Methods in Molecular Biology
ISBN 978-1-0716-3030-3 ISBN 978-1-0716-3028-0 (eBook)
https://doi.org/10.1007/978-1-0716-3028-0

This Humana imprint is published by the registered company Springer Science+Business Media, LLC, part of Springer Nature.
The registered company address is: 1 New York Plaza, New York, NY 10004, U.S.A.

Preface

Since Paul W. K. Rothemund reported the very first examples of DNA origami in 2006, the field has grown exponentially, attracting an increased interest from different multidisciplinary areas of research, including but not limited to chemistry, biology, and physics. Conceptually, DNA origami represents one of the most extraordinary paradigms of efficient molecular self-assembly, where hundreds of DNA molecules (staples) are programmed to fold into a desired nanostructure guided by a long scaffold that provides the thermodynamic stability necessary to recruit and nucleate all the different staples during DNA origami formation. After only 16 years, the DNA origami technology has shown a small part of its immense potential for building complex nanoarchitectures, nanomechanical devices, and molecular computing systems that can operate and perform highly complex tasks in biological environments such as delivery of drugs, nanolithography, template-directed synthesis, immunomodulation, and nanolocomotion, among others. Nowadays, the shared consensus in the field is that sky is the limit and that the future of biomolecular origami (this also includes the use of RNA and proteins as foldable biopolymers) and implications in other research areas are enormous. Talking about the future, I would also like to acknowledge the pioneering work and pay tribute to Nadrian C. "Ned" Seeman (1945–2021) whose revolutionary ideas and seminal contribution sparked the DNA nanotechnology and origami fields.

This MiMB volume comprehensively describes diverse methodological approaches towards the assembly and applications of nucleic acid (DNA and RNA) origami assemblies. In particular, different synthetic and computational methods as well as the isolation and structural characterization of 2D and 3D DNA and RNA origami nanoarchitectures will be discussed. Multidisciplinary applications of these nanostructures in the fields of nanophotonics, drug delivery, biophysics, and synthetic biology, among others, will be described. Moreover, alternative approaches towards the assembly of other complex DNA and RNA nanoarchitectures will be reviewed. Overall, this book aims to serve as a guideline describing the current state-of-the-art assembly methodologies and applications of nucleic acid origami nanostructures, fostering and inspiring their potential applicability in other arenas at the interface between physics, chemistry, and biology.

Aarhus, Denmark ***Julián Valero***

Contents

Part IV Applications of DNA and RNA Origami

Contributors

ALEKSEI AKSIMENTIEV • *Department of Physics and Beckman Institute for Advanced Science and Technology, University of Illinois Urbana-Champaign, Urbana, IL, USA*

EBBE SLOTH ANDERSEN • *Interdisciplinary Nanoscience Center, Aarhus University, Aarhus, Denmark*

MAARTJE M. C. BASTINGS • *Programmable Biomaterials Laboratory, EPFL, EPFL-STI-IMX-PBL MXC 340 Station 12, Lausanne, Switzerland*

ERIK BENSON • *Department of Physics, Clarendon Laboratory, University of Oxford, Oxford, UK*

PETER E. BESHAY • *Department of Mechanical and Aerospace Engineering, The Ohio State University, Columbus, OH, USA*

JOAKIM BOHLIN • *Department of Physics, Clarendon Laboratory, University of Oxford, Oxford, UK*

CARLOS E. CASTRO • *Department of Mechanical and Aerospace Engineering, The Ohio State University, Columbus, OH, USA; Biophysics Graduate Program, The Ohio State University, Columbus, OH, USA*

JIE CHAO • *Key Laboratory for Organic Electronics and Information Displays, Jiangsu Key Laboratory for Biosensors, Institute of Advanced Materials, National Synergetic Innovation Center for Advanced Materials, Nanjing University of Posts and Telecommunications, Nanjing, China*

HUYEN DINH • *Institute of Advanced Energy, Kyoto University, Uji, Kyoto 611-0011, Japan*

JONATHAN P. K. DOYE • *Physical and Theoretical Chemistry Laboratory, Department of Chemistry, University of Oxford, Oxford, UK*

MASAYUKI ENDO • *Department of Chemistry, Graduate School of Science, Kyoto University, Sakyo-ku, Kyoto, Japan; Institute for Integrated Cell-Material Sciences, Kyoto University, Sakyo-ku, Kyoto, Japan; Organization for Research and Development of Innovative Science and Technology, Kansai University, Suita, Osaka, Japan*

MEGAN C. ENGEL • *School of Engineering and Applied Sciences, Harvard University, Cambridge, MA, USA*

CHUNHAI FAN • *School of Chemistry and Chemical Engineering, Frontiers Science Center for Transformative Molecules and National Center for Translational Medicine, Shanghai Jiao Tong University, Shanghai, China*

HANNAH FOWLER • *Physical and Theoretical Chemistry Laboratory, Department of Chemistry, University of Oxford, Oxford, UK*

HENRI G. FRANQUELIM • *Max Planck Institute of Biochemistry, Munich, Germany; Interfaculty Centre for Bioactive Matter (b-ACTmatter), Leipzig University, Leipzig, Germany*

KRISTEN FROEHLICH • *Department of Materials Science and Engineering, North Carolina State University, Raleigh, NC, USA*

CODY GEARY • *Interdisciplinary Nanoscience Center, Aarhus University, Aarhus, Denmark*

WEI GUO • *Microfluidics and Soft Matter Group, Department of Mechanical Engineering, Faculty of Engineering, The University of Hong Kong, Pokfulam, Hong Kong SAR, China*

MICHAEL HAYDELL • *Chemical Biology and Medicinal Chemistry Unit, Life and Medical Sciences (LIMES) Institute, University of Bonn, Bonn, Germany*

HEINI IJÄS • *Nanoscience Center, Department of Biological and Environmental Science, University of Jyväskylä, Jyväskylä, Finland; Biohybrid Materials, Department of Bioproducts and Biosystems, Aalto University, Aalto, Finland; LIBER Center of Excellence, Aalto University, Aalto, Finland; Ludwig-Maximilians-University, Munich, Germany*

METTE JEPSEN • *Interdisciplinary Nanoscience Center, Aarhus University, Aarhus, Denmark*

DANYANG JI • *Department of Chemistry and State Key Laboratory of Marine Pollution, City University of Hong Kong, Kowloon Tong, Hong Kong SAR, China; Shenzhen Research Institute of City University of Hong Kong, Shenzhen, China*

JOSHUA A. JOHSON • *Biophysics Graduate Program, The Ohio State University, Columbus, OH, USA*

HIMANSHU JOSHI • *Department of Physics and Beckman Institute for Advanced Science and Technology, University of Illinois Urbana-Champaign, Urbana, IL, USA; Department of Biotechnology, Indian Institute of Technology Hyderabad, Kandi, Sangareddy, Telangana, India*

CARSON KEY • *Department of Materials Science and Engineering, North Carolina State University, Raleigh, NC, USA*

ALENA KHMELINSKAIA • *Max Planck Institute of Biochemistry, Munich, Germany; Institute of Protein Design, University of Washington, Seattle, WA, USA; Life & Medical Sciences Institute (LIMES), University of Bonn, Bonn, Germany*

ANDREW B. KINGHORN • *School of Biomedical Sciences, Li Ka Shing Faculty of Medicine, The University of Hong Kong, Pokfulam, Hong Kong SAR, China*

JØRGEN KJEMS • *Interdisciplinary Nanoscience Centre (iNANO), Department of Molecular Biology and Genetics, Aarhus University, Aarhus C, Denmark*

ENZO KOPPERGER • *Physics Department – E14, TU Munich, Garching, Germany*

MAURI A. KOSTIAINEN • *Biohybrid Materials, Department of Bioproducts and Biosystems, Aalto University, Aalto, Finland; LIBER Center of Excellence, Aalto University, Aalto, Finland*

ABHICHART KRISSANAPRASIT • *Department of Materials Science and Engineering, North Carolina State University, Raleigh, NC, USA*

CHUN KIT KWOK • *Department of Chemistry and State Key Laboratory of Marine Pollution, City University of Hong Kong, Kowloon Tong, Hong Kong SAR, China; Shenzhen Research Institute of City University of Hong Kong, Shenzhen, China*

THOMAS H. LABEAN • *Department of Materials Science and Engineering, North Carolina State University, Raleigh, NC, USA*

JENNY V. LE • *Biophysics Graduate Program, The Ohio State University, Columbus, OH, USA*

CHEN-YU LI • *Department of Physics and Beckman Institute for Advanced Science and Technology, University of Illinois Urbana-Champaign, Urbana, IL, USA*

DI LI • *School of Chemistry and Molecular Engineering, East China Normal University, Shanghai, China*

QIAN LI • *School of Chemistry and Chemical Engineering, Frontiers Science Center for Transformative Molecules and National Center for Translational Medicine, Shanghai Jiao Tong University, Shanghai, China*

PENG LIN • *Institute of Advanced Energy, Kyoto University, Uji, Kyoto 611-0011, Japan*

VEIKKO LINKO • *Biohybrid Materials, Department of Bioproducts and Biosystems, Aalto University, Aalto, Finland; LIBER Center of Excellence, Aalto University, Aalto, Finland; Institute of Technology, University of Tartu, Tartu, Estonia*

JONATHAN LIST • *Physics Department – E14, TU Munich, Garching, Germany*
ARD A. LOUIS • *Rudolf Peierls Centre for Theoretical Physics, University of Oxford, Parks Road, Oxford, UK*
YINZHOU MA • *Chemical Biology and Medicinal Chemistry Unit, Life and Medical Sciences (LIMES) Institute, University of Bonn, Bonn, Germany; Department of Mechanics and Engineering Science, College of Engineering, Peking University, Beijing, China; Beijing Innovation Center for Engineering Science and Advanced Technology, Peking University, Beijing, China*
MICHAEL MATTHIES • *School of Molecular Sciences and Center for Molecular Design and Biomimetics, The Biodesign Institute, Arizona State University, Tempe, AZ, USA*
XIAOYONG MO • *Department of Chemistry, CAS-HKU Joint Laboratory of Metallomics on Health and Environment, The University of Hong Kong, Pokfulam, Hong Kong SAR, China*
TAKASHI MORII • *Institute of Advanced Energy, Kyoto University, Uji, Kyoto 611-0011, Japan*
EIJI NAKATA • *Institute of Advanced Energy, Kyoto University, Uji, Kyoto 611-0011, Japan*
NIBEDITA PAL • *Indian Institute of Science Education and Research (IISER) Tirupati, Tirupati, Andhra Pradesh, India*
ERIK POPPLETON • *School of Molecular Sciences and Center for Molecular Design and Biomimetics, The Biodesign Institute, Arizona State University, Tempe, AZ, USA*
DOMEN PREŠERN • *Physical and Theoretical Chemistry Laboratory, Department of Chemistry, University of Oxford, Oxford, UK*
FLAVIO ROMANO • *Dipartimento di Fisica, Sapienza Universitá di Roma, Rome, Italy*
LORENZO ROVIGATTI • *Dipartimento di Fisica, Sapienza Universitá di Roma, Rome, Italy*
JOHN S. SCHRECK • *Computational and Information Systems Laboratory, National Center for Atmospheric Research (NCAR), Boulder, USA*
PETRA SCHWILLE • *Max Planck Institute of Biochemistry, Munich, Germany*
SIMON CHI-CHIN SHIU • *School of Biomedical Sciences, Li Ka Shing Faculty of Medicine, The University of Hong Kong, Pokfulam, Hong Kong SAR, China*
ANDERSON HO CHEUNG SHUM • *Microfluidics and Soft Matter Group, Department of Mechanical Engineering, Faculty of Engineering, The University of Hong Kong, Pokfulam, Hong Kong SAR, China*
FRIEDRICH C. SIMMEL • *Physics Department – E14, TU Munich, Garching, Germany*
LIANE S. SLAUGHTER • *Division of Life Science, Hong Kong University of Science and Technology, Clear Water Bay, Hong Kong SAR, China*
BENEDICT E. K. SNODIN • *Department of Philosophy, Future of Humanity Institute, University of Oxford, Oxford, UK*
RASMUS S. SØRENSEN • *Interdisciplinary Nanoscience Centre (iNANO), Department of Molecular Biology and Genetics, Aarhus University, Aarhus C, Denmark*
HIROSHI SUGIYAMA • *Department of Chemistry, Graduate School of Science, Kyoto University, Sakyo-ku, Kyoto, Japan; Institute for Integrated Cell-Material Sciences, Kyoto University, Sakyo-ku, Kyoto, Japan*
PETR ŠULC • *School of Molecular Sciences and Center for Molecular Design and Biomimetics, The Biodesign Institute, Arizona State University, Tempe, AZ, USA*
LELE SUN • *School of Chemistry and Molecular Engineering, East China Normal University, Shanghai, China*

YUKI SUZUKI • *Frontier Research Institute for Interdisciplinary Sciences, Tohoku University, Aoba-ku, Sendai, Japan; Department of Chemistry for Materials, Graduate School of Engineering, Mie University, Tsu, Mie, Japan*

JULIAN A. TANNER • *School of Biomedical Sciences, Li Ka Shing Faculty of Medicine, The University of Hong Kong, Pokfulam, Hong Kong SAR, China*

RASMUS P. THOMSEN • *Interdisciplinary Nanoscience Centre (iNANO), Department of Molecular Biology and Genetics, Aarhus University, Aarhus C, Denmark*

NGOC CHAU TRAN • *School of Biomedical Sciences, Li Ka Shing Faculty of Medicine, The University of Hong Kong, Pokfulam, Hong Kong SAR, China*

EDMUND CHUN MING TSE • *Department of Chemistry, CAS-HKU Joint Laboratory of Metallomics on Health and Environment, The University of Hong Kong, Pokfulam, Hong Kong SAR, China; HKU Zhejiang Institute of Research and Innovation, Zhejiang, China*

NÉSTOR SAMPEDRO VALLINA • *Interdisciplinary Nanoscience Center, Aarhus University, Aarhus, Denmark*

NILS G. WALTER • *Single Molecule Analysis Group, Department of Chemistry, University of Michigan, Ann Arbor, MI, USA*

LIN WANG • *School of Biomedical Sciences, Li Ka Shing Faculty of Medicine, The University of Hong Kong, Pokfulam, Hong Kong SAR, China*

PENGFEI WANG • *Institute of Molecular Medicine, Renji Hospital, School of Medicine, Shanghai Jiao, Tong University, Shanghai, China*

CHAK KUI WONG • *Physical and Theoretical Chemistry Laboratory, Department of Chemistry, University of Oxford, Oxford, UK*

FAN XU • *Institute of Molecular Medicine, Renji Hospital, School of Medicine, Shanghai Jiao, Tong University, Shanghai, China*

AN YAN • *School of Chemistry and Molecular Engineering, East China Normal University, Shanghai, China*

DONGLEI YANG • *Institute of Molecular Medicine, Renji Hospital, School of Medicine, Shanghai Jiao, Tong University, Shanghai, China*

HONGLU ZHANG • *School of Biomedical Sciences and Engineering, National Engineering Research Center for Tissue Restoration and Reconstruction, Key Laboratory of Biomedical Materials and Engineering of the Ministry of Education, South China University of Technology, Guangzhou, China*

Part I

Background and Design of Nucleic Acid Origami

Chapter 1

DNA Origami: Recent Progress and Applications

Michael Haydell and Yinzhou Ma

Abstract

This chapter explores the basic concept of DNA origami and its various types. By showing the progress made in structural DNA nanotechnology during the last 15 years, the chapter draws attention to the capability of DNA origami to construct complex structures in both 2D and 3D level. As well as looking at a few examples of dynamic DNA nanostructures, the chapter also explores the possible applications of DNA origami in different fields, such as biological computing, nanorobotics, and DNA walkers.

Key words DNA origami, Structural DNA nanotechnology, Dynamic DNA nanotechnology, DNA nanorobots, DNA nanomachines, DNA drug delivery

1 Introduction

DNA origami is a revolutionary fabrication method that allows one to create a vast array of two- or three-dimensional, nanoscale structures either as static structural elements or controllable, dynamic nanomachines. Applications for this field are continuously being expanded and include nanofabrication [1], nanorobotics [2–4], nanomedicine [5–7], biological computing [8], and basic science [9, 10]. Since the advent of DNA origami in 2006, the complexity of possible structures has increased dramatically. The field is still fairly young and ever more interesting and useful devices continue to be reported in top tier journals. The potential for this kind of technology has yet to be determined, but the trend implies that we have only explored the tip of the iceberg. The following is a crash course of sorts into the history of DNA origami and an overview of some of the capabilities and applications for the field. The papers discussed herein are merely a small fraction of the projects utilizing DNA origami and are not meant as a comprehensive review, merely a starting point to enable further research.

Julián Valero (ed.), *DNA and RNA Origami: Methods and Protocols*, Methods in Molecular Biology, vol. 2639,
https://doi.org/10.1007/978-1-0716-3028-0_1,

2 History and Development

The field of DNA nanotechnology involves using DNA as a structural material instead of, or sometimes in addition to, an information storage material. The field began when Nadrian Seeman demonstrated that DNA can be used to build nanoscale shapes [11] precisely and controllably but was revolutionized by the invention of DNA origami [12] in 2006. Paul Rothemund showed that one can control the folding of a long, biologically sourced, circular DNA "scaffold" strand by allowing it to hybridize with specifically designed short, synthetic DNA "staple" strands. The staples bring distant parts of the scaffold together which causes the scaffold to take the desired shape through folding in a method reminiscent of paper origami. The scaffold will take different shapes depending on the sequence of the staple strands used (shown in Fig. 1). Lastly, the

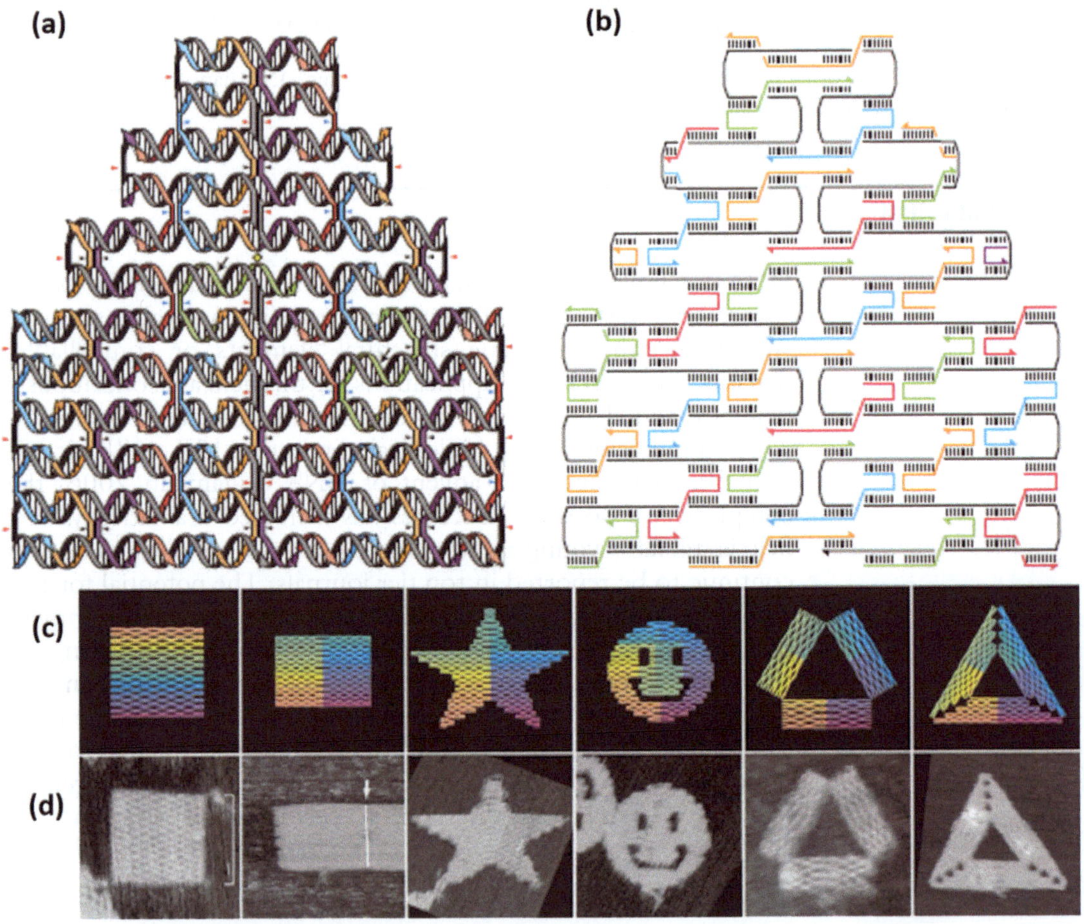

Fig. 1 DNA origami. (**a**) Helical depiction of how staple strands (colored) cross over to and cause a scaffold strand (black) to fold. (**b**) Schematic view equivalent to (**a**). (**c**) Schematic designs for the origamis shown in (**d**), AFM images (165 nm × 165 nm) of folded origamis. (Figure adapted with permission from Ref. [1] © 2006 Nature Publishing Group)

process for assembling the origami is fairly simple. The staples are added to the scaffold in excess (typically 4–10 equivalents) in buffer solution, heated to 90 °C, and slowly cooled to room temperature over several hours.

2.1 Structural DNA Origami

Structures made with DNA origami can be much larger and more rigid than those previously produced with DNA nanotechnology [11, 13, 14]. They can also be programmed to tile together to form supramolecular structures with megadalton [12] weights. Impressive as they were, Rothemund's origamis were unfortunately constrained to two dimensions. Other than six-helix-bundle nanotubes [15], three-dimensional DNA origami was first demonstrated by Andersen et al. in May of 2009 by forcing planar DNA origami squares to fold into a hollow box with a controllable lid [16]. Later that month, William Shih reported 3D origamis [17] with bulk honeycomb cross-sections that could be formed into various shapes. Both methods relied on the staple strands "crossing-over" to distant locations of the scaffold strand, but in the first (box with lid), the crossovers occur at every full turn, much in the same way as Rothemund's 2D origamis. The Shih method, however, induced crossovers after 7 base pairs (bp) (every 2/3 of a helical turn) between one strand and each of its neighboring strands (Fig. 2)

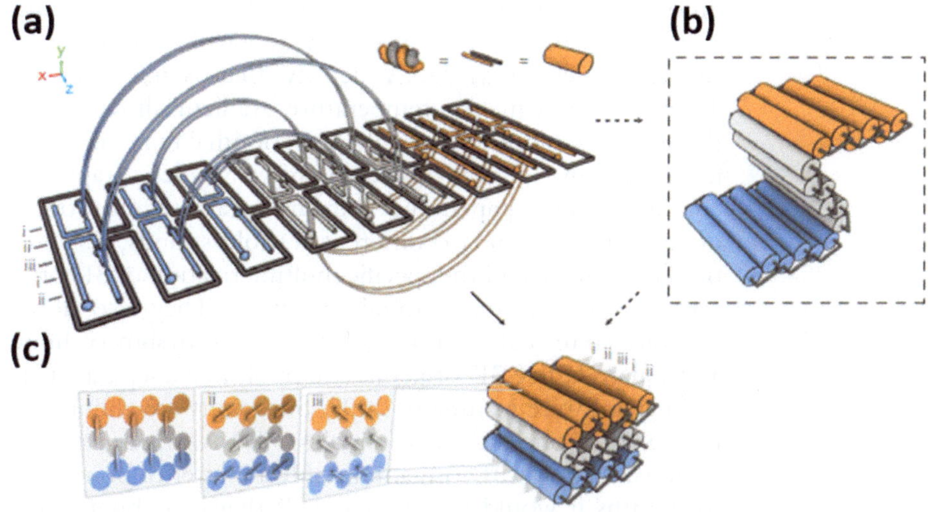

Fig. 2 Forming three-dimensional DNA origami. (**a**) Crossovers (blue and orange arcs above and below origami plane) between different parts of the scaffold (black) cause a 2D origami to fold as in (**b**). (**c**) Cross-sectional slices showing that crossovers between neighboring helices occur every 7 bp, which corresponds to 2/3 of a helical turn. Therefore, interactions occur at 240° angles resulting in a honeycomb shape, and crossovers between the same two helices occur every 21 bp. (Figure adapted with permission from Ref. [17] © 2009 Nature Publishing Group)

allowing two adjacent crossovers to form an angle of 240°. As such, crossovers from one strand to another repeat every 21 bp (two full turns), and the DNA will fold into a 3D structure with a honeycomb lattice. The Shih group reported several different nanostructures that were impossible to fabricate with traditional nanofabrication technologies such as a square tube with hollow interior, a bridge with rails, a slotted cross where one beam passes through the other, and a stacked cross again with hollow components. Furthermore, not only does DNA origami produce well-defined nanostructures, but the self-assembly nature facilitates the assembly of billions of such structures in a single reaction, with no extra effort on the researcher's part and with far less harmful ingredients than traditional nanofabrication.

One consideration when assembling 3D lattice origamis compared to 2D planar origamis is the annealing time. Planar origamis can fold in a matter of hours with a fairly high yield. In contrast, the Shih group reported that 3D origamis require days to fold. The reason for this is that the DNA strands do most of their hybridization at a certain temperature and that temperature can change depending on the complexity of the structure and the composition of the buffer solution (divalent cations were shown to increase successful folding over monovalent cations) and at different points during the folding process. The authors explain this as being due to the structures needing to overcome more challenging kinetic traps due to requiring a higher density of crossovers and folding and unfolding between different layers of the origami during the annealing process. Also, a 3D origami will have a much higher (negatively charged) DNA density than single-layer origamis. Therefore, slowing the temperature gradient allows each part of the structure more time to fold at its optimal temperature, thereby having a higher chance of overcoming any kinetic traps.

In the same paper, the Shih group also demonstrated a 3D origami made from two different scaffold strands, polymerization of 3D origamis, and the specific multimerization of 3D origamis to assemble into a larger overall structure. They also developed a software suite (caDNAno) to facilitate the design of honeycomb and square-lattice 3D origamis and published a separate paper [18] on that software in June of 2009. Such software eliminates repetitive, error-prone tasks and enables a DNA origami researcher to design a complex 3D origami in only a matter of hours, compared to months it would take by hand. Perhaps the biggest impact of caDNAno is that it opens the door for more researchers to get involved with DNA origami research and push the field even closer to the physical limit.

The Shih group also reported the ability to make square-lattice 3D DNA origami [19]. By inducing crossovers every 8 bp (or 76% of a helical turn), the angle between two adjacent crossovers can be set to 274° allowing the helices to interact at almost right angles to each other. Due to the angle between two adjacent crossovers not being exactly 90°, the torsional strain on the DNA helices will unwind the DNA to 10.67 bp/turn, resulting in a global right-handed twist to the structure. This torsional strain can be mitigated by removing base pairs in every helix at strategic locations throughout the origami in order to bring the average base pair density closer to the 10.5 bp/turn in regular B-form DNA.

In designs shown above, the structures assembled with DNA origami employ crossovers equidistant from each other. This results in minimal strain on the structure and therefore the whole structure remains straight. In 2009, Hendrik Dietz, while working in the Shih group, reasoned that adding or removing base pairs between crossovers (as described in Fig. 3) should cause helical strain, which would result in a twisted origami [20]. Specifically, removing base pairs between two adjacent crossovers should cause a left-handed torque between two crossovers which result in a left-handed twist. Conversely, adding base pairs between two adjacent crossovers should cause a right-handed torque between two crossovers which result in a right-handed twist. Curves can be induced by removing base pairs between crossovers on one side of the origami (resulting in inward tensile forces) and adding base pairs between crossovers on the other side (resulting in outward compressive forces). Dietz experimentally demonstrated that DNA origamis can be curved up to 180°, with a radius of curvature of 6 nm. Moreover, the degree of curvature can be tuned by changing the number of base pairs per turn on either side of the origami.

Other research groups have also achieved tremendous success in developing complicated structures from DNA origami. One example is the research on how to induce complex curvature in origamis [21] in order to produce almost any arbitrary shape from the Yan group which was published in 2011. This method builds upon Dietz' method of inducing curvature by changing the number of base pairs between crossovers in order to either create concentric ring structures or induce out-of-plane curvature in normally planar structures. Out-of-plane curvature is induced by changing the number of base pairs between crossovers connecting three adjacent DNA helices (A, B, and C). In planar origami the crossovers occur every turn, but if the crossover from B to C occurs at a base pair with a different position along the helical axis than the crossover from A to B, then the plane defined by helices BC will be at an angle to the plane defined by helices AB. Because base pairs are discrete, however, the minimum angular resolution is $360°/10.5 = 34.3°$, though the authors mention the possibility of using non-B-form DNA to fine-tune the angle as well as the fact that DNA is flexible enough to bend to most angles required by various structures. The authors reported creating both hemispheres and full spheres with hollow interiors as well as irregularly shaped hollow objects such as an ellipsoid and a nanoflask.

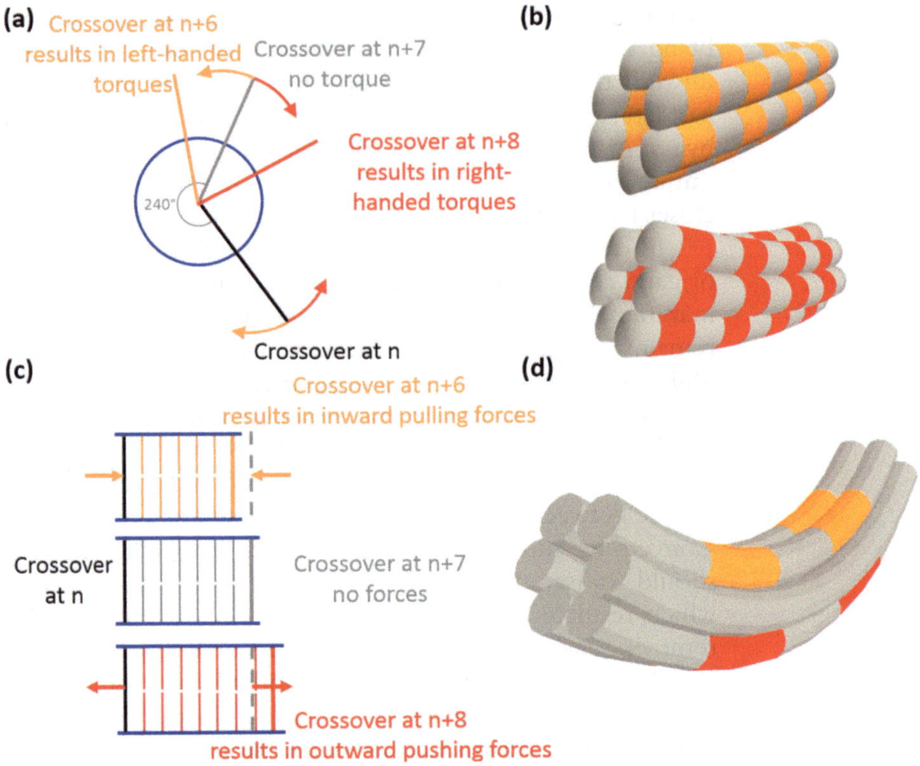

Fig. 3 Inducing twists and curves in DNA origami. (**a**) Crossovers separated by 7 bp (gray) do not experience any torsional strain about the DNA helix (blue circle) compared to the reference crossover (black). Reducing the number of base pairs between crossovers (in this case by one, orange) results in left-handed torques. Increasing the number of base pairs between crossovers (in this case by one, red) results in right-handed torques. (**b**) Global left-handed twists (top) can be induced by reducing the number of base pairs between crossovers (orange) and right-handed twists (bottom) induced by increasing the number of base pairs between crossovers (red). (**c**) Reducing the number of base pairs between crossovers (in this case by one, orange) results in pulling forces between the reference crossover (black) and the "no strain" crossover location (gray dashes). Crossovers separated by 7 bp (gray) do not experience any pulling or pushing forces. Increasing the number of base pairs between crossovers (in this case by one, red) results in pushing forces between the reference crossover (black) and the "no strain" crossover location (gray dashes). (**d**) Curves can be induced by introducing pulling (<7 bp between crossovers, orange) forces on one side of the origami (gray) and pushing (>7 bp between crossovers, red) forces on the other

Although most origami assemblies require divalent cations such as Mg^{2+} to coordinate the negatively charged DNA and promote folding, Mg^{2+} can be undesirable as a buffer component in some situations. An example is when working with DNA-wrapped carbon nanotubes where Mg^{2+} seems to cause faster nanotube aggregation [1]. In 2012, Dietz and coworkers, at his lab in Munich, published a paper where they report origami folding conditions independent of divalent cations [22]. The researchers assembled several 42-helix bundle origamis according to 4 different rules for how many base pairs away from a crossover the staple should end. The origamis were assembled in various buffer solutions with varying concentrations of sodium and magnesium. The highest yields were achieved in origamis where staples were on average 42 nucleotides long with ends no more than 4 bp from a crossover with crossovers only 7 bp apart. Staples were never ended between crossovers more than 7 bp away from each other. Structures following this rule also either formed the fastest (1.5 days in 2.4 M NaCl) or with the lowest concentration of sodium (1 M NaCl over 12 days).

Another challenge with DNA origami is procuring starting materials. Staples strands are synthesized and can be performed by numerous companies. The scaffold strand, however, is biologically sourced and can sometimes be difficult to acquire, especially in large quantities. Therefore, researchers working in the Yin group at Harvard devised a way to create nanoscale 2D [23] and 3D [24] shapes using only synthetic DNA. Instead of using a long scaffold strand, the strategy was to use oligodeoxynucleotides to form a "single-stranded tile" (SST). The SSTs consist of four domains, and each domain specifically interacts with a complementary domain on another SST. For 2D structures, the domains are 10–11 bp long, ensuring that all interactions occur in the same plane. For 3D structures, what the researchers called "DNA bricks," domains are 8 bp long so that the SSTs always hybridize perpendicular to each other. The SSTs were designed so that if the full library is used, they will bind and grow into a square or cube. This square or cube was abstracted into a molecular "canvas" similar to a block of marble. When sculpting a statue, an artist removes pieces of the block. Analogously, to create a 2D or 3D structure from these SSTs, simply leave corresponding SSTs out of the solution. As such, the researchers reported that the same SST library could be used to create over 100 different nanostructures. Compared to scaffolded DNA origami, DNA brick nanostructures are easier to design, are cheaper to procure, tile easier into multimeric structures, and are expected to induce little to no immune response when used in vivo (not tested) due to the lack of biological DNA. DNA bricks, however, take more time to fold than scaffolded origamis and are not expected to be as stiff or stable as their scaffold-based counterparts.

2.2 Dynamic DNA Origami

By now it should be clear that DNA origami is a powerful method for fabricating diverse nanoscale structural elements via bottom-up self-assembly. In order to create nanomachines, however, static structures are not enough. The first dynamic 3D origami was actually also the first 3D origami from Andersen et al., the nanoscale box with a controllable lid [16] made from the planar DNA origami squares. The motion in this case is rotational, but many macroscale machines benefit from the conversion of rotational motion into linear motion or vice versa. In January of 2015, researchers in the Carlos Castro lab at Ohio State University created a DNA origami crank-slider device [25] that replicates macroscale mechanics at the nanoscale. It consists of a six-helix bundle origami piston that slides through a hollow tube origami. The tube and piston are both connected by a flexible hinge origami that is free to open and close along a wide range of angles. As Brownian motion opens the hinge (rotational motion), the piston retracts into the tube (linear motion). When the hinge closes, the piston extends out from the tube. The expected piston extension distance can be predicted for a given hinge angle through simple geometry. The researchers used transmission electron microscopy to image the assembled origamis and measured the angles and extension distances of 56 samples. They found that the measurements matched their predictions. Such a finding shows that DNA origami nanomachines behave in a similar manner as their macroscale counterparts.

Another set of dynamic origamis [3] was reported 2 months later by the Dietz group. These devices were actually composed of numerous, separately assembled, origamis and later joined together based on shape complementarity similar to RNAse-P tRNA recognition, instead of base pairing. One origami had a pocket that was shaped to accept a protrusion on another origami. If the protrusion fits into the pocket, stacking interactions between terminal base pairs on the protrusion and pocket will further stabilize the joining. This principle enabled construction of homo- or hetero-multimeric structures on a micrometer scale and, in a later report, structures up to a gigadalton in mass [26] as well as lattices and reconfigurable structures. Furthermore, unlike the crank-slider which was stochastically actuated by Brownian motion, the conformation of the reconfigurable structures could be controlled based on cation concentration or temperature. One of the devices reported was a humanoid-shaped "nanorobot" whose arms would raise in low cation concentration and lower in high cation concentration.

The shape complementarity principle was used in a different study to assemble a rotary device [27] from three tight fitting components, two clamp units, and one rotor unit. The clamp units contained pockets that accept the protrusion on the rotor. Once the clamps surround the rotor, the clamp lids were secured through addition of complementary DNA strands, thereby mechanically interlocking the rotor to the clamps. This means the rotor should have a high rotational degree of freedom, but limited translational degree of freedom from the clamp units. The free rotation of the rotor was confirmed using total internal reflection fluorescence microscopy, but such rotation was only due to Brownian motion. Thus, the directionality of the rotor was not controllable. It was, however, a proof of concept that dynamic machines can be assembled with high precision from separate origamis.

Another interesting development to come out of Dietz' lab is the use of bacteria and bacteriophage DNA to increase the scale of assembled origami [28] by three orders of magnitude over that possible with chemical synthesis. This method is ideal for mass producing the same origami structure. The strategy was to encode the staples into a phagemid strand with two Zn^{2+}-dependent self-cleaving DNAzymes flanking every staple. The scaffold is added as a helper plasmid, and after uptake the bacteria continuously replicates both the staple phagemid and the scaffold through rolling circle amplification. The origami is assembled by isolating the phagemid particle and incubating in $ZnCl_2$ to liberate the staples from the phagemid. The origami can be assembled through conventional means after the $ZnCl_2$ is removed from the buffer. The self-cleaving DNAzymes do leave short overhangs on the staples which can be utilized as toeholds to replace specific staples with labeled variants. The researchers demonstrated the ability to produce 163 mg of origami nanorods using this method. They expect these scales will make possible new applications for DNA origami as many applications are economically precluded at scales of micrograms.

DNA origami is capable of creating complex nanoscale machines and in increasingly large amounts, but control of such machines has been limited to strand displacement (which can pollute the system, thereby decreasing further efficiencies), changing the buffer composition/concentration (which can be complicated and imprecise), or photoswitches (which are not easily attainable and are limited to only a few switching light wavelengths). Electric fields are the ideal means of control since computers are capable of very quickly and precisely controlling the strength and direction of such fields. It is known that DNA's negative charge allows it to be separated by gel electrophoresis, so researchers in Friedrich Simmel's lab attempted to use this property to control the orientation of a DNA origami arm [2] in relation to an attached origami plate. The arm was a six-helix bundle origami and was attached to the square plate with two noncomplementary single-stranded DNA strands. The movement of the arm was tracked via fluorescence and was shown to be controllably actuated with external electric fields, through several full rotations, in either direction, at speeds up to 25 Hz. They also reported the ability for the arm to push a gold nanoparticle cargo. The paper also discusses that future computer-controlled origami nanomachines can be specifically placed through a combination of lithographic techniques and self-assembly and actuated by nanoscale electrodes to create a rudimentary nanoscale production factory.

3 Applications

While most DNA origami research is a "proof of concept" for future applications, some specific applications for DNA origami have already been published as prototypes or already demonstrated as useful in basic science. Those that are still in the "prototype" phase typically can be classed as either drug delivery systems or self-assembled biocomputing systems, and representative examples are shown in Fig. 4. The final use applications primarily exist as tools for basic research and Fig. 5 presents some examples.

3.1 Prototype Applications

Concerning prototype DNA origami applications, one exciting area of research is in drug delivery systems. Many different designs have been presented in literature but they all follow a basic premise (presented in Fig. 4a). The DNA origami itself acts as a nanoscale container for a small payload, namely, pharmaceuticals. The container is able to change conformation between an open and a closed state. In the closed state, the medication is sterically prevented from interacting with its target or other molecules. Aptamers [29–31] are the preferred method for controlling the conformation changes. When the aptamer interacts with a specific molecule, the origami will open and release its payload. The aforementioned DNA origami box with controllable lid [16] is a one such drug delivery system, but another approach was reported by Shawn Douglas, Ido Bachelet, and George Church as a logic-gated nanorobot that releases a molecular payload [5] under certain conditions. This "robot" consisted of a clamshell-like origami that would surround a molecular payload (e.g., an anticancer drug), thereby preventing the payload from interacting with off-target molecules. The clamshell robot would open when exposed to the target molecule, releasing the drug, which subsequently performs its pharmacological function. Different aptamers can be used to respond to different targets, and origamis have even been employed to create artificial membrane ion channels [32], which theoretically enable one to deliver payloads into the cell.

Using the shape complementary principle for origami assembly, Sigl et al. in 2021 reported a DNA origami structure that is able to bind to and encapsulate viruses [33], thereby helping to neutralize viral infections. They created DNA origami triangular subunits that can be programmed with virus-binding molecules, and these subunits use shape complementary principle to assemble into octahedral, icosahedral, or larger containers up to 925 MDa. They demonstrated this method on hepatitis-B core particles and adeno-associated viruses. They were able to prevent the hepatitis-B core particle interactions in vivo, and for human cells exposed to AAV2, origami half shells were able to prevent infection.

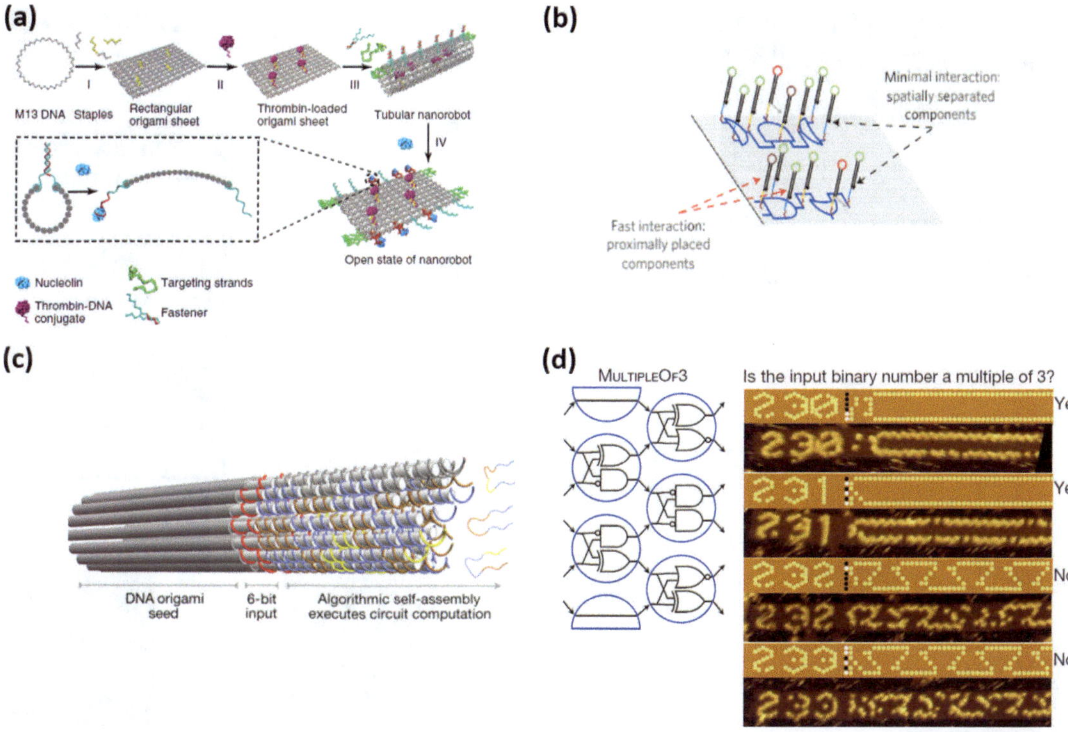

Fig. 4 Prototype DNA origami applications. (**a**) Schematic for DNA origami drug delivery systems. When closed, the origami (gray) acts as a container to prevent undesired interactions with the pharmaceutical payload (in this case thrombin, purple). Aptamers (green and blue/red) both bind the origami to a cell surface (green targeting strands) and fasten the origami together (blue/red). In the presence of nucleolin (blue), the fastening aptamers will preferentially bind to the nucleolin, thereby allowing the origami to open and thus releasing the pharmaceutical payload (Figure reprinted with permission from Ref. [6] © 2018 Nature Publishing Group). (**b**) Schematic for a DNA-based logic circuit. The DNA origami (gray square) serves as breadboard on which the logic elements (various colored DNA hairpins) can be placed. Elements readily interact with other nearby hairpin elements, while interference between different circuits is mitigated through spatial separation (Figure reprinted with permission from Ref. [34] © 2017 Nature Publishing Group). (**c**) Graphic representation for iterative DNA computing. A DNA origami seed (gray) programs a six-bit input (red) which interacts with different SSTs (blue/yellow/brown) to self-assemble into a nanotube in a one-pot reaction. (**d**) Left: Different logic circuits (in this case MulipleOf3) can be selected by using different SSTs. Right: Four different iterations of the MultipleOf3 logic circuit. Top images are schematics showing expected results. Bottom images are AFM images of unwrapped nanotubes. The circuit will output computed bits (yellow) based on the input (white dots = 1, black dots = 0). Note that the base-10 numerals are the experimental label, not the base-10 conversion of the base-2 inputs. (For (**c**) and (**d**) Figure adapted with permission from Ref. [8] © 2019 Nature Publishing Group)

Another promising application is in biological computing. There already exist various origami-based computing systems, and a comprehensive list is beyond the scope of this introduction. It is worth mentioning that the origami itself is typically used as a substrate for precisely positioning the logic elements [1, 34] or, as in a recent paper from Erik Winfree's lab at the California Institute of Technology, a seed element that encodes the input bits for

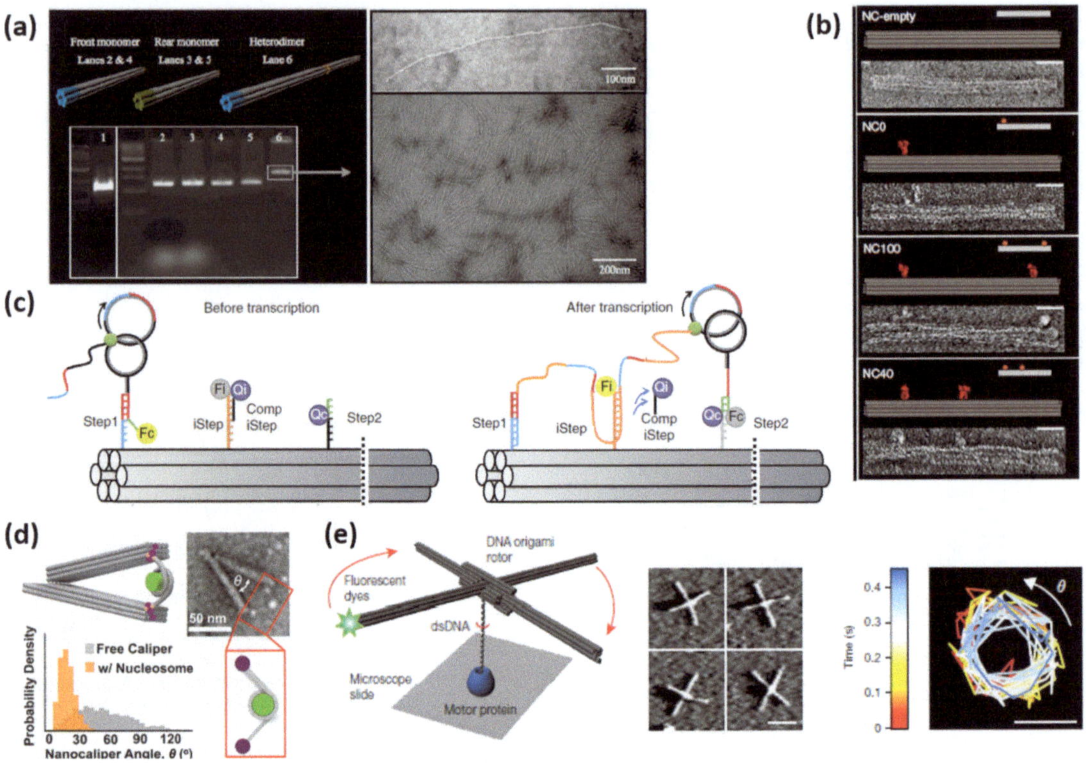

Fig. 5 DNA origami as final applications. (**a**) Using DNA origami nanorods to enhance NMR signals of membrane proteins. The nanorods are made by combining two monomers into a heterodimer (upper left). The gel (lower left) shows the results of heterodimer assembly (lane 6). The rods are approximately 800 nm long as can be seen in the TEM image of a single nanorod (upper right) which causes them to tend to align (TEM image lower right). Membrane proteins that collide with these nanorods will therefore also temporarily align with the nanorod, thereby enhancing the NMR signal (Figure adapted with permission from Ref. [9] © 2019 Nature Publishing Group). (**b**) DNA origami nanorods used to measure ligand/receptor distance-based interaction in human breast cancer cells. The origami rods (dark gray) can be used to controllably separate the ligands (red) as verified by the TEM images that accompany each schematic (scale bars, 20 nm) (Figure a-dapted with permission from Ref. [37] © 2019 Nature Publishing Group). (**c**) Schematic for a DNA catenane nanoengine walker on a DNA origami track. The nanoengine walker consists of two catenated DNA rings (black circles), one of which encodes a sequence complementary to Step1 (red and blue), and an RNA polymerase (green dot) attached to the catenane via a fused zinc finger. As the nanoengine transcribes RNA, the RNA can displace the fluorophore (Fc) labeled leg from Step1 allowing the leg to bind to the quencher (Qc) labeled Step2, quenching Fc fluorescence. The RNA also binds to a fluorophore (Fi) labeled iStep, displacing the quencher (Qi) labeled Comp-iStep, thereby activating Fi fluorescence (Figure adapted with permission from Ref. [39] © 2019 Nature Publishing Group). (**d**) DNA origami hinged nanocaliper used for studying DNA/histone binding. Upper left: Schematic of nanocaliper (gray) with histone (green). Upper right: TEM image of assembled caliper and histone (enhanced in lower right). Lower left: Histogram of measured nanocaliper angles either with or without nucleosome (Figure adapted with permission from Ref. [10] © 2019 Nature Publishing Group). (**e**) DNA origami rotary device for measuring motor protein rates of actuation. Leftmost: Schematic of device operation. Four blades extend from a central hub. A double-stranded DNA (dsDNA) strand extends from the bottom and is unwound by a surface-attached motor protein. As the DNA strand unwinds, the blades rotate and can be tracked by the fluorescent dye attached to one of the blades. Middle: TEM images of four assembled rotary devices (scale bar, 100 nm). Rightmost: Fluorescent tracking of the rotor blades. The different colors represent time; scale bar, 100 nm. (Figure adapted with permission from Ref. [42] © 2019 Nature Publishing Group)

iterative computation [8]. Considering the former, origami is an ideal material for building biological computing systems because of its self-assembling nature, ability to tile, and the level of precision and specificity with which elements can be located on the origamis. As mentioned previously, 2D origamis can tile together to create very large structures, and the self-assembly means they do so without any extra input during the assembly process. Furthermore, the tiles can be programmed to hybridize with logic elements that are then located at specific places on the origami (Fig. 4b), much how conventional transistors are specifically located on silicon chips. Concerning the latter, an origami was used to precisely encode input bits that serve as a growth seed for SSTs, which add iteratively to the growing structure based on how each SST is "programmed" (different DNA sequences) in relation to the input. The resulting structure (Fig. 4c) was read with an atomic force microscope (Fig. 4d) and shown to reliably output values as designed. This origami/SST computing system was able to perform several complex logic operations including Turing complete computations.

Biological computing is not the only type of processing technology that DNA origami can enable. It can also be used to assist in making next-generation computers using carbon nanotubes (CNTs). Maune et al. reported on the use of 2D DNA origami ribbons to align CNTs into a cross [1] and then used electron beam lithography to attach electrodes and demonstrate a limited field effect. Ten years later, a similar principle used 3D DNA tile origamis to arrange CNTs into equally spaced lines [35] with pitches from 25 down to 10 nm. This method was used to fabricate high-performance CNT-based field effect transistors [36] whose position could be spatially controlled to enable centimeter scale alignment.

3.2 Expanding the Toolbox of Applications

The above examples are considered still in the "prototype" phase as they cannot yet be employed as final applications. DNA origami has, however, been used in final application for several basic research studies. One such example is DNA nanorods that were used to enhance the nuclear magnetic resonance signal for membrane proteins [9]. The idea is that the nanorods (Fig. 5a) are long and align in the same direction. Proteins in solution that collide with the nanorods will also temporarily align, and this alignment enhances the NMR signal for such proteins. Moreover, the nanorods are detergent resistant, which is a prerequisite for working with membrane proteins.

Another interesting feature of DNA origami is its ability to precisely control the position of moieties attached to the origami. In one study (Fig. 5b), DNA origami nanorods were used to controllably separate ephrin-A5 ligands [37] to show that the distribution of such ligands results in different levels of EphA2 activation in human breast cancer cells as well as the cell's invasive characteristics. Such studies can lead to a better understanding of membrane receptor-mediated signaling and provide more knowledge to use in the fight against cancer. In another study, a research group led by Prof. Högberg used a similar strategy to study antibody-binding affinities for an antibody to two antigens when the antigens are spatially separated [38]. The antigens were placed at distances ranging from 3 to 43 nm on a DNA origami platform. They found that the ideal separation distance for immunoglobulin G is 16 nm and drastically decreases at 17 nm, whereas immunoglobulin M is able to bind well to antigens separated by as much as 29 nm. Such insights can provide a better understanding of how the immune system functions which itself can lead to more effective treatment and vaccines.

Origami nanorods or 2D plates can also serve as tracks for DNA walkers [39]. The increased stiffness of DNA origami over double-stranded DNA is advantageous for walkers as an origami track (Fig. 5c) is less likely to fold, allowing the walker to skip steps. Also, origami rods can be very long, and the distance between steps can be easily changed just by exchanging certain staples. Origami plates can provide more advanced environments for walking [40] or stochastic cargo sorting [41].

Another study where origamis were employed was in the determination of nucleosome stability [10] by Le et al. in the Castro lab. The researchers used a DNA origami hinge (Fig. 5d), similar to the hinge used in the previously discussed crank-slider mechanism, as a nanocaliper that would close to certain angles when bound to a DNA-wrapped histone protein by "linker" DNA strands of differing lengths (6, 26, 51, and 75 nm). As expected, the shorter the linker strand, the more tension would be induced on the histone, thereby causing it to unwrap. The researchers found that linker strands of 75 nm do not cause histone unwrapping and could determine histone-DNA interaction strength based on the results.

More recently, researchers used a DNA origami rotary device (Fig. 5e) to measure the rate of RecBCD (a DNA repair helicase) actuation as it unwound DNA [42]. Such measurements were not possible with conventional methods due to insufficient time resolution. The origami rotor had four 80-nm-long blades extending from a central hub. This profile provides a low hydrodynamic drag, and the central hub provided a high torsional stiffness to mitigate the effects of Brownian motion on the rotor. Initial rotor characterization experiments showed that 20 ms was long enough to resolve single base pair rotations with a high signal-to-noise ratio, whereas conventional methods require up to an hour to achieve the same resolution [43, 44]. A 52 bp double-stranded DNA segment extended from the bottom of the central hub perpendicular to the blades. One of the blades is modified with a fluorescent dye which can be tracked via microscopy as a surface-bound RecBCD unwinds the 52-bp-long strand. This same setup, dubbed origami-rotor-based imaging and tracking (ORBIT), was also used to measure the step size of RNA polymerase (RNAP) during transcription. Previously, optical tweezers measured single bp steps for RNAP, but only under applied force [45, 46]. The ORBIT method was able to confirm such single bp steps under more realistic circumstances, namely, no applied forces.

4 Conclusion

The capabilities for DNA origami have progressed considerably in the 15 years since its founding. Not only have researchers repurposed DNA to be a building material, something never to be expected from nature, but the methods used to fold and contort DNA are both simple and ingenious. Software programs exist to simplify the origami design process, thereby allowing people with almost no experience to quickly design and construct origamis for their own purposes. Furthermore, the possible applications for the field are diverse, and more will continue to be developed as more and more people become aware of the capabilities of DNA origami. We hope you enjoy reading the work presented in this book and learn something that could be useful for your own research.

References

1. Maune HT, Han SP, Barish RD, Bockrath M, Goddard WA III, Rothemund PW, Winfree E (2010) Self-assembly of carbon nanotubes into two-dimensional geometries using DNA origami templates. Nat Nanotechnol 5(1): 61–66. https://doi.org/10.1038/nnano.2009.311
2. Kopperger E, List J, Madhira S, Rothfischer F, Lamb DC, Simmel FC (2018) A self-assembled nanoscale robotic arm controlled by electric fields. Science 359(6373):296–301. https://doi.org/10.1126/science.aao4284
3. Gerling T, Wagenbauer KF, Neuner AM, Dietz H (2015) Dynamic DNA devices and assemblies formed by shape-complementary, non--base pairing 3D components. Science 347(6229):1446–1452. https://doi.org/10.1126/science.aaa5372
4. Muscat RA, Bath J, Turberfield AJ (2011) A programmable molecular robot. Nano Lett 11(3):982–987. https://doi.org/10.1021/nl1037165
5. Douglas SM, Bachelet I, Church GM (2012) A logic-gated nanorobot for targeted transport of molecular payloads. Science 335(6070): 831–834. https://doi.org/10.1126/science.1214081
6. Li SP, Jiang Q, Liu SL, Zhang YL, Tian YH, Song C, Wang J, Zou YG, Anderson GJ, Han JY, Chang Y, Liu Y, Zhang C, Chen L, Zhou GB, Nie GJ, Yan H, Ding BQ, Zhao YL (2018) A DNA nanorobot functions as a cancer therapeutic in response to a molecular trigger in vivo. Nat Biotechnol 36(3):258. https://doi.org/10.1038/nbt.4071
7. Ma WJ, Zhang YX, Zhang YX, Shao XR, Xie XP, Mao CC, Cui WT, Li Q, Shi JY, Li J, Fan CH, Lin YF (2019) An intelligent DNA nanorobot with in vitro enhanced protein lysosomal degradation of HER2. Nano Lett 19(7): 4505–4517. https://doi.org/10.1021/acs.nanolett.9b01320
8. Woods D, Doty D, Myhrvold C, Hui J, Zhou F, Yin P, Winfree E (2019) Diverse and robust molecular algorithms using reprogrammable DNA self-assembly. Nature 567(7748): 366–372. https://doi.org/10.1038/s41586-019-1014-9
9. Bellot G, McClintock MA, Chou JJ, Shih WM (2013) DNA nanotubes for NMR structure determination of membrane proteins. Nat Protoc 8(4):755–770. https://doi.org/10.1038/nprot.2013.037

10. Le JV, Luo Y, Darcy MA, Lucas CR, Goodwin MF, Poirier MG, Castro CE (2016) Probing nucleosome stability with a DNA origami nanocaliper. ACS Nano 10(7):7073–7084. https://doi.org/10.1021/acsnano.6b03218
11. Chen JH, Seeman NC (1991) Synthesis from DNA of a molecule with the connectivity of a cube. Nature 350(6319):631–633
12. Rothemund PWK (2006) Folding DNA to create nanoscale shapes and patterns. Nature 440(7082):297–302
13. Goodman RP, Schaap IA, Tardin CF, Erben CM, Berry RM, Schmidt CF, Turberfield AJ (2005) Rapid chiral assembly of rigid DNA building blocks for molecular nanofabrication. Science 310(5754):1661–1665. https://doi.org/10.1126/science.1120367. 310/5754/1661 [pii] 1126/science.1120367
14. Zhang YW, Seeman NC (1994) Construction of a DNA-truncated octahedron. J Am Chem Soc 116(5):1661–1669
15. Mathieu F, Liao SP, Kopatscht J, Wang T, Mao CD, Seeman NC (2005) Six-helix bundles designed from DNA. Nano Lett 5(4):661–665. https://doi.org/10.1021/nl050084f
16. Andersen ES, Dong M, Nielsen MM, Jahn K, Subramani R, Mamdouh W, Golas MM, Sander B, Stark H, Oliveira CL, Pedersen JS, Birkedal V, Besenbacher F, Gothelf KV, Kjems J (2009) Self-assembly of a nanoscale DNA box with a controllable lid. Nature 459(7243):73–76. https://doi.org/10.1038/nature07971
17. Douglas SM, Dietz H, Liedl T, Hogberg B, Graf F, Shih WM (2009) Self-assembly of DNA into nanoscale three-dimensional shapes. Nature 459(7245):414–418. https://doi.org/10.1038/nature08016. nature08016 [pii] 1038/nature08016
18. Douglas SM, Marblestone AH, Teerapittayanon S, Vazquez A, Church GM, Shih WM (2009) Rapid prototyping of 3D DNA-origami shapes with caDNAno. Nucleic Acids Res 37(15):5001–5006. https://doi.org/10.1093/nar/gkp436
19. Ke YG, Douglas SM, Liu MH, Sharma J, Cheng AC, Leung A, Liu Y, Shih WM, Yan H (2009) Multilayer DNA origami packed on a square lattice. J Am Chem Soc 131(43):15903–15908. https://doi.org/10.1021/ja906381y
20. Dietz H, Douglas SM, Shih WM (2009) Folding DNA into twisted and curved nanoscale shapes. Science 325(5941):725–730. https://doi.org/10.1126/science.1174251. 325/5941/725 [pii] 1126/science.1174251
21. Han DR, Pal S, Nangreave J, Deng ZT, Liu Y, Yan H (2011) DNA origami with complex curvatures in three-dimensional space. Science 332(6027):342–346. https://doi.org/10.1126/science.1202998
22. Martin TG, Dietz H (2012) Magnesium-free self-assembly of multi-layer DNA objects. Nat Commun 3:1103. https://doi.org/10.1038/ncomms2095
23. Wei B, Dai M, Yin P (2012) Complex shapes self-assembled from single-stranded DNA tiles. Nature 485(7400):623–626. https://doi.org/10.1038/nature11075
24. Ke Y, Ong LL, Shih WM, Yin P (2012) Three-dimensional structures self-assembled from DNA bricks. Science 338(6111):1177–1183. https://doi.org/10.1126/science.1227268
25. Marras AE, Zhou LF, Su HJ, Castro CE (2015) Programmable motion of DNA origami mechanisms. Proc Natl Acad Sci U S A 112(3):713–718. https://doi.org/10.1073/pnas.1408869112
26. Wagenbauer KF, Sigl C, Dietz H (2017) Gigadalton-scale shape-programmable DNA assemblies. Nature 552(7683):78–83. https://doi.org/10.1038/nature24651
27. Ketterer P, Willner EM, Dietz H (2016) Nanoscale rotary apparatus formed from tight-fitting 3D DNA components. Sci Adv 2(2):e1501209. https://doi.org/10.1126/sciadv.1501209
28. Praetorius F, Kick B, Behler KL, Honemann MN, Weuster-Botz D, Dietz H (2017) Biotechnological mass production of DNA origami. Nature 552(7683):84. https://doi.org/10.1038/nature24650
29. Robertson DL, Joyce GF (1990) Selection in vitro of an RNA enzyme that specifically cleaves single-stranded-DNA. Nature 344(6265):467–468. https://doi.org/10.1038/344467a0
30. Tuerk C, Gold L (1990) Systematic evolution of ligands by exponential enrichment – RNA ligands to bacteriophage-T4 DNA-polymerase. Science 249(4968):505–510. https://doi.org/10.1126/science.2200121
31. Ellington AD, Szostak JW (1990) In vitro selection of RNA molecules that bind specific ligands. Nature 346(6287):818–822. https://doi.org/10.1038/346818a0
32. Langecker M, Arnaut V, Martin TG, List J, Renner S, Mayer M, Dietz H, Simmel FC (2012) Synthetic lipid membrane channels formed by designed DNA nanostructures. Science 338(6109):932–936. https://doi.org/10.1126/science.1225624

33. Sigl C, Willner EM, Engelen W, Kretzmann JA, Sachenbacher K, Liedl A, Kolbe F, Wilsch F, Aghvami SA, Protzer U, Hagan MF, Fraden S, Dietz H (2021) Programmable icosahedral shell system for virus trapping. Nat Mater. https://doi.org/10.1038/s41563-021-01020-4
34. Chatterjee G, Dalchau N, Muscat RA, Phillips A, Seelig G (2017) A spatially localized architecture for fast and modular DNA computing. Nat Nanotechnol 12(9):920. https://doi.org/10.1038/Nnano.2017.127
35. Sun W, Shen J, Zhao Z, Arellano N, Rettner C, Tang J, Cao T, Zhou Z, Ta T, Streit JK, Fagan JA, Schaus T, Zheng M, Han SJ, Shih WM, Maune HT, Yin P (2020) Precise pitch-scaling of carbon nanotube arrays within three-dimensional DNA nanotrenches. Science 368(6493):874–877. https://doi.org/10.1126/science.aaz7440
36. Zhao M, Chen Y, Wang K, Zhang Z, Streit JK, Fagan JA, Tang J, Zheng M, Yang C, Zhu Z, Sun W (2020) DNA-directed nanofabrication of high-performance carbon nanotube field-effect transistors. Science 368(6493): 878–881. https://doi.org/10.1126/science.aaz7435
37. Shaw A, Lundin V, Petrova E, Fordos F, Benson E, Al-Amin A, Herland A, Blokzijl A, Hogberg B, Teixeira AI (2014) Spatial control of membrane receptor function using ligand nanocalipers. Nat Methods 11(8):841–846. https://doi.org/10.1038/Nmeth.3025
38. Shaw A, Hoffecker IT, Smyrlaki I, Rosa J, Grevys A, Bratlie D, Sandlie I, Michaelsen TE, Andersen JT, Hogberg B (2019) Binding to nanopatterned antigens is dominated by the spatial tolerance of antibodies. Nat Nanotechnol 14(4):398–398. https://doi.org/10.1038/s41565-019-0404-3
39. Valero J, Pal N, Dhakal S, Walter NG, Famulok M (2018) A bio-hybrid DNA rotor-stator nanoengine that moves along predefined tracks. Nat Nanotechnol 13(6):496–503. https://doi.org/10.1038/s41565-018-0109-z
40. Lund K, Manzo AJ, Dabby N, Michelotti N, Johnson-Buck A, Nangreave J, Taylor S, Pei R, Stojanovic MN, Walter NG, Winfree E, Yan H (2010) Molecular robots guided by prescriptive landscapes. Nature 465(7295):206–210. https://doi.org/10.1038/nature09012
41. Thubagere AJ, Li W, Johnson RF, Chen ZB, Doroudi S, Lee YL, Izatt G, Wittman S, Srinivas N, Woods D, Winfree E, Qian LL (2017) A cargo-sorting DNA robot. Science 357(6356):eaan6558. https://doi.org/10.1126/science.aan6558. ARTN eaan65581126/science.aan6558
42. Kosuri P, Altheimer BD, Dai MJ, Yin P, Zhuang XW (2019) Rotation tracking of genome-processing enzymes using DNA origami rotors. Nature 572(7767):136. https://doi.org/10.1038/s41586-019-1397-7
43. Lipfert J, Wiggin M, Kerssemakers JWJ, Pedaci F, Dekker NH (2011) Freely orbiting magnetic tweezers to directly monitor changes in the twist of nucleic acids. Nat Commun 2: 439. https://doi.org/10.1038/ncomms1450. Artn 4391038/Ncomms1450
44. Lebel P, Basu A, Oberstrass FC, Tretter EM, Bryant Z (2014) Gold rotor bead tracking for high-speed measurements of DNA twist, torque and extension. Nat Methods 11(4): 456. https://doi.org/10.1038/nmeth.2854
45. Righini M, Lee A, Canari-Chumpitaz C, Lionberger T, Gabizon R, Coello Y, Tinoco I, Bustamante C (2018) Full molecular trajectories of RNA polymerase at single base-pair resolution. Proc Natl Acad Sci USA 115(6): 1286–1291. https://doi.org/10.1073/pnas.1719906115
46. Abbondanzieri EA, Greenleaf WJ, Shaevitz JW, Landick R, Block SM (2005) Direct observation of base-pair stepping by RNA polymerase. Nature 438(7067):460–465. https://doi.org/10.1038/nature04268

Chapter 2

Design, Assembly, and Function of DNA Origami Mechanisms

Peter E. Beshay, Joshua A. Johson, Jenny V. Le, and Carlos E. Castro

Abstract

This chapter provides an overview of the common procedures used in making functional DNA origami devices. These procedures include the design, assembly, purification, and characterization of the DNA origami structures, with a focus on dynamic devices.

Key words DNA origami, DNA nanotechnology, Self-assembly, Dynamic nanodevices, Biomolecular nanotechnology, Gel electrophoresis, Transmission electron microscopy

1 Introduction

Structural DNA nanotechnology [1, 2] involves the self-assembly of nanoscale 2D and 3D structures using DNA as building blocks via Watson-Crick base pairing [3]. Scaffolded DNA origami [4], in particular, enables the fabrication of nanostructures and dynamic nanodevices with unprecedented control of structure, mechanical, and dynamic properties. DNA origami structures are comprised of a long, usually ~7000–8000 nucleotides, "scaffold" strand [5, 6], and many shorter "staple" strands. A staple is an oligonucleotide generally 20–60 bases long, with the constraint on the longer end generally imposed by production costs and the constraint on the shorter end to ensure stable binding of staple strands at room temperature. The scaffold is typically a circular single-stranded DNA, often derived from the M13 bacteriophage genome, with a known sequence that is used as the foundation of a DNA origami nanostructure [5, 7]. Double helices are formed by base pairing between staples and scaffold, and the structure is held together via crossovers, similar to naturally occurring Holliday junctions [8], that connect neighboring helices together at regular intervals along their length. DNA origami has become a powerful toolset to solve current and future problems, with exciting potential for

Julián Valero (ed.), *DNA and RNA Origami: Methods and Protocols*, Methods in Molecular Biology, vol. 2639,
https://doi.org/10.1007/978-1-0716-3028-0_2,

applications in fields including drug delivery [9], biophysics [10], molecular sensing [11], plasmonics [12], and cellular studies [13]. In this chapter, we will provide some common protocols and tips for designing, assembling, purifying, and characterizing DNA structures, with a focus on dynamic devices. Other useful resources for designing and assembling DNA origami structures can be found in [14–16].

2 Materials

2.1 DNA Origami Self-Assembly

1. Scaffold: Common scaffolds used in DNA origami include M13mp18 bacteriophage plasmid variants, which include lengths such as 7249 nt, 7308 nt, 7560 nt, 7704 nt, and 8064 nt [5]. Other scaffolds have been used [17–20]. The scaffold can be produced as previously described [14] or purchased from commercial vendors such as New England Biolabs, Tilibit Nanosystems, and Guild Biosciences. It is possible to generate your own scaffold through a variation on giga-preps if you already have an aliquot of scaffold [21], and recent studies have demonstrated new methods for preparation of single-stranded DNA [22, 23].
2. Staple oligos: Can be purchased from various vendors, such as Integrated DNA Technologies (IDT), Eurofins, Bioneer, and Millipore Sigma.
3. Tris(hydroxymethyl)aminomethane (Tris base).
4. Ethylenediaminetetraacetic acid (EDTA).
5. Sodium chloride.
6. Magnesium chloride hexahydrate.
7. 96-well plates.
8. Pipette basins.
9. 10x FOB: 50 mM of TRIS-HClpH 7,4, 50 mM of NaCl, and 10 mM of EDTA.

2.2 Purification

2.2.1 Gel Electrophoresis

1. Agarose.
2. TBE (Tris-borate-EDTA) buffer.
3. Magnesium chloride hexahydrate.
4. Intercalating agent (e.g., ethidium bromide (EtBr), SYBR gold, SYBR safe). EtBr, stock solution is at 10 mg/mL.
5. Gel loading dye.
6. 1 kb DNA ladder.
7. Running buffer: 0.5× TBE + 11 mM $MgCl_2$.

2.2.2 Centrifugation in Polyethylene Glycol (PEG Solution)

1. Polyethylene glycol (PEG), 8000 Da.
2. Tris(hydroxymethyl)aminomethane (Tris base).
3. Ethylenediaminetetraacetic acid (EDTA).
4. Magnesium chloride hexahydrate.
5. Sodium chloride.

2.2.3 Molecular Weight Cutoff (MWCO) Centrifugal Filters

1. 100 kDa or 50 kDa MWCO centrifugal filter.
2. Tris(hydroxymethyl)aminomethane (Tris base).
3. Ethylenediaminetetraacetic acid (EDTA).
4. Magnesium chloride hexahydrate.
5. Prewet buffer: 5 mM TRIS, 1 mM EDTA, and 10 mM $MgCl_2$.

2.3 Transmission Electron Microscopy (TEM) Imaging and Verification Methods

1. TEM grids: Formvar/carbon-coated copper grids are the typical choices for imaging DNA origami structures.
2. Uranyl formate.
3. 10 mL syringe.
4. 0.22 μm syringe filter.
5. Sodium hydroxide.
6. Filter paper.

3 Methods

3.1 DNA Origami Design

The design process for a simple rigid beam is as follows (*see* **Note 1**):

3.1.1 Simple Structures

1. Choose cross-section lattice: the most common is to use square [24] or honeycomb [5] lattice cross-section although there are other variations [25].
2. Select circles in the left-hand side slice panel (circles represent the cross-section of a DNA helix) to closely approximate the cross-section of the desired shape. This should also populate the path panel with numbered short blue lines which represent the scaffold strand (*see* Fig. 1a).
3. Extend the work area and scaffold strands to the approximate length of the intended shape by selecting and dragging the scaffold. It is worth noting that caDNAno has a number of shortcuts, which are also useful and can be explored in video tutorials that can be found via the caDNAno website.
4. Using the clickable numbers in the path view, create crossovers at the left- and rightmost ends of the scaffold. These are referred to as external crossovers.

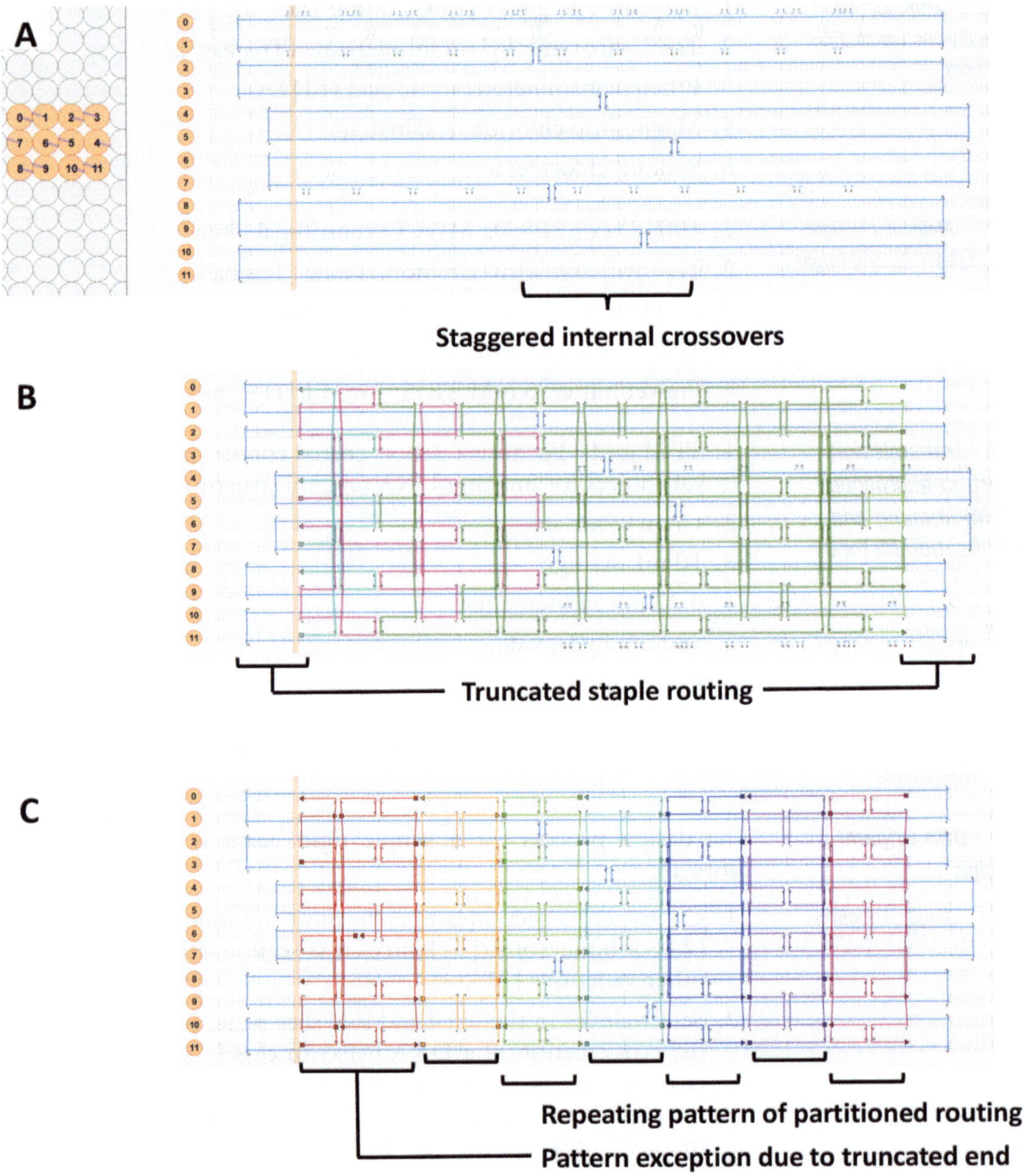

Fig. 1 Example of scaffold and staple routing in caDNAno for a 3 by 4 square lattice DNA origami design. (**a**) The scaffold routing must form a continuous loop which is typically done using nearly aligned external crossovers and staggered internal crossovers. (**b**) Using the Autostaple function, easily fill the design with staples that already have all possible crossovers. Staples must be truncated at the edges to create single-stranded scaffold loops which help prevent blunt end stacking between structures. (**c**) Staples should be broken into segments which are less than 60 bp long and preferably less than 50 bp. It is recommended to break staples such that they can be partitioned into segments with similar routing motifs. In some edge cases, it may not be possible to maintain this repeating pattern

5. Create crossovers in a staggered fashion which connect bundles that are not already connected on the ends. These are referred to as internal crossovers. Appropriate locations for crossovers according to the helical geometry of DNA are highlighted in caDNAno (*see* Fig. 1a). This process should result in one continuous scaffold strand that is routed throughout the entire desired shape.
6. Adjust the scaffold ends until the total length is equal to the scaffold you intend to use for experiments (typically m13MP18 which is 7249 nucleotides long). If you do not use the entire scaffold length, the structure will contain a loop of scaffold DNA, which may or may not be an issue depending on the size of the loop, the structure, and the intended application.
7. Insert staple strands using shift + right click or with the pencil tool and extend to the edges of the structure (*see* **Note 2**).
8. Add staple crossovers by clicking on all available locations unless it would be within seven bases of a scaffold crossover between the same two helices.
9. Use the break tool to cut staple lines into segments that are between ~20 and 60 bases long, which usually consist of 3–5 sections that bind to the scaffold in a piecewise manner. One typical rule of thumb is to avoid having shorter sections sandwiched between longer sections, since the longer sections may anneal first and inhibit annealing of the shorter section in the middle. Exact positions of staple breaks can influence overall yield [26]. Ideally into a periodic pattern which can be grouped into distinct sections (*see* **Note 3**).
10. It is recommended to color code grouped sections of staples using the paint tool.
11. Create a single break within the scaffold loop. This can be anywhere. If you do not use the entire scaffold, the remaining scaffold will form a ssDNA loop that will be positioned at this break point where you start/end the scaffold.
12. Use the Add Sequence tool and click on the 5′ end of the scaffold (denoted by the box) at the break point. It is not necessary to create a break in order to add the scaffold sequence, but it is a useful practice to mark where the sequence is started. If using caDNAno 2, you may click anywhere on the scaffold.
13. Click export to create a csv document containing all necessary staple sequences.
14. It is recommended to sort these sequences in a spreadsheet by assigned color and checking for any unassigned bases, which will appear as a question mark. Unassigned bases may be a sign

that the scaffold is not one single complete loop. In caDNAno 2 this can also be checked directly as the sequence appears in the routing diagram window after it is added. If the structure has overhangs, these will also appear as question marks on the exported sequence list.

3.1.2 Design of Overhangs (See Note 4)

1. To add binding locations along the surface of the structure, click on bundles in the slice panel that are adjacent to the desired binding location. For example, in Fig. 2a, if an overhang is desired on the top of the structure sticking out of helix 2, then you would click on the circle just above helix 2 to add a reference helix there (*see* **Notes 5** and **6**).
2. Add a staple segment in the path panel without adding a scaffold segment (i.e., by right-clicking while holding down the shift key). Find a crossover location between this new staple segment and a staple within the structure at the desired position along the length of the structure.
3. In some cases, it may be necessary to create additional breaks in staples within the structure to add a crossover to an overhang (*see* Fig. 2b). However, this is not recommended if creating a break to accommodate an overhang results in a staple strand that is very short (i.e., less than 14 bp) as shown in Fig. 2e. Convenient locations for adding crossovers to overhangs are ones which are already located at staple breaks (*see* Fig. 2c, d). These locations will also allow the user to choose between overhangs which start with a 5′ or 3′ end (*see* Fig. 2f, g).
4. Adjust the nearby staple routing if necessary to ensure that all staples are still ~20–60 bases long. Additionally, it is recommended to adjust the staple routing to ensure there are not many staple breaks or short binding regions in close proximity to one another (*see* Fig. 3).
5. If using a square lattice, it is necessary to use skips to counteract an accumulation of strain that results in a global twist of the structure. Add one skip for every six tokens (48 bp) (*see* Fig. 4).
6. Verify that the structure has been appropriately corrected using CanDo [14, 27], which can be used through a convenient open online interface (cando-dna-origami.org). Make sure to adjust the total scaffold length to match the scaffold you intend to use.
7. Export the staple list which will now have question marks on the staples with overhang ends. A sample condensed staple list is depicted in Fig. 5.
8. Replace question marks with the desired overhang sequence or with a newly generated sequence (*see* **Notes** 7 and **8**).

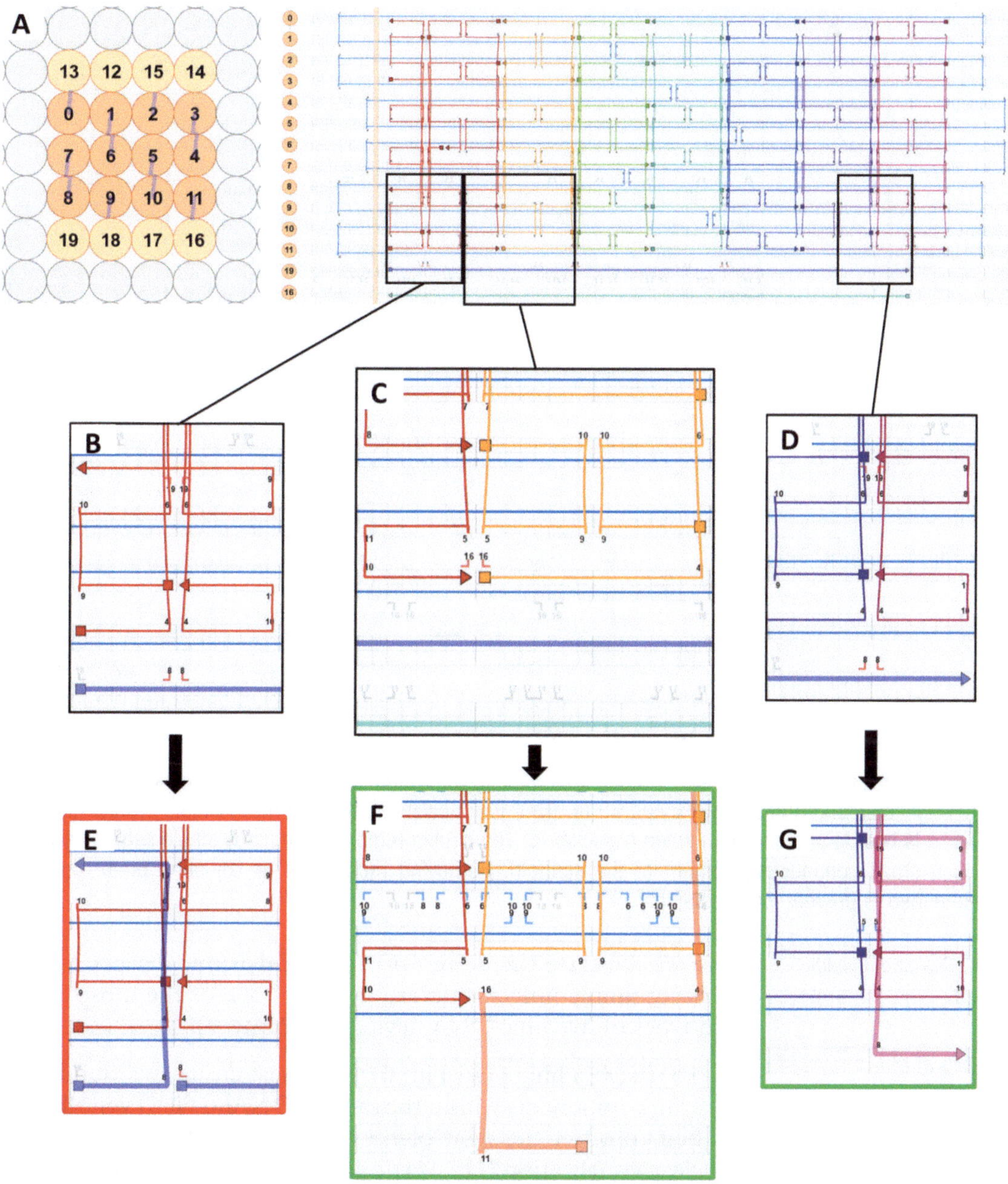

Fig. 2 Example of how to add overhangs. (**a**) Cross-sectional view shows new bundles added above and below the main structure. Staple segments spanning the entire structure (purple and cyan) are added to show all locations where it is possible to add an overhang. (**b**) Due to the staple routing along the edge of this structure, adding an overhang between bundles 8 and 19 is not recommended. (**c**) Due to the repeated pattern of partitioned staple routing, there are more likely to be convenient points to add overhangs. Here a break between the red and orange sections also allows for crossover to overhang strands without needing to adjust other staples. (**d**) Although the magenta section is located along a structure edge, it still conforms to the repeating pattern of staple routing, and therefore, overhangs can be added closer to the structure end more conveniently. (**e**) Adding an overhang to this location would create a staple that only has one token bound to the scaffold and is not recommended. (**f**) Example of a good overhang using the 5′ end of a staple. (**g**) Example of a good overhang using the 3′ end of a staple

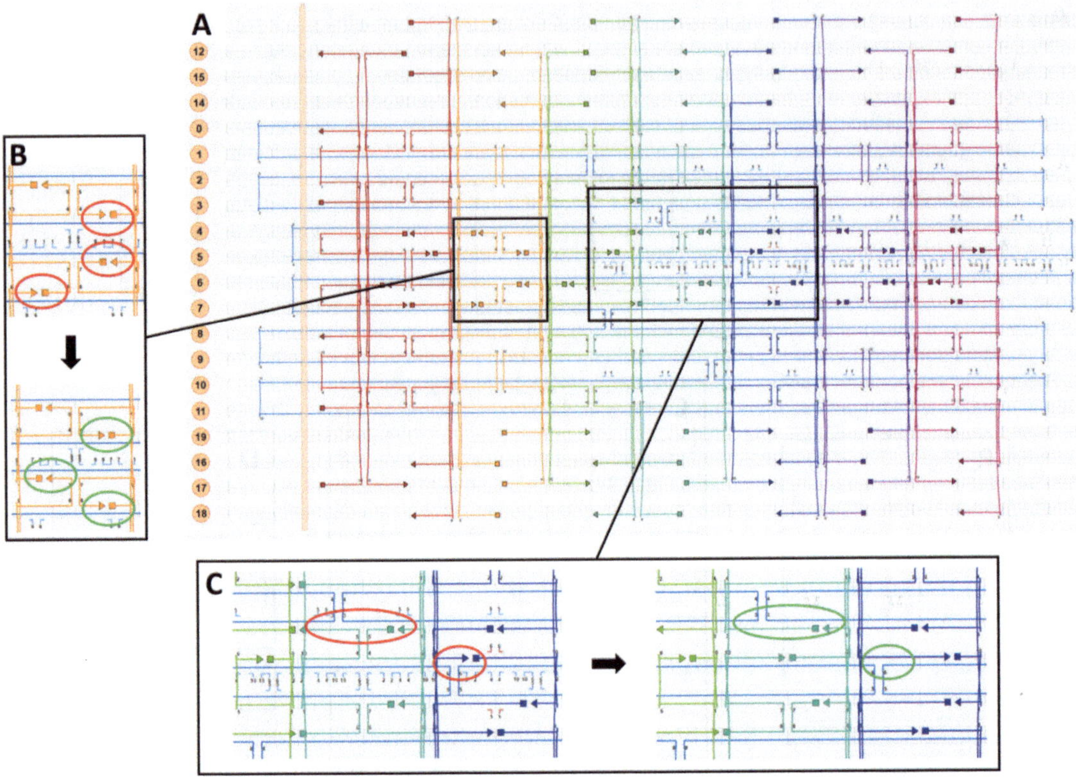

Fig. 3 Example of corrections to staple routing after adding overhangs. (**a**) Initial pass of breaking up staples to ensure their total length does not exceed 60 nt. (**b**) If it is necessary to have many staple breaks near one another, it is best to stagger them when possible. (**c**) The circled regions have many staple segments which bind to a small continuous segment of the scaffold. Removing crossovers or adjusting break points is recommended to prevent local instability

3.1.3 Design of Dynamic Origami

Here we illustrate the design process for dynamic devices using a relatively simple hinge structure, which is one of the most widely used types of dynamic DNA devices (*see* **Note 9**):

1. To design a hinge (*see* Fig. 6), one must create two bundle, or arm, components within the same caDNAno file. Typically, it is best to offset the rigid components within the slice panel for clear visualization (i.e., vertical separation distance in Fig. 6b).
2. The scaffolds of the separate bodies must be joined to create one continuous scaffold strand that is routed through the entire structure. Scaffold lengths connecting hinge arm components must alternate between short and long to account for the helical alignment of connection points.
3. The shorter connections that occur between the bottom of the top arm and the top of the bottom arm are often two nucleotides long to allow for some rotational flexibility but still effectively constrain the translational motion or rotations in other directions.

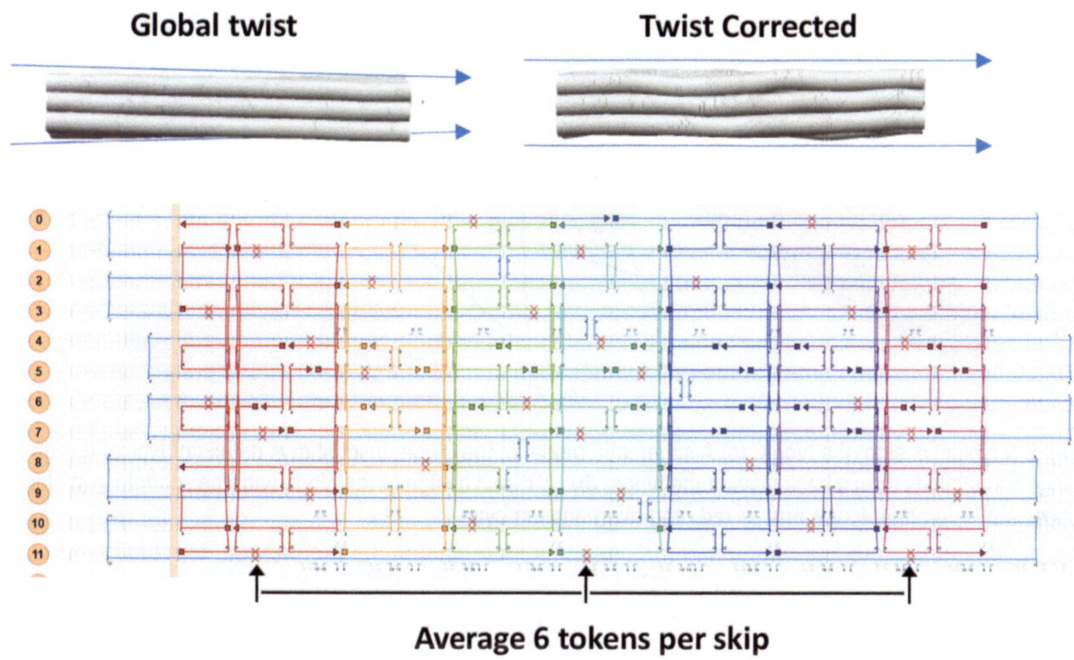

Fig. 4 Correcting for global twist in square lattice designs. CanDo simulations show the example structure before (top left) and after (top right) adding skips

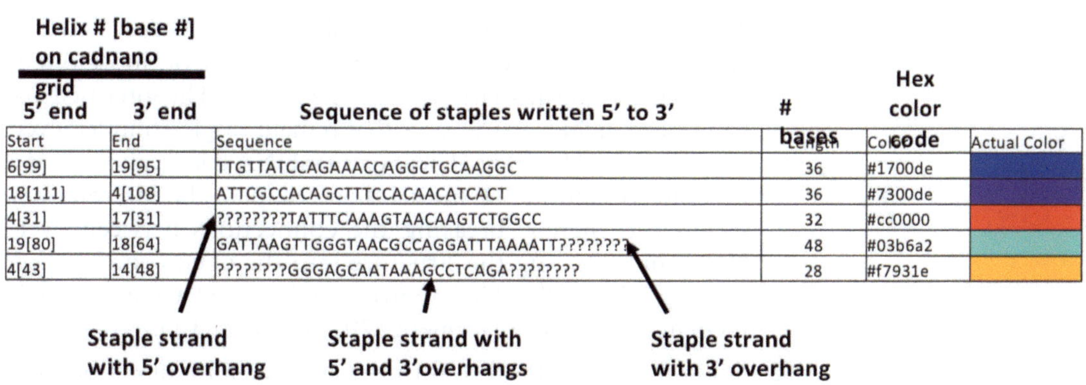

Start	End	Sequence	Length	Color	Actual Color
6[99]	19[95]	TTGTTATCCAGAAACCAGGCTGCAAGGC	36	#1700de	
18[111]	4[108]	ATTCGCCACAGCTTTCCACAACATCACT	36	#7300de	
4[31]	17[31]	????????TATTTCAAAGTAACAAGTCTGGCC	32	#cc0000	
19[80]	18[64]	GATTAAGTTGGGTAACGCCAGGATTTAAAATT????????	48	#03b6a2	
4[43]	14[48]	????????GGGAGCAATAAAGCCTCAGA????????	28	#f7931e	

Fig. 5 Sample staple list exported from caDNAno with labels added to describe entries. The entries in the table and top row of table headings are all exported directly from caDNAno except for the column labeled "Actual Color." The Actual Color column was added for reference (colors are approximate)

4. The longer scaffold connections must span twice the diameter of the DNA helices; hence these must be made longer, at least ~15 nucleotides, and likely at least ≳20 nucleotides to obtain a relatively flexible hinge.

5. The inset in Fig. 6a illustrates short and long ssDNA scaffold connections that form a hinge joint. While the relation between these design parameters and the resulting hinge properties (i.e.,

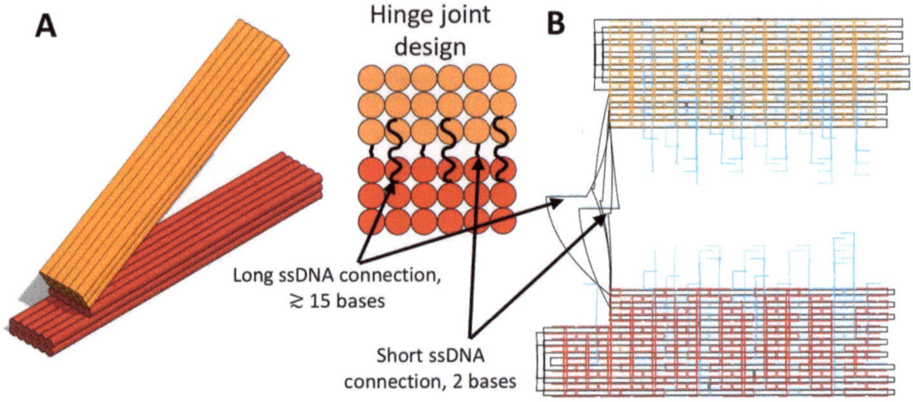

Fig. 6 A typical design of a hinge. (**a**) A schematic model of the hinge, inset shows details of typical design of hinge connections from a back view of the hinge. (**b**) caDNAno design of the hinge with the scaffold is in black, upper arm in orange, lower arm in red, and overhangs in cyan

torsional stiffness and equilibrium angle) are not well-understood, we previously showed the length of the longer connections can be used to tune the torsional stiffness of the hinge.

3.1.4 Design of Actuation Methods

A widely used standard actuation method involves (*see* **Note 10**):

1. Two or more overhangs that can form a connection between two or more rigid components either directly or, more often, through an intermediary strand (*see* Fig. 7a). The intermediary strand can include extra bases that remain single-stranded even after the initial binding. These extra ssDNA bases can serve as a toehold for another strand to then compete the intermediary strand off, thereby breaking the connections between overhangs and releasing the structure into the open configuration (*see* Fig. 7b) [28]. This approach is referred to as toehold-mediated strand displacement [29] (*see* **Note 11**).
2. Similarly, any method of forming and breaking direct binding between overhanging strands can be used as an actuation method. Methods to disrupt this binding between overhangs (*see* Fig. 7c) include changing buffer conditions to modify pH [30, 31], ionic concentration [32–34], temperature [32, 35], or light [36, 37]. For pH-based actuation schemes, it is necessary to have overhang sequences which can form pH-sensitive triplex or quadruplex structures, such as an i-motif structure that can form in C-rich sequences [38].
3. Alternatively, base-stacking interactions can be similarly stabilized or destabilized by temperature changes or changes in ion concentrations (*see* Fig. 7d) to achieve actuation of shape complementary components [32].

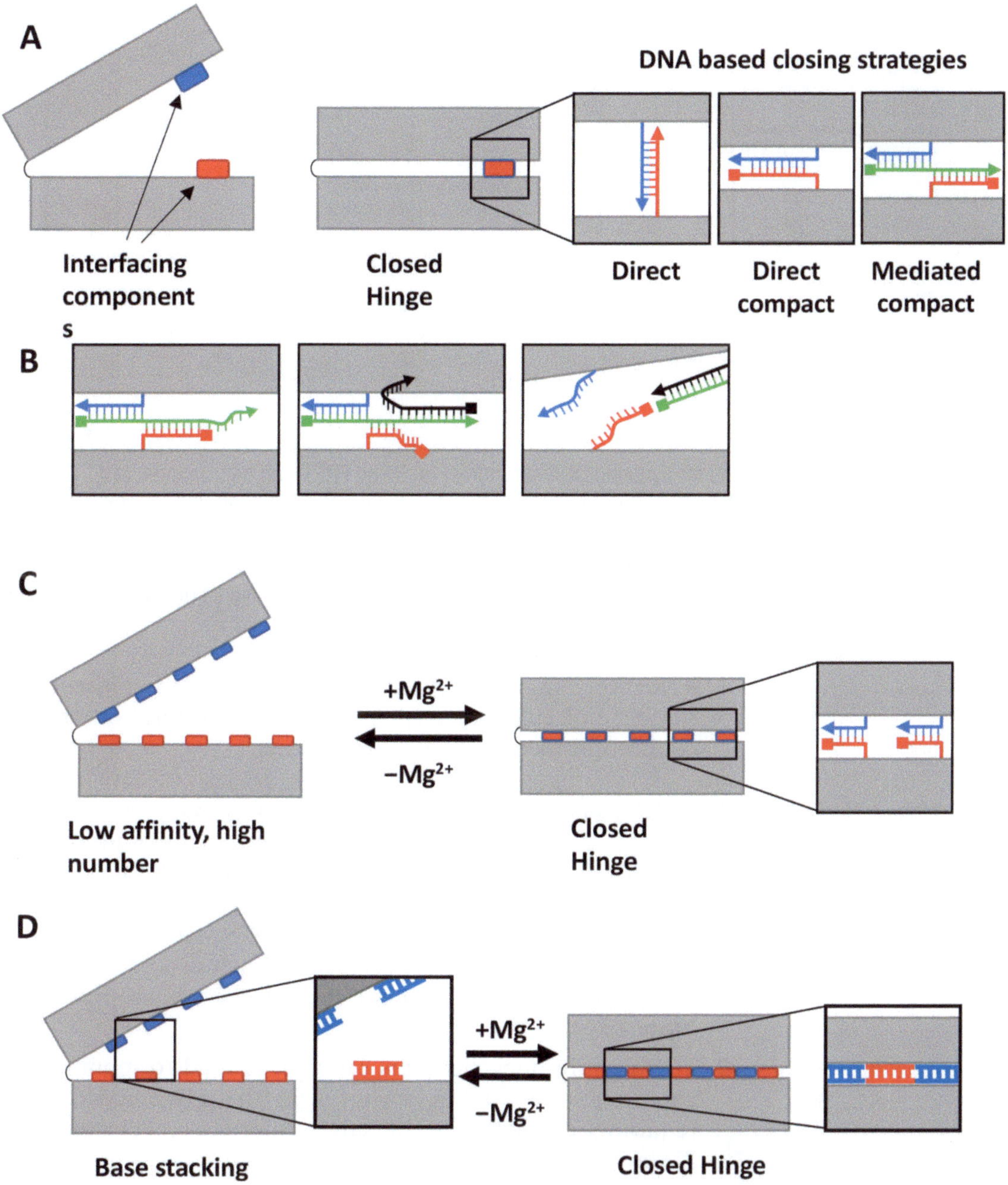

Fig. 7 Methods for reconfiguration of dynamic DNA origami. (**a**) The most commonly used actuation method is direct binding between two overhangs or mediated by a "closing strand" (green in rightmost panel). (**b**) The closing strand, or strands, can be removed through toehold-mediated strand displacement where an invader strand (black) binds to a single-stranded toehold on the closing strand and ultimately competes it off the structure. (**c**) Another approach for reversible actuation is to stabilize or destabilize direct binding, for example, by changing ion conditions or changing temperature. (**d**) An alternative approach to actuation leverages base-stacking interactions between blunt ends of helices, which are also sensitive to ion conditions or temperature, to achieve binding between shape complementary components

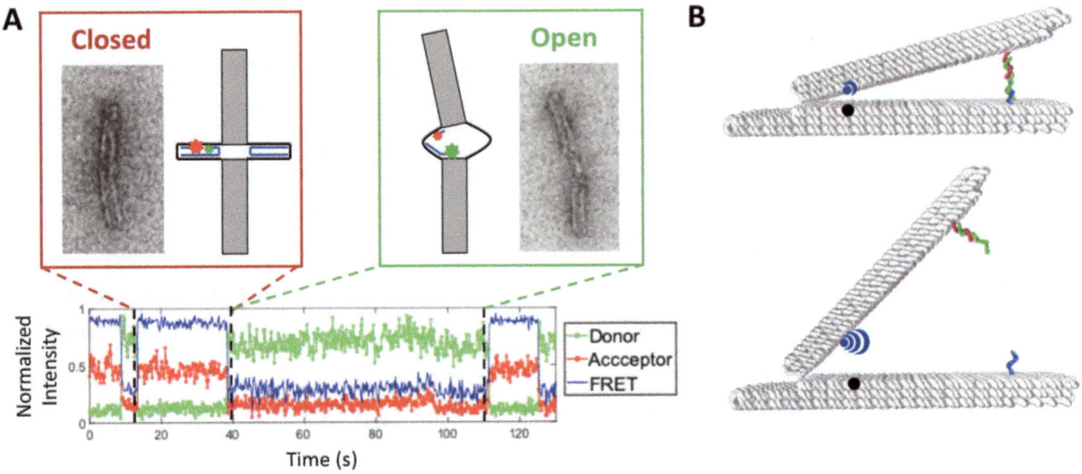

Fig. 8 Common readout methods for dynamic DNA structures. (**a**) Using FRET-pair fluorophores (e.g., Cy3 and Cy5) to monitor the conformational change of a dynamic DNA origami structure, where, when exciting the donor molecule, high acceptor fluorophore emission indicates the structure is closed and high donor fluorophore emission indicates the structure is open. (**b**) A fluorophore-quencher pair can similarly be used to monitor the conformational change of a hinge structure, where the fluorophore signal gets quenched when the hinge is closed

4. Actuation can also be used as a readout for the presence of specific molecules when aptamers are used to form or displace connections between rigid components.

*3.1.5 Design of Readout Methods (See **Note 12**)*

1. FRET pairs can be included on the ends of staple strands such that they protrude out of the surface. As such, it is important to keep in mind the orientation of the DNA at the location where a fluorophore is desired. Additionally, certain fluorophores can be quenched by neighboring guanines [39, 40], and as such it may be necessary to include spacers, such as additional thymines, between staple strand and attached fluorophore (*see* **Note 13**).
2. Fluorophore-quencher pairs can also be used to monitor the conformation of DNA devices with the advantage of only requiring measurement of single fluorescence channel (*see* **Notes 14** and **15**) (Fig. 8).

3.2 DNA Origami Assembly (See Note 16) (Depending on Applications, See Notes 25, 26, and 27)

3.2.1 Prestocks and Working Stocks and Folding Reactions

1. Start by briefly centrifuging the commercially ordered oligo plates to get any liquid off the lid. Usually centrifuging for 1–2 min at 2000 g is sufficient.
2. Pipette staples from your oligo plates into the respective well of a prestock plate, a 96-well deep well plate, according to your prestock sheet (*see* Fig. 9a–c and **Notes 17** and **18**).
3. When finished with prestocks, seal the prestock plate, and then vortex and centrifuge as mentioned above.

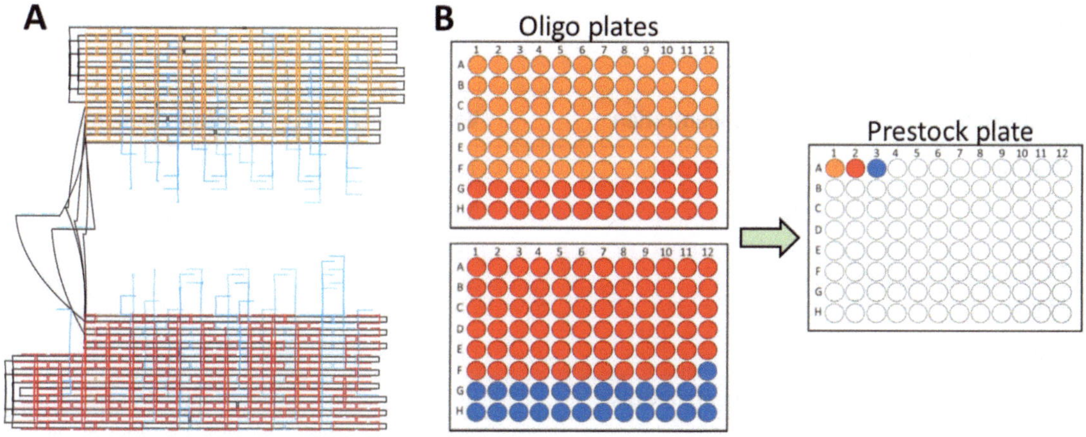

C

Pre-Stock Location	Oligo Plate	Plate Location Start	Plate Location End	Description	# of Oligos
A1	1	A1	F9	Upper arm	69
A2	1	F10	H12	Lower arm	98
	2	A1	F11		
A3	2	F12	H12	Overhangs	25

D

Pre-Stock Location	Description	# of Oligos	# of Working Stocks	Volume (µl)
A1	Upper arm	69	1	69
A2	Lower arm	98	1	98
A3	Overhangs	25	1	25
			ddH_2O	8
			Total	200

Fig. 9 Converting a given (**a**) DNA origami design into (**b**) prestocks by consolidating the staple strands corresponding to a certain region, or design module, into one well according to a prestock sheet similar to (**c**). (**d**) shows an example of a working stock sheet

4. For working stocks, unseal the prestock plate, and pipette the respective prestocks needed into a 1.5 mL tube according to the working stock sheet (*see* Fig. 9d). This involves pipetting a volume proportional to the number of staples in the prestock. For example, if a prestock contains 98 staples, you would pipette 98 μL of that prestock into the working stock. Typically, the setup of working stocks has the equivalence of adding 1 μL of each relevant oligo at 100 μM into the working stock.
5. For structures with less than 200 staples, dilute to 500 nM by adding volume to fill the working stock up to 200 μL. For structures with more than 200 staples, leave as is. Since the staples are diluted by a factor of 200 (i.e., 1 μL in a total of

Table 1
Typical folding reaction recipe

Component	Stock concentration	Volume [μL]	Final concentration
Scaffold	100 nM	10	20 nM
Staples	500 nM	20	200 nM
FOB	10x	5	1x
H_2O		5	
$MgCl_2$	10x	5	1x
	Resulting structures	50	≲20 nM

200 μL of working stock), each staple strand will be at a concentration of 500 nM in the working stock solution. For structures with more than 200 staples, each strand will be slightly more dilute than 500 nM, which means the staples will be in less than tenfold excess in the folding reaction. While tenfold excess is that standard, we have found reducing down to fivefold excess of staple strands has no noticeable effect on the folding results.

6. Briefly vortex and centrifuge the working stock.
7. Folding reactions are typically done in 50 μL aliquots of 20 nM structure within eight-tube strips (*see* Table 1). This folding recipe can be scaled for a 100 μL reaction, which is the limit for many thermocycler machines. The reaction could also be scaled to larger volume [15] if desired, and scaled folding methods will be discussed in the notes section (*see* **Note 19**).
8. Briefly vortex and centrifuge.

3.2.2 DNA Origami Thermal Annealing Protocols (See ***Note 20****)*

1. Most structures fold under a 2.5-day annealing ramp (*see* Fig. 11a), which is as follows (*see* **Note 21**):
   Melting temperature: 65 °C, for 30 min.
   Folding temperatures:
   (a) 65–62 °C, stepping down at 1°/1 h.
   (b) 61–59 °C, stepping down at 1°/2 h.
   (c) 58–46 °C, stepping down at 1°/3 h.
   (d) 45–40 °C, stepping down at 1°/1 h.
   (e) 39–25 °C, stepping down at 1°/30 min.
   (f) 24–4 °C, stepping down at 1°/min.
   Cooling temperature 4 °C, for at least 30 min
2. Key parameters to vary in order to determine ideal folding conditions for specific DNA origami structures are primarily salt, time, and temperature. When first characterizing DNA

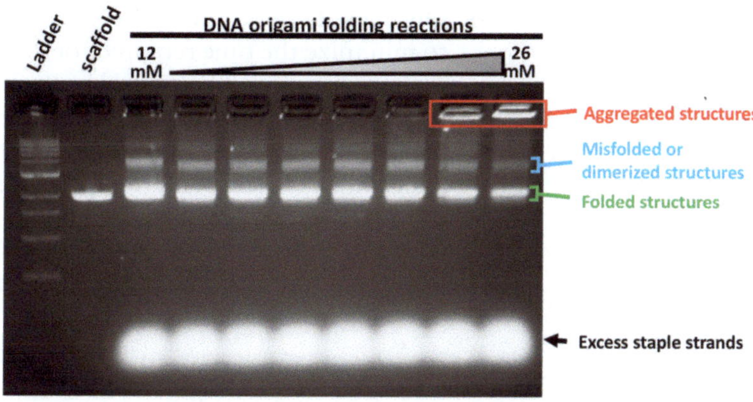

Fig. 10 An example of a salt screen gel electrophoresis for a DNA origami structure. The first lane is a ladder, followed by a scaffold. The following lanes are DNA origami folding reactions carried out at 12 mM to 26 mM $MgCl_2$. The fact that the structures folded at 12 mM run slightly slower than the other conditions suggests they are likely not as well folded. The higher salt conditions result in some amount of structures that form larger aggregates that get stuck in the well. These qualitative results are typical of DNA origami folding (poor folding at lower salt and aggregation at higher salt)

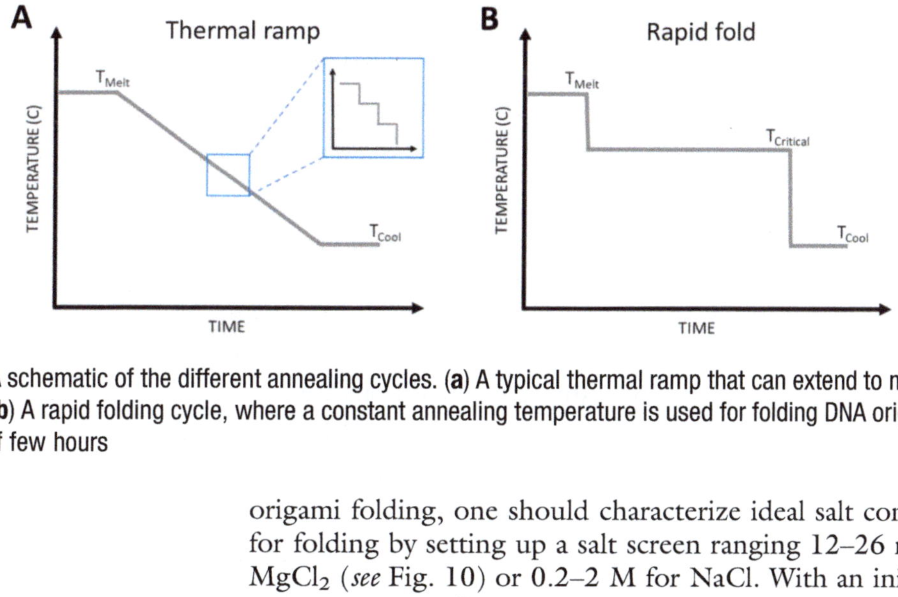

Fig. 11 A schematic of the different annealing cycles. (**a**) A typical thermal ramp that can extend to more than 2 days. (**b**) A rapid folding cycle, where a constant annealing temperature is used for folding DNA origami in a matter of few hours

origami folding, one should characterize ideal salt conditions for folding by setting up a salt screen ranging 12–26 mM for $MgCl_2$ (*see* Fig. 10) or 0.2–2 M for NaCl. With an initial salt screen, we typically use an extended folding reaction like the 2.5-day annealing process to ensure some degree of folding.

3. Quantify the salt screen via gel electrophoresis (*see* Subheading 3.3). There should be a shift in gel mobility, generally transition from a spread to one or two distinct bands. A secondary change could also be discrepancies in band intensities if there is not an obvious gel shift. A stronger band intensity indicates a greater yield.

4. Once a salt condition has been established, it is often desirable to minimize the time required for folding reactions, especially if the fabrication will be carried out repeatedly. This can be done using a constant temperature annealing approach [41], which consists of a high-temperature melting phase followed by annealing at a constant temperature for ~1–4 h and finally cooling to 4 °C (*see* **Note 22**).

3.2.3 Rapid Folding

The suitable folding temperature can be found as follows (*see* **Notes 23** and **24**):

1. 60–40 °C gradient for 4 h:
   - (a) Annealing conditions are:
     - (i) 65 °C melt time for 15 min.
     - (ii) Gradient annealing temperatures for 4 h (i.e., each tube sits at a different constant temperature).
     - (iii) 4 °C cooling time for 15 min.
     - (iv) Store in a refrigerator at 4 °C.
   - (b) Quantify via gel electrophoresis and TEM.
   - (c) If a mobility shift into a well-folded population is not apparent:
     - (i) Incrementally increase gradient folding time.
     - (ii) It may be necessary to use longer folds with annealing over a range of temperatures.
2. If it is desired to more precisely determine the appropriate range of annealing temperature, another constant temperature annealing reaction may be carried out with a Δ4 °C gradient:
   - (a) Based on where the gel shifts, determine a Δ4 °C gradient range to more precisely identify the range of annealing temperatures or optimal annealing temperature.
   - (b) Annealing conditions are:
     - (i) 65 °C melt time for 15 min.
     - (ii) Δ4 °C gradient folding temperature (e.g., 51 °C–55 °C) for 4 h.
     - (iii) 4 °C cooling time for 15 min.
   - (c) Quantify via gel electrophoresis and TEM.
   - (d) The temperature corresponding to the first identified band shift is the highest annealing temperature, which we refer to as a critical folding temperature, to fold DNA origami structures. We typically round down the critical folding temperature to the nearest whole number.

3. Critical folding temperature:
   (a) Annealing conditions are:
      (i) 65 °C melt time for 15 min.
      (ii) Critical folding temperature for 4 h or most viable range.
      (iii) 4 °C cooling time for 15 min.
   (b) Quantify via gel electrophoresis and TEM.
4. If it is desired to further reduce the folding time, a series of decreasing annealing times can be tested using the critical folding temperature approach. It is not unusual that 1 h of folding or even less is sufficient to fold many DNA origami structures.

3.3 Purification

There are several purification methods that exist for DNA origami. Most of these methods are size based, where folded DNA origami structures can be separated from excess staples by using gel electrophoresis, polyethylene glycol (PEG), or spin columns. Here we are going to show how each of these methods can be utilized for purifying DNA origami structures.

3.3.1 Gel Electrophoresis

Below is a protocol (based on protocol presented in [14]) for making a small 2% agarose gel that has a volume of approximately 50 mL (*see* **Note 28**):

1. Weigh 1 g of agarose in a glass beaker.
2. Fill to 49.6 g with 0.5× TBE buffer.
3. Stir the beaker briefly and microwave for 1 min.
4. Fill the beaker back up to 49.6 g with water.
5. Add 0.4 mL of 1.375 M $MgCl_2$.
6. Add 2 μL of EtBr.
7. Cast the gel mixture into the gel rig.
8. Add the comb and leave it for 20–30 min to solidify (*see* **Note 29**).
9. While the gel is solidifying, start preparing your samples to be loaded into gel wells as follows (*see* **Note 30**):
   (a) For DNA origami structures, mix 15 μL of sample with 3 μL of a loading dye.
   (b) For scaffold, mix 15 μL of scaffold at 20 nM with 3 μL of a loading dye.
10. Remove the comb and flip the gel so that the wells point toward the negative electrode.
11. Cover the gel with the running buffer.

12. Load the samples into wells starting with the ladder, followed by the scaffold and the rest of the samples (*see* **Note 31**).
13. Connect the rig to a power supply and run it at 70–90 V for 1.5–2 h (*see* **Note 32**).
14. To protect the gel from overheating, surround the rig with water/ice.

3.3.2 Polyethylene Glycol (PEG) Purification

The protocol for PEG purification is as follows (*see* **Note 33**):

1. In 1.5 mL tube, add the volume of your structure. Then add the same volume of 15% PEG8000 (*see* **Note 34**), i.e., if you have 150 μL of structure, add 150 μL of PEG (*see* **Note 35**). Then pipet up and down to mix.
2. Spin in the tabletop centrifuge at 16,000 g for 30 min (*see* **Notes 36** and **37**).
3. Remove supernatant and resuspend in 1x FOB + 20 mM $MgCl_2$, or alternative desired buffer (*see* **Notes 38** and **39**).

3.3.3 Molecular Weight Cutoff (MWCO) Filters

The general procedure is as follows (based on protocol presented in [16] where more details can be found) (*see* **Note 40**):

1. Prewet the filter by adding prewet buffer (*see* **Note 41**).
2. Spin the filter for 8 min at 5000 g (*see* **Note 42**).
3. Add at least 50 μL of DNA origami structures to the tube and fill with the buffer, mentioned above, to 500 μL.
4. Spin the filter for 8 min at 5000 g.
5. Wash the filter by adding 500 μL of buffer and spinning for 8 min at 5000 g.
6. The washing step can be repeated three times.
7. Recover DNA origami structures by inverting the inner tube and spinning at 10000 g for 2 min.

3.4 Imaging and Verification Methods

The most common imaging and direct visualization methods include TEM, AFM, and gel electrophoresis. Gel electrophoresis is the most cost-effective method to determine folding of a structure, often appearing as a single, tight band.

3.4.1 Grid Preparation for Transmission Electron Microscopy (TEM) (See ***Notes 43*** *and* ***44****)*

The protocols below are based on protocols provided in [14], and sample TEM images of DNA origami mechanisms, first presented in [28], are depicted in Fig. 12:

1. First, prepare the uranyl formate solution (UFo): For 2% UFo (*see* **Note 45**), add 5 mL of boiling ddH_20 to 100 mg of depleted UFo (SPI) in a 15 mL falcon tube.
2. Protect from light by wrapping the tube in foil. Vortex for 10 min.

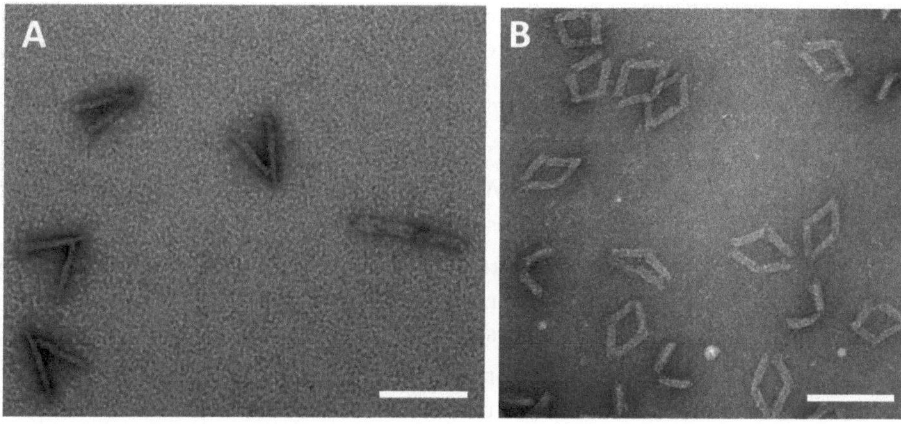

Fig. 12 Sample TEM images of A) DNA origami hinges and B) DNA origami Bennett linkages (modified from [35]). Scale bars are 100 nm

3. Pour solution into a 10 mL syringe with an attached syringe filter (0.22 μm pore size). Filter into a new 15 mL falcon tube. Replace the filter if it pops or breaks.
4. Make 200 μL aliquots. Spin down and store at –20 °C. Each aliquot stains six grids.
5. For grid preparation, first thaw an aliquot of UFo (*see* **Note 46**).
6. Use fine-tip tweezers to place one to six copper grids, dark shiny side up, on a parafilm-covered glass slide. Grab only the edge of the grids with the tweezers. It is helpful to use negative-action-type tweezers that require a force to open them, so the tweezers can hold the grid without need to apply force.
7. Place in a plasma cleaner, set to 25 mAmps for 40 s to make the grid surface hydrophilic.
8. Stick a section of parafilm about 6 inches long onto the workbench with ddH_2O.
9. Add 1 μL of 5 M NaOH to the UFo tube to bring to 25 mM NaOH in UFo. Vortex for 2 min. Since the NaOH can cause crystallization, it is helpful to add the NaOH to the side of the tube, so it is only mixed upon vortexing.
10. Centrifuge at max speed for 3–4 min.
11. Remove grids from plasma cleaner. Use tweezers to hold grids in place above the parafilm, dark shiny side up. Pipette 4 μL of sample to the grid, and let adsorb for 4 min.
12. In the last ~30 s, pipette a 10 μL and 20 μL UFo droplet onto the parafilm. The less time the UFo spends on the parafilm, the less likely it will crystallize (*see* **Notes 47** and **48**).

13. Turn tweezers such that the grid is matte side up. Wick away excess sample on grid with the filter paper (touch filter paper to sample solution orthogonal to the grid surface). Remove filter paper once diffusion stops.
14. Pick up a 10 μL UFo droplet with the shiny side (upside down) of the grid. Wick off right away. Remove filter paper when diffusion stops.
15. Pick up a 20 μL UFo droplet with shiny side (upside down) of grid. Stain for 40 s and wick off. Remove filter paper when diffusion stops.
16. Place in storage or let air dry covered to avoid any contamination.
17. Let dry for at least 15 min before imaging.

4 Notes

1. Design of DNA origami nanostructures is most often carried out in a program called caDNAno [42]. However, other programs have been developed which specialize in different design styles. In particular, recent efforts have developed automated design tools for specific classes of static structures [43–47]. We will focus on design principles as they apply to caDNAno, specifically caDNAno 2, since it is the most widely used design tool, especially within the DNA origami community.
2. You can use the Autostaple feature for basic designs to get started. This populates the scaffold routing with staples that have all possible crossovers (except near internal scaffold crossovers) but without any staple breaks.
3. It is possible to use the autobreak feature but recommended to do this manually.
4. DNA-binding locations are often needed along the surface of a DNA origami for functionalization with other biomolecules, labeling with fluorophores, connections to other DNA origami components or structures, etc. The most common means of creating binding locations is to extend staples with segments that are not complementary to any scaffold location such that it protrudes out of the structure. These single-stranded portions of staples are commonly referred to as ssDNA overhangs, or just overhangs.
5. Although overhangs may not be necessary everywhere on the structure surface, it is recommended to further adjust staple routings to accommodate as many foreseeable overhang positions as possible. This can help avoid troublesome changes to designs and pipetting procedures for fabrication in the future.

6. Deliberate twisted or curved structures may be desired, and design principles for various degrees of twist or curvature can be found in more detail elsewhere [48].
7. Typically, a randomly generated sequence is fine; however, as an added precaution, it is recommended to check the resulting staple for any undesired interactions with itself, such as hairpin formation, or other staples.
8. We recently demonstrated an approach [35] for overhang sequence design that minimizes potential complementarity with the scaffold or other staples, which could lead to problems during folding.
9. Dynamic, or reconfigurable, DNA origami devices are of great interest for applications such as sensors, measurement tools, or triggered delivery. Our laboratory and others have demonstrated approaches to design, fabricate, and actuate dynamic devices with programmed 1D, 2D, and 3D motion, tunable mechanical properties, and conformational dynamics. In these designs the flexibility that facilitates motion is typically achieved by strategically incorporating single-stranded DNA components (i.e., leaving part of the scaffold single-stranded in specific locations).
10. Actuation of DNA origami generally refers to any change in structure configuration which typically involves forming or breaking connections between multiple rigid components within a single DNA origami nanostructure. Typically, these can be triggered by some change in the local environment or an externally applied field.
11. A major advantage of this approach is the sequence specificity of DNA base pairing. This allows for selective actuation of particular motions within the structure, for example, to actuate a series of joints in sequence to fold DNA origami waterbomb base into different final configurations [49].
12. Typically, dynamic DNA origami can be characterized in terms of the distribution of configurations that are observed through negative stain transmission electron microscopy (TEM). However, for measuring real-time changes in solution, other methods must be employed to read out the configuration of a structure. FRET pairs can be readily used to determine when one rigid component of a nanostructure is close to another.
13. Our laboratory and collaborators have effectively employed Cy3/Cy5 FRET pairs to measure DNA origami conformations (*see* Fig. 8a) [33, 50].
14. We have previously employed Alexa488 and Blackhole Quencher 1 as an effective pair to monitor DNA origami conformations (*see* Fig. 8b) [28, 35]. However, FRET has the

benefit of an internal control (i.e., correlation between donor and acceptor fluorophores) to ensure changes in fluorescence are due to conformation changes and not photobleaching or some other photophysical effect. This is especially useful for single-molecule measurements, but appropriate controls can be done in any case to ensure fluorescence changes are due to conformation changes in either case.

15. A critical advantage of DNA origami is the ability to design many more locations which can have fluorescent molecules for readout. Although we show examples containing only a single pair of fluorophores or a single fluorophore and quencher pair, it is possible to include multiple sets of these for enhanced signal output [51].

16. A common practice in DNA origami design is to divide the design into modules. This is particularly important to facilitate the fabrication of different versions of the structure. Since each structure comprises ~150–200 strands, the initial pipetting is tedious. Hence, it is highly advantageous to divide the design into modules where it is easy to substitute different modules. For example, for the hinge shown in Fig. 6a, a typical approach would be to divide the design into staples that comprise the top arm and bottom arm. In addition, staples that contain overhangs for actuation are often considered a separate module, since it may often be necessary to have different versions of the device that either include or do not include an overhang. In versions that do not include the overhang, a replacement strand, which comprises the same staple sequence minus the overhang, must be included. In addition to separating strands that make up different components within the structure and overhangs (e.g., for actuation, attachment to a surface, or functionalization), another typical module might include staple strands near the edges of a structure, for example, to promote or avoid connection to another device. Because of this design modularity, the DNA origami folding protocols are divided into three main steps: prestocks, working stocks, and folding reactions [14, 16]. A single prestock is a pool of staples that would comprise a specific module of a DNA origami structure as described above. A working stock is a pool of staples that would comprise a full version of a DNA origami structure with its desired modifications. The working stock made here is sufficient to make 200 μL of working stock solution. Assuming one follows a typical folding reaction recipe (*see* Table 1), this 200 μL is sufficient for ten 50 μL folding reaction, although folding reactions are typically carried out in standard PCR tube strip with eight tubes. A folding reaction comprises of a working stock (i.e., staple strands), scaffold, and relevant salt buffers, all of which comprise the solution that will be subjected to the thermal annealing process.

17. As a rule of thumb, pipette 10 μL per oligo for wells containing more than ten oligos, 15 μL for three to nine oligos per well, and 20 μL for one to two oligos in a prestock. For more than ~30 oligos, it is helpful to use a multichannel pipette and a basin for ease.
18. Scaffold concentration can typically between 10–40 nM, with working stocks varying two- to tenfold excess of scaffold. $MgCl_2$ or NaCl are typically used in DNA origami folding reactions, varying from 12 to 26 mM and 0.2 to 2 M, respectively. Subheading 3.2 goes into further detail on determining ideal salt conditions for specific DNA origami structures.
19. With an established annealing ramp for the DNA origami, one can scale up the folding process from 50 μL aliquots within a thermocycler to tens of milliliters in a water bath. Specific details can be found in [15, 52].
20. Once the folding reaction is ready, it can now undergo an annealing ramp, typically carried out in a thermocycler, but simpler versions may be carried out in heated water baths [15]. An annealing ramp comprises of a melting temperature, folding temperature(s), and a cooling temperature. The melting temperature brings the folding reaction to at least 65 °C for 15–60 min to melt any DNA-binding interactions. Many protocols use higher temperatures for this melt phase. DNA origami structures self-assemble during the folding temperature step. The folding temperatures required can vary significantly across multiple designs from a single temperature for approximately 10 min to a range of temperatures over the course of days. Optimization is generally required. Although to our knowledge, there is no clear evidence the final cooling below room temperature is required, reactions are often terminated by cooling to 4 °C for convenience until the solution can be transferred to refrigerated storage.
21. For DNA origami that cannot fold in a 2.5-day ramp, the origami structure needs a higher melting temperature or begins folding at higher temperatures. The 2.5-day ramp does not accommodate long folding periods at temperatures above 58 °C. One can modify the rapid fold experiments to check the temperature range in which the structure folds.
22. The appropriate range of annealing temperatures may vary from structure to structure. However, 52 °C (or ~ 51–53 °C) seems to be a robust annealing temperature that works for many structures; however, the yield and speed (i.e., time required) may be optimized by testing a variety of annealing temperatures. This can be done conveniently in some thermocyclers by setting a spatial temperature gradient across the heat block to allow testing of several constant annealing

temperatures at once. Note that the temperature discrepancy between each row of the thermal block is typically not homogenous.

23. Although typical folding reactions can take more than 2 days (*see* Fig. 11a), most DNA structures can rapidly fold, in few hours, when annealed at a constant temperature. Hence, doing long thermal ramps is not a necessity, once the appropriate folding temperature is found (*see* Fig. 11b). However, identifying appropriate annealing temperatures may require screening.
24. For DNA origami that does not fold during the rapid fold experiments, it is possible that the structure needs a longer folding time. One can incrementally increase the folding time to determine the length of time required. Alternatively, one can create a mini-ramp, a shortened temperature range. One can iterate through different ranges while maintaining a constant step-down rate (i.e., 1 °C/3 h).
25. Low salt: Physiological salt condition for experiments is on the order of ~1 mM $MgCl_2$ and ~ 200 mM NaCl. Not all structures will survive low salt conditions. Most structures will survive down to at least ~5 mM $MgCl_2$. Structure design is key to maintain low salt stability. This includes introducing porous designs with a relatively high surface area-to-volume ratio and using square lattice cross-sections.
26. Cell culture media: Typical cell culture conditions include 10% fetal bovine serum (FBS), which contains growth factors and nucleases that could degrade the DNA origami nanostructures. For cell culture experiments containing DNA origami, aim for a low surface area-to-volume ratio. Also, the media may be supplemented with low levels of $MgCl_2$ (~1 mM) to improve stability. Higher levels may disrupt cells. Using heat-inactivated serum will also help improve DNA origami structure stability.
27. Endotoxins: For biological studies, it is desirable for DNA origami scaffold to be endotoxin-free. This is essential when considering any experiments using DNA origami in animal models. This can be done via endotoxin-free removal kits (Qiagen), but this will also reduce scaffold production yields.
28. Agarose gel electrophoresis is often the first method of purification used to quantify folding of DNA origami structures. Well-folded DNA structures are typically more compact than incompletely folded structures; thus they usually run faster than their corresponding scaffold in an agarose gel. Agarose gel electrophoresis very effectively separates structures, but recovery of structures from a gel will often leave agarose residue in the solution and may perturb the structure due to the presence of dyes (e.g., ethidium bromide). In some cases running through the gel may disrupt molecules attached to the

surface of the structure (i.e., proteins bound to DNA origami). Unlike the standard DNA gel electrophoresis protocols, DNA origami gel electrophoresis requires salt in its gel and running buffer, typically 10–11 mM $MgCl_2$, but any suitable salt condition for the DNA origami structure is sufficient. Include an intercalating dye for imaging if the structures are not fluorescently labeled.

29. To prevent bubbles from interfering with the running samples, you can use a pipette tip to move any bubbles to the bottom of the gel before it solidifies.

30. Incubating samples at 37 °C for 5–10 min before loading into gels can help prevent aggregation.

31. Controls for DNA origami gels should include a 1 kb ladder and scaffold, the latter being at the same concentration as the folding reaction. Run the gel 70 V for 90–120 min for a small gel (~50–70 mL) and 180–240 min for a large gel (~130 mL). The gel rig should be immersed in ice or an ice bath. After imaging, bands can be cut out and filtered out via spin columns to reclaim the purified sample. Sample concentrations are typically 1–5 nM.

32. The gel can be run at higher voltage for a shorter amount of time, but one must determine the structural integrity of the DNA origami when proceeding.

33. Based on [53], PEG centrifugation pellets out DNA origami structures while excess staples remain in the supernatant. PEG centrifugation removes ~90% of excess staples but can leave PEG residue and aggregated or misfolded structures in solution. Specifically, 15% w/v PEG 8000 solution is used. Salt is optionally included but encouraged. It varies based on preference for experiments, but buffer conditions include 5 mM Tris, 1 mM EDTA, and 500 mM NaCl.

34. The PEG solution, while usable for at least 2–3 months, can be of suspect for low yields. Remake it if low yields persist.

35. Use at least 20 nM of 100 μL of unpurified structures to a 1.5 mL tube.

36. When spinning in a tabletop centrifuge, the pellet may not be visible.

37. Face the hinge part of tube outward in the centrifuge to identify the location of the pellet.

38. If the concentration after purification is extremely low or nonexistent, the pellet may have been lost. Spinning down the supernatant can sometimes reclaim the lost pellet.

39. After resuspending the pellet, pulse vortex the sample.

40. Amicon and spin column filters can purify out excess staples leaving the folded structures. This approach, however, does not remove misfolded or aggregated structures. For purifying DNA origami structures, there is a departure from the typical manufacturer-recommended spin column procedures.
41. In some cases, salt conditions need to be dropped to ~5 mM $MgCl_2$ in order to minimize interference with the filter.
42. When using spin columns, do not exceed 13,000 g.
43. Transmission electron microscopy is often used to obtain high-resolution images of DNA origami structures. In this technique, a heavy metal (e.g., uranyl acetate or uranyl formate) is used to negatively stain DNA origami structures, resulting in high-contrast images. It is especially useful for imaging of structures for bulk measurements.
44. Typical tools to analyze TEM micrographs include ImageJ, EMAN2, and Relion. ImageJ has been reliable as a basic image analysis toolbox. EMAN2 and Relion are open-source program originally created for cryo-EM images. Many aspects of these two programs are great for TEM bulk analysis and image averaging. EMAN2 requires a LINUX/UNIX or MAC OS. EMAN2 can run on Windows OS but with limited functionality.
45. Uranyl formate tends to produce sharper images with higher contrast compared to that stained with uranyl acetate.
46. Crystallization on UFo will prevent good negative staining and imaging of TEM grids. Items that increase crystallization or positive staining include high concentration (>5 nM DNA origami), prolonged staining times (>40 s), high salt, and/or poor wicking.
47. A common approach for troubleshooting poor staining is to reduce staining times.
48. Any UFo remaining after wicking off stain from TEM grids will continue to stain and thus crystallize.

References

1. Seeman NC (1982) Nucleic acid junctions and lattices. J Theor Biol 99(2):237–247
2. Seeman NC, Sleiman HF (2018) DNA nanotechnology. Nat Rev Mater 3(1):17068
3. Watson JD, Crick FH (1953) Molecular structure of nucleic acids; a structure for deoxyribose nucleic acid. Nature 171(4356): 737–738. https://doi.org/10.1038/171737a0
4. Rothemund PWK (2006) Folding DNA to create nanoscale shapes and patterns. Nature 440(7082):297–302. https://doi.org/10.1038/nature04586
5. Douglas SM, Dietz H, Liedl T, Hogberg B, Graf F, Shih WM (2009) Self-assembly of DNA into nanoscale three-dimensional shapes. Nature 459(7245):414–418. https://doi.org/10.1038/nature08016
6. Douglas SM, Marblestone AH, Teerapittayanon S, Vazquez A, Church GM, Shih WM (2009) Rapid prototyping of 3D DNA-origami shapes with caDNAno. Nucleic

Acids Res 37(15):5001–5006. https://doi.org/10.1093/nar/gkp436

7. Rothemund PW (2006) Folding DNA to create nanoscale shapes and patterns. Nature 440(7082):297–302. https://doi.org/10.1038/nature04586
8. Holliday R (1964) A mechanism for gene conversion in fungi. Genet Res 5(2):282–304. https://doi.org/10.1017/S0016672300001233
9. Jiang Q, Liu S, Liu J, Wang ZG, Ding B (2019) Rationally designed DNA-origami nanomaterials for drug delivery in vivo. Adv Mater 31(45):e1804785. https://doi.org/10.1002/adma.201804785
10. Castro CE, Dietz H, Högberg B (2017) DNA origami devices for molecular-scale precision measurements. MRS Bull 42(12):925–929. https://doi.org/10.1557/mrs.2017.273
11. Liu Y, Kumar S, Taylor RE (2018) Mix-and-match nanobiosensor design: logical and spatial programming of biosensors using self-assembled DNA nanostructures. Wiley Interdiscip Rev Nanomed Nanobiotechnol 10(6):e1518. https://doi.org/10.1002/wnan.1518
12. Liu N, Liedl T (2018) DNA-assembled advanced plasmonic architectures. Chem Rev 118(6):3032–3053. https://doi.org/10.1021/acs.chemrev.7b00225
13. Mishra S, Feng Y, Endo M, Sugiyama H (2019) Advances in DNA origami-cell interfaces. Chembiochem 21:33. https://doi.org/10.1002/cbic.201900481
14. Castro CE, Kilchherr F, Kim DN, Shiao EL, Wauer T, Wortmann P, Bathe M, Dietz H (2011) A primer to scaffolded DNA origami. Nat Methods 8(3):221–229. https://doi.org/10.1038/nmeth.1570
15. Halley PD, Patton RA, Chowdhury A, Byrd JC, Castro CE (2019) Low-cost, simple, and scalable self-assembly of DNA origami nanostructures. Nano Res 12(5):1207–1215
16. Wagenbauer KF, Engelhardt FA, Stahl E, Hechtl VK, Stömmer P, Seebacher F, Meregalli L, Ketterer P, Gerling T, Dietz H (2017) How we make DNA origami. Chembiochem 18(19):1873–1885
17. Engelhardt FAS, Praetorius F, Wachauf CH, Bruggenthies G, Kohler F, Kick B, Kadletz KL, Pham PN, Behler KL, Gerling T, Dietz H (2019) Custom-size, functional, and durable DNA origami with design-specific scaffolds. ACS Nano 13(5):5015–5027. https://doi.org/10.1021/acsnano.9b01025
18. Marchi AN, Saaem I, Vogen BN, Brown S, LaBean TH (2014) Toward larger DNA origami. Nano Lett 14(10):5740–5747. https://doi.org/10.1021/nl502626s
19. Nafisi PM, Aksel T, Douglas SM (2018) Construction of a novel phagemid to produce custom DNA origami scaffolds. Synth Biol (Oxf) 3(1). https://doi.org/10.1093/synbio/ysy015
20. Majikes JM, Nash JA, LaBean TH (2016) Competitive annealing of multiple DNA origami: formation of chimeric origami. New J Phys 18(11):115001
21. Kick B, Praetorius F, Dietz H, Weuster-Botz D (2015) Efficient production of single-stranded phage DNA as scaffolds for DNA origami. Nano Lett 15(7):4672–4676. https://doi.org/10.1021/acs.nanolett.5b01461
22. Minev D, Guerra R, Kishi JY, Smith C, Krieg E, Said K, Hornick A, Sasaki HM, Filsinger G, Beliveau BJ, Yin P, Church GM, Shih WM (2019) Rapid in vitro production of single-stranded DNA. Nucleic Acids Res 47(22):11956–11962. https://doi.org/10.1093/nar/gkz998
23. Shepherd TR, Du RR, Huang H, Wamhoff EC, Bathe M (2019) Bioproduction of pure, kilobase-scale single-stranded DNA. Sci Rep 9(1):6121. https://doi.org/10.1038/s41598-019-42665-1
24. Ke Y, Douglas SM, Liu M, Sharma J, Cheng A, Leung A, Liu Y, Shih WM, Yan H (2009) Multilayer DNA origami packed on a square lattice. J Am Chem Soc 131(43):15903–15908. https://doi.org/10.1021/ja906381y
25. Ke Y, Voigt NV, Gothelf KV, Shih WM (2012) Multilayer DNA origami packed on hexagonal and hybrid lattices. J Am Chem Soc 134(3):1770–1774. https://doi.org/10.1021/ja209719k
26. Ke Y, Bellot G, Voigt NV, Fradkov E, Shih WM (2012) Two design strategies for enhancement of multilayer-DNA-origami folding: underwinding for specific intercalator rescue and staple-break positioning. Chem Sci 3(8):2587–2597. https://doi.org/10.1039/C2SC20446K
27. Kim DN, Kilchherr F, Dietz H, Bathe M (2012) Quantitative prediction of 3D solution shape and flexibility of nucleic acid nanostructures. Nucleic Acids Res 40(7):2862–2868. https://doi.org/10.1093/nar/gkr1173
28. Marras AE, Zhou L, Su HJ, Castro CE (2015) Programmable motion of DNA origami mechanisms. Proc Natl Acad Sci U S A 112(3):713–718. https://doi.org/10.1073/pnas.1408869112

29. Yurke B, Turberfield AJ, Mills AP, Simmel FC, Neumann JL (2000) A DNA-fuelled molecular machine made of DNA. Nature 406(6796): 605–608. https://doi.org/10.1038/35020524
30. Kuzuya A, Watanabe R, Yamanaka Y, Tamaki T, Kaino M, Ohya Y (2014) Nanomechanical DNA origami pH sensors. Sensors (Basel) 14(10):19329–19335. https://doi.org/10.3390/s141019329
31. Urban MJ, Dutta PK, Wang P, Duan X, Shen X, Ding B, Ke Y, Liu N (2016) Plasmonic toroidal metamolecules assembled by DNA origami. J Am Chem Soc 138(17):5495–5498
32. Gerling T, Wagenbauer KF, Neuner AM, Dietz H (2015) Dynamic DNA devices and assemblies formed by shape-complementary, non-base pairing 3D components. Science 347(6229):1446–1452. https://doi.org/10.1126/science.aaa5372
33. Marras AE, Shi Z, Lindell MG 3rd, Patton RA, Huang CM, Zhou L, Su HJ, Arya G, Castro CE (2018) Cation-activated avidity for rapid reconfiguration of DNA Nanodevices. ACS Nano 12(9):9484–9494. https://doi.org/10.1021/acsnano.8b04817
34. Bruetzel LK, Walker PU, Gerling T, Dietz H, Lipfert J (2018) Time-resolved small-angle X-ray scattering reveals millisecond transitions of a DNA origami switch. Nano Lett 18(4): 2672–2676. https://doi.org/10.1021/acs.nanolett.8b00592
35. Johnson JA, Dehankar A, Winter JO, Castro CE (2019) Reciprocal control of hierarchical DNA origami-nanoparticle assemblies. Nano Lett 19:8469. https://doi.org/10.1021/acs.nanolett.9b02786
36. Liu M, Jiang S, Loza O, Fahmi NE, Sulc P, Stephanopoulos N (2018) Rapid Photoactuation of a DNA nanostructure using an internal photocaged trigger strand. Angew Chem 57(30):9341–9345. https://doi.org/10.1002/anie.201804264
37. Suzuki Y, Endo M, Yang Y, Sugiyama H (2014) Dynamic assembly/disassembly processes of photoresponsive DNA origami nanostructures directly visualized on a lipid membrane surface. J Am Chem Soc 136(5): 1714–1717. https://doi.org/10.1021/ja4109819
38. Modi S, Swetha M, Goswami D, Gupta GD, Mayor S, Krishnan Y (2009) A DNA nanomachine that maps spatial and temporal pH changes inside living cells. Nat Nanotechnol 4(5):325
39. Mao H, Luo G, Zhan Y, Zhang J, Yao S, Yu Y (2018) The mechanism and regularity of quenching the effect of bases on fluorophores: the base-quenched probe method. Analyst 143(14):3292–3301. https://doi.org/10.1039/c8an00116b
40. Noble JE, Wang L, Cole KD, Gaigalas AK (2005) The effect of overhanging nucleotides on fluorescence properties of hybridising oligonucleotides labelled with Alexa-488 and FAM fluorophores. Biophys Chem 113(3): 255–263. https://doi.org/10.1016/j.bpc.2004.09.012
41. Sobczak J-PJ, Martin TG, Gerling T, Dietz H (2012) Rapid folding of DNA into nanoscale shapes at constant temperature. Science 338(6113):1458–1461
42. Douglas SM, Marblestone AH, Teerapittayanon S, Vazquez A, Church GM, Shih WM (2009) Rapid prototyping of 3D DNA-origami shapes with caDNAno. Nucleic Acids Res 37(15):5001–5006
43. Jun H, Shepherd TR, Zhang K, Bricker WP, Li S, Chiu W, Bathe M (2019) Automated sequence design of 3D polyhedral wireframe DNA origami with honeycomb edges. ACS Nano. https://doi.org/10.1021/acsnano.8b08671
44. Jun H, Zhang F, Shepherd T, Ratanalert S, Qi X, Yan H, Bathe M (2019) Autonomously designed free-form 2D DNA origami. Sci Adv 5(1):eaav0655. https://doi.org/10.1126/sciadv.aav0655
45. Linko V, Kostiainen MA (2016) Automated design of DNA origami. Nat Biotechnol 34(8):826–827. https://doi.org/10.1038/nbt.3647
46. Veneziano R, Ratanalert S, Zhang K, Zhang F, Yan H, Chiu W, Bathe M (2016) Designer nanoscale DNA assemblies programmed from the top down. Science 352(6293):1534. https://doi.org/10.1126/science.aaf4388
47. Benson E, Mohammed A, Gardell J, Masich S, Czeizler E, Orponen P, Hogberg B (2015) DNA rendering of polyhedral meshes at the nanoscale. Nature 523(7561):441–444. https://doi.org/10.1038/nature14586
48. Dietz H, Douglas SM, Shih WM (2009) Folding DNA into twisted and curved nanoscale shapes. Science 325(5941):725–730. https://doi.org/10.1126/science.1174251
49. Zhou L, Marras AE, Huang CM, Castro CE, Su HJ (2018) Paper origami-inspired design and actuation of DNA nanomachines with complex motions. Small 14(47):e1802580. https://doi.org/10.1002/smll.201802580
50. Hudoba MW, Luo Y, Zacharias A, Poirier MG, Castro CE (2017) Dynamic DNA origami device for measuring compressive depletion

forces. ACS Nano 11(7):6566–6573. https://doi.org/10.1021/acsnano.6b07097

51. Selnihhin D, Sparvath SM, Preus S, Birkedal V, Andersen ES (2018) Multifluorophore DNA origami beacon as a biosensing platform. ACS Nano 12(6):5699–5708
52. Praetorius F, Kick B, Behler KL, Honemann MN, Weuster-Botz D, Dietz H (2017) Biotechnological mass production of DNA origami. Nature 552(7683):84–87. https://doi.org/10.1038/nature24650
53. Stahl E, Martin TG, Praetorius F, Dietz H (2014) Facile and scalable preparation of pure and dense DNA origami solutions. Angew Chem Int Ed Engl 53(47):12735–12740. https://doi.org/10.1002/anie.201405991

Chapter 3

Computer-Aided Design and Production of RNA Origami as Protein Scaffolds and Biosensors

Néstor Sampedro Vallina, Cody Geary, Mette Jepsen, and Ebbe Sloth Andersen

Abstract

RNA nanotechnology is able to take advantage of the modularity of RNA to build a wide variety of structures and functional devices from a common set of structural modules. The RNA origami architecture harnesses the property of RNA to fold as it is being enzymatically synthesized by the RNA polymerase and enables the design of single-stranded devices that integrate multiple structural and functional RNA motifs. Here, we provide detailed procedures on how to design and characterize RNA origami structures. The process is illustrated by two examples: one that forms lattices and another example that acts as biosensors.

Key words RNA nanotechnology, RNA origami, RNA aptamers, AFM, FRET, Scaffolding, RNA lattices

1 Introduction

Nucleic acid nanotechnology enables the creation of custom shapes and patterns [1–7] by taking advantage of the sequence-dependent programmability of the molecules. Compared to DNA, RNA nanostructures can utilize a wider variety of naturally occurring tertiary motifs that can function as structural modules [8]. Furthermore, the RNA nanostructures can be enzymatically synthesized and expressed inside cells [9], and in addition have a low cost of production [7, 10]. Rationally designed RNA nanostructures have the potential to be applied as, e.g., scaffolds for enzymes in synthetic biology or diagnostic biosensors in medicine [11].

The RNA folding starts just as soon as transcription from the DNA template begins, and the structure gradually compacts into helices as the strand grows longer. The RNA origami architecture benefits from this property and allows for the design of a wide range of shapes and sizes of RNA devices [7] that can be produced in vitro and in vivo [10]. RNA devices are inherently modular and can

Julián Valero (ed.), *DNA and RNA Origami: Methods and Protocols*, Methods in Molecular Biology, vol. 2639, https://doi.org/10.1007/978-1-0716-3028-0_3, © Springer Science+Business Media, LLC, part of Springer Nature 2023

contain multiple functionalities while maintaining their folding. The RNA origami architecture can be functionalized with different RNA motifs such as RNA aptamers for binding proteins [12], fluorogenic RNA aptamers for RNA tracking in vitro and in vivo [10], intermolecular-interacting motifs for lattice growth [13, 14], siRNAs, riboswitches, and ribozymes [15].

RNA origami uses structural RNA motifs that are present in nature as building blocks, and the rational combination of structural modules can produce single-stranded RNA molecules, that for example can self-assemble into multimeric lattices. This architecture, which is folded from a single strand [7, 16] (Fig. 1), is comprised of hairpins (RNA folded into double helices with terminal loops [17]), double crossovers (analogous to their counterparts in DNA nanotechnology [18]) that connect double helices, and kissing loop (KL) interactions that consist of two hairpins linked by base pairing [19].

Programmable RNA origami lattices have been created using KLs positioned at the end of helices, which are responsible for intermolecular interaction between individual RNA origami structures. For example, the use of 120° external KLs results in hexagonal lattices [7, 14]. Furthermore, Di Liu and colleagues have modified the KL interaction into a synthetic RNA motif termed the branched kissing loop (bKL). In order to do so, they substituted the two adenines on the 5′-side of one KL with a double helix stem. This new motif provides the possibility to obtain 3D designs or to introduce functional RNA elements such as protein-binding motifs at the positions of the KLs within the RNA origami tile (Fig. 2). The RNA origami principles have also been useful to create biosensors. Jepsen and colleagues [10] incorporated the fluorescent RNA aptamers Mango and Spinach [20, 21] on a 2-helix origami tile to produce a Förster resonance energy transfer (FRET)-based sensor capable of detecting specific RNA strands and the small molecule *S*-adenosylmethionine (SAM) (Fig. 3).

In this tutorial, we will guide the user on how to design a 2-helix antiparallel even (even number of helical half turns between crossovers) RNA origami tile, and how to design programmable hexagonal lattices based on 2-helix tile systems. We will step-by-step incorporate different RNA elements into the origami such as the bKL and the Mango and Spinach fluorogenic aptamers [20, 21] to build FRET-based systems that can sense single-stranded RNA inputs or the small molecule SAM [22]. We will guide the readers through the software design pipeline that comprises the text-based incorporation of RNA motifs into origami structural blueprints, and the design of RNA sequences matching the specified constraints. Finally, stepwise-protocols are provided for the experimental in vitro RNA production and characterization through atomic force microscopy (AFM) and FRET (Fig. 1).

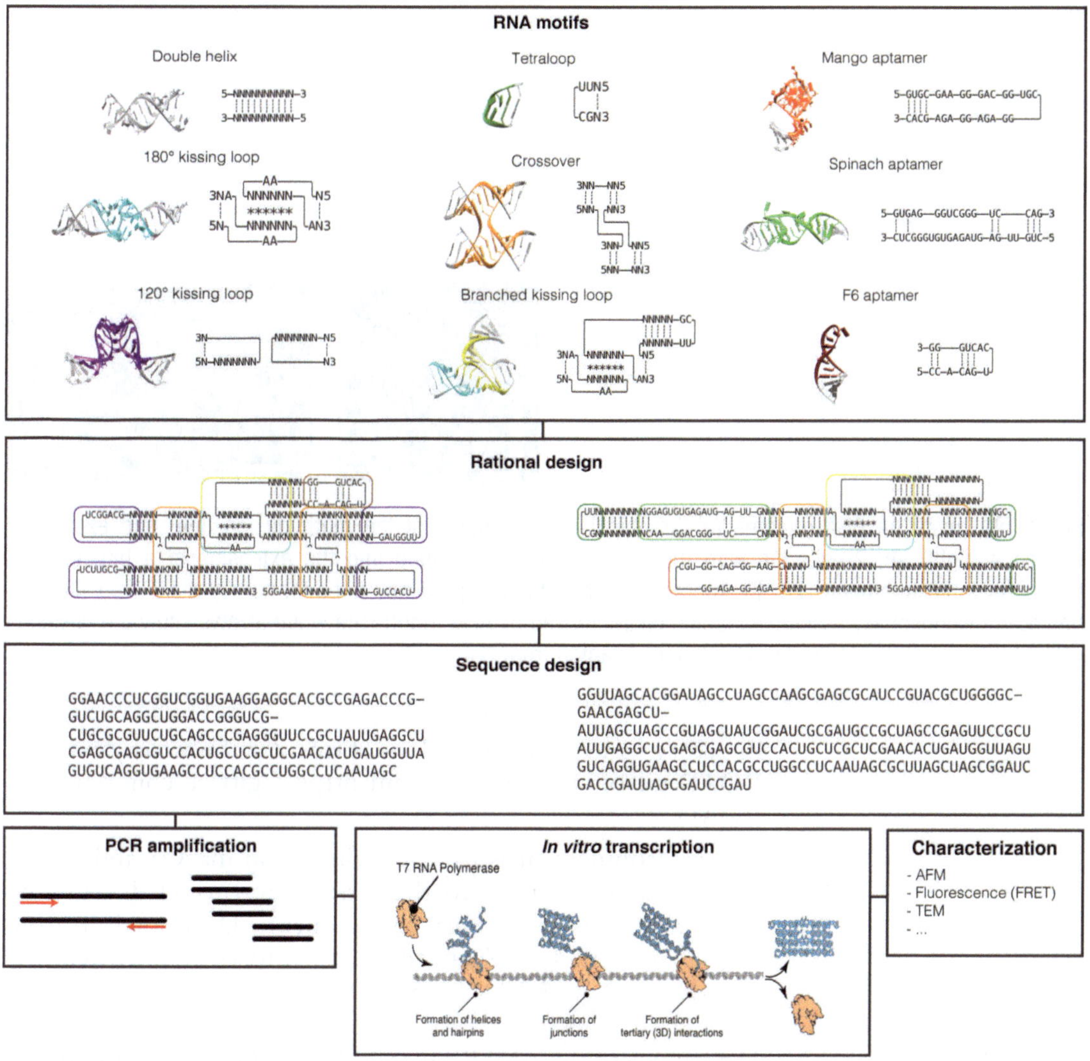

Fig. 1 RNA origami workflow. The chosen natural or synthetic RNA motifs are merged by rational design onto a 2D text-based blueprint following the RNA origami design principles. The blueprint is used as an input for the NUPACK sequence design algorithm. The DNA templates are amplified and then used for in vitro transcription, where after the different RNA devices can be characterized according to their structure or function

2 Materials

2.1 Software

- Text editor software that can copy/paste rows/columns of fixed-width text, such as TextEdit on MacOS or Sublime Text 3 for MacOS and Windows.
- The RNA ROAD package comprises a group of scripts used to design RNA Origami structures [23]. To obtain the software package, open a Terminal window, navigate to your working

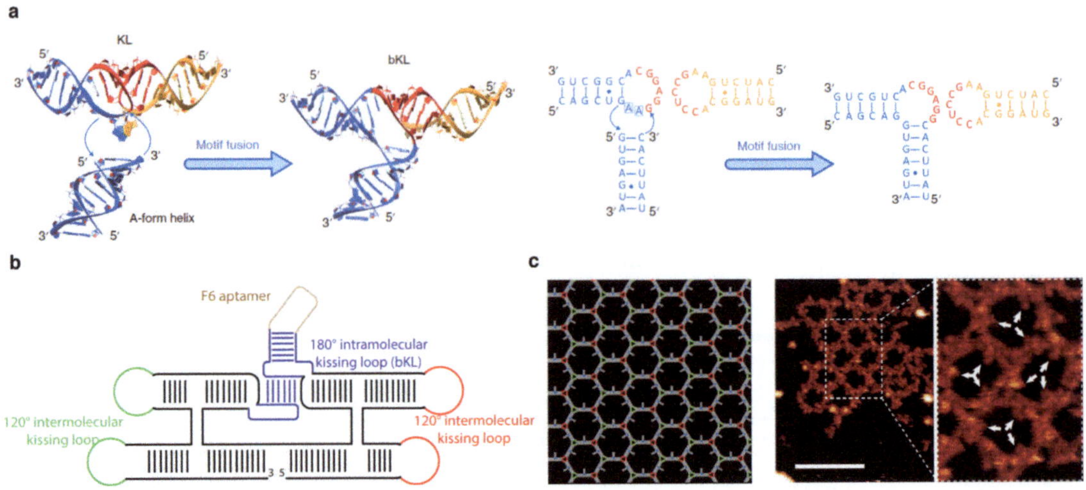

Fig. 2 bKL motif on RNA origami. (**a**) Graphical representation of the motif design (left) and exemplary sequence (right). (**b**) 2-helix tile with the F6 aptamer incorporated into the extension of a bKL forming hexagonal lattices through 120° KL interactions. (**c**) 2D lattice predicted to be formed by the 2-helix tile (left) and co-transcriptionally assembled lattice characterized by AFM (right). The white arrows show the positions of the F6 aptamer. Scale bar 50 nm. (Figure adapted from [14])

directory, and type: git clone https://github.com/esa-lab/ROAD.git or download it from https://github.com/esa-lab/ROAD.

- Perl environment, it is usually included on macOS and Linux, but the user might not have the latest version installed. Windows users need to install this environment. Download it from https://www.perl.org/get.html. This tutorial was written and tested with Perl v5.18.4 for MacOS.

2.2 Wet Lab Materials

2.2.1 DNA Template Amplification and Purification

- Deoxynucleotide (dNTP) mix: dATP, dCTP, dGTP, and dTTP (10 mM of each).
- Q5 DNA polymerase kit (New England Biolabs) (*see* **Note 1**).
- DNA primers and double-stranded DNA gBlock templates. These can be ordered from a DNA synthesis company like Integrated DNA Technologies or Twist Biosciences.
- PCR clean-up kit. In our laboratory, we use the one manufactured by Macherey-Nagel.
- Thermocycler.
- NanoDrop spectrophotometer.

2.2.2 *In Vitro* Transcription of RNA

- T7 RNA polymerase, can be ordered from a standard manufacturer or produced in-house.

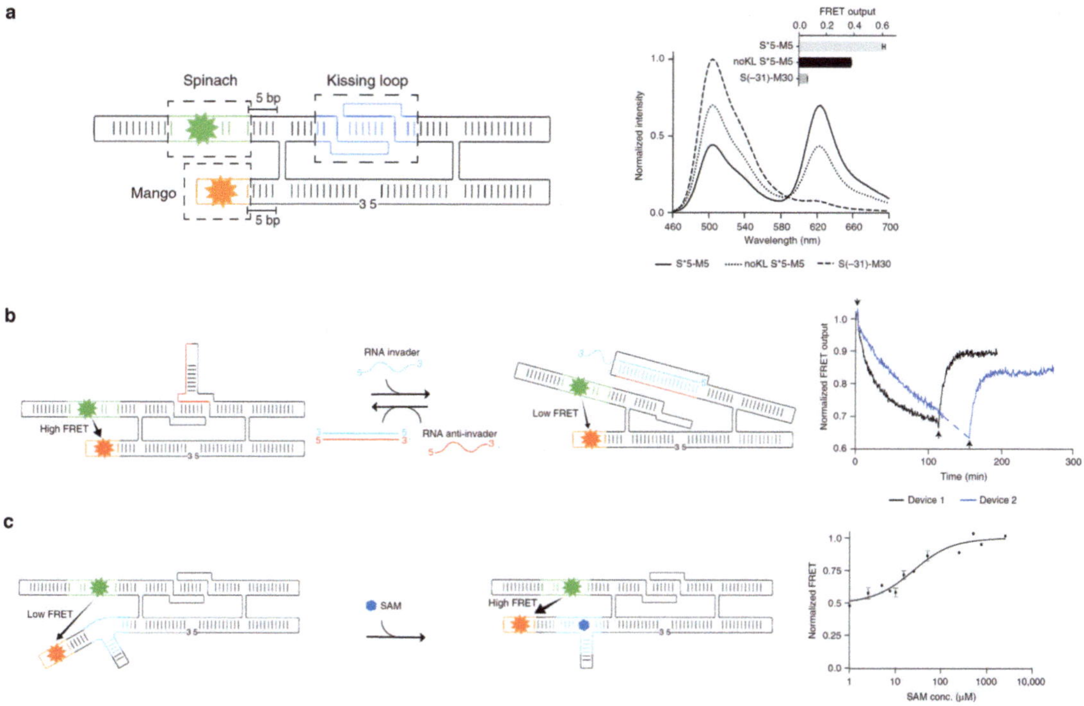

Fig. 3 RNA origami FRET tiles. (**a**) 2-helix AE apta-FRET tile with Spinach (green) and Mango (orange) aptamers incorporated and a 180° KL (blue) locking the structure (left). FRET spectra and output of apta-FRET constructs after excitation of DFHBI-1T (right). The solid line is the S*5-M5 construct with an intact KL expected to give high FRET, the dotted line is the S*5-M5 construct designed without a KL expected to give lower FRET, and the dashed line is the S(-31)-M30 construct designed to have no FRET. The inset shows calculated FRET outputs of the three constructs. Error bars indicate standard deviations calculated using triplicate measurements. (**b**) Apta-FRET sensor against specific ssRNA sequences. The reversible strand-displacement and conformational change resulting in a FRET change are illustrated on the left and the FRET output of two different devices that detect two different RNA inputs are shown on the right. (**c**) SAM sensor. The structure of the riboswitch changes its conformation upon sensing of the target (left) and increases the FRET output (right). (Figure adapted from [10])

- Nucleotide triphosphate (NTP) mix: ATP, UTP, GTP, and CTP 25 mM each.
- 100 mM 1,4-Dithiothreitol (DTT).
- 10× transcription buffer 1: 60 mM $Mg(OAc)_2$, 400 mM NaOAc, 400 mM KCl, and 500 mM Tris-OAc, pH 7.8.
- 10× transcription buffer 2: 150 mM $Mg(OAc)_2$, 500 mM Tris-OAc pH 7.8, 400 mM NaCl, 1% Tween20.
- DNase I (ThermoFisher Scientific).
- Aluminum block incubator.

2.2.3 Denaturing Acrylamide Gel Electrophoresis

- SequaGel-UreaGel System (UreaGel Concentrate and Urea Gel Diluent, national diagnosis).
- Tetramethylethylenediamine (TEMED).
- 10% ammonium persulfate (APS).
- 10× Tris-borate-Ethylenediaminetetraacetic acid (EDTA) buffer (TBE).
- Denaturing loading buffer: 95% formamide, 18 mM EDTA, 0.025% SDS, bromophenol blue, xylene cyanol.
- Glass plates, spacers, comb and clamps for gel casting.
- Power source and vertical gel chamber.
- "Freeze n' squeeze" DNA gel extraction Spin columns (Bio-Rad).
- 3M NaOAc pH 5.2, 96% and 70 % Ethanol.

2.2.4 Folding RNA Origami by Heat-Annealing Procedure

- 5× assembly buffer: 5X Tris-Borate, 625 mM KCl, 5 mM $MgCl_2$.
- Thermocycler and freezer (−20 °C).

2.2.5 Fluorescence Spectroscopy

- Fluorometer (FluoroMax 4 from Horiba, Jobin Yvon).
- Fluorometer cuvettes.
- YO3-biotin fluorophore.
- DFHBI-1T fluorophore.

2.2.6 Atomic Force Microscopy

- Silicon nitride probes.
- Digital Instruments Multimode AFM with a Nanoscope IIIA controller and either an E or J-scanner.
- AFM Buffer: Tris-Borate 1X (pH 8.1), 2 mM $Mg(OAc)_2$, 50 mM KCl, 50 mM NaCl.

3 Methods

3.1 Design and Characterization of RNA Origami Programmable Lattices with bKLs

Hexagonal and rectilinear lattices have previously been shown to assemble co-transcriptionally using 120° and 180° external KLs [7]. A new promising motif that allows the functionalization of RNA tiles and lattices at specific positions is the bKL [14]. In the bKL, the double adenine bulge on the 5′-end of the KL is replaced with a hairpin loop (Fig. 2), giving each tile at least one additional point of functionalization, depending on the number of internal KLs in the tile. In the following section, we will guide the users on how to conceptually build this type of tiles based on a 2D, text-based blueprint. Furthermore, we will explain how to design RNA sequences that match the specified constraints and describe how to produce and characterize the RNA lattices.

3.1.1 RNA Origami Design

In the following, we will go through the RNA origami design process step-by-step.

1. The first step involves drawing a 2D blueprint in a text file. The 2D motifs can either be typed by hand (*see* **Note 2**) or copy-pasted from files included with the trace.pl script on Github (https://github.com/esa-lab/trace). Initially, an RNA hairpin is created by typing two lines of an equal number of "N"s (random nucleotides), adding base pairs in between with ":" symbols, and joining together the double helix using a UUCG terminal loop motif on one of its ends. You will need to indicate the positions of the 5′ and 3′-ends with a "5" and a "3," respectively, in the blueprint diagram. Each letter on the strand indicates a nucleotide position and can be A, U, C, or G. Positions marked with an "N" are the positions that are specified to be designed by an inverse folding algorithm (such as NUPACK [24]). As seen below, the secondary structure can either be written with the symbols (\/|:-) or using monospace font or with more advanced symbols (╮ ╯ | ┊ ─) using, e.g., the Menlo font type. For the purposes of this tutorial, we will use the Menlo font type.

```
5-NNNNNNNNNNN-UU\        5─NNNNNNNNNNN─UU╮
  :::::::::::    |   =     ┊┊┊┊┊┊┊┊┊┊┊    │
3-NNNNNNNNNNN-GC/        3─NNNNNNNNNNN─GC╯
```

2. Extend two terminal loops into a large double helix, closed on both sides, as shown as follows:

```
╭UUN5    3NGC╮
│  ┊      ┊  │
╰CGN3    5NUU╯
```

```
╭UUNNNNNNNNNN─NNNNNNNNNNN──NNNNNNNNNNN─NNNNNNNNNNGC╮
│   ┊┊┊┊┊┊┊┊┊┊ ┊┊┊┊┊┊┊┊┊┊┊   ┊┊┊┊┊┊┊┊┊┊┊ ┊┊┊┊┊┊┊┊┊┊   │
╰CGNNNNNNNNNN─NNNNNNNNNNN3 5NNNNNNNNNNN─NNNNNNNNNNUU╯
```

3. Stack multiple helices on top of each other and align them where you want to merge them together. Start by drawing two complete double helices (*see* **Note 3**):

```
╭UUNNNNNNNNNN─NNNNNNNNNNN──NNNNNNNNNNN─NNNNNNNNNNGC╮
│   ┊┊┊┊┊┊┊┊┊┊ ┊┊┊┊┊┊┊┊┊┊┊   ┊┊┊┊┊┊┊┊┊┊┊ ┊┊┊┊┊┊┊┊┊┊   │
╰CGNNNNNNNNNN─NNNNNNNNNNN3 5NNNNNNNNNNN─NNNNNNNNNNUU╯

╭UUNNNNNNNNNN─NNNNNNNNNNN──NNNNNNNNNNN─NNNNNNNNNNGC╮
│   ┊┊┊┊┊┊┊┊┊┊ ┊┊┊┊┊┊┊┊┊┊┊   ┊┊┊┊┊┊┊┊┊┊┊ ┊┊┊┊┊┊┊┊┊┊   │
╰CGNNNNNNNNNN─NNNNNNNNNNN3 5NNNNNNNNNNN─NNNNNNNNNNUU╯
```

Now, connect them introducing the crossovers at multiples of 11 bp distance (1 helical turn in A-form RNA) (*see* **Note 4**):

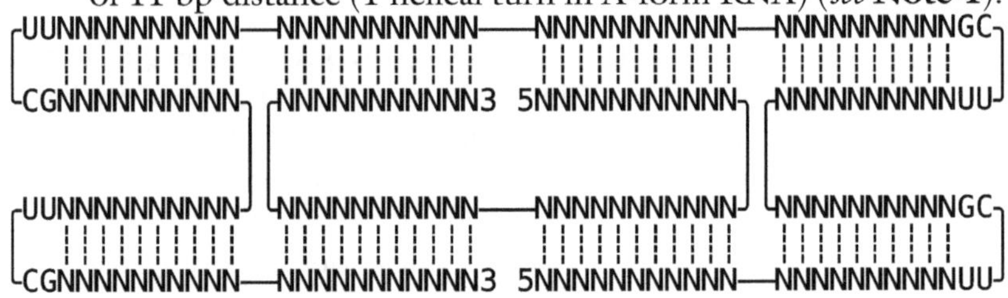

4. Kissing loops are now introduced to achieve a single-stranded RNA structure. The kissing loop structural module has a length equivalent to 8 bp of A-form helix:

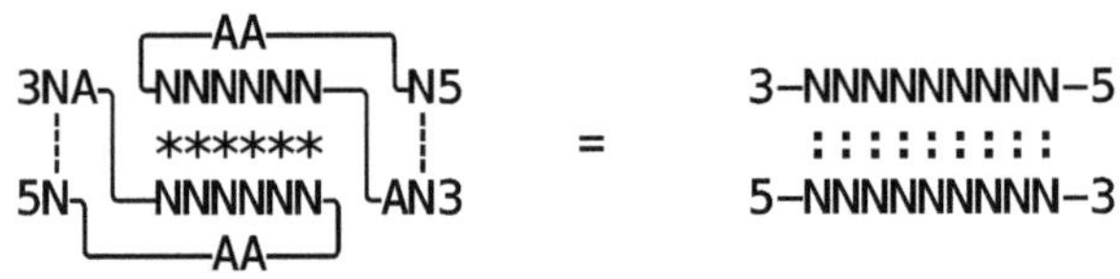

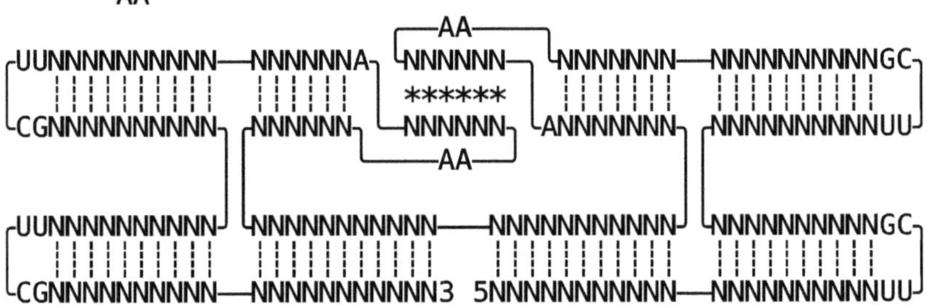

5. Add sequence constraints. Insert GU wobbles to avoid >8 bp stems in DNA for more efficient DNA synthesis and PCR amplification. This can be done by inserting the letter K instead of N in the two partners of a base pair. The 5′-end of the sequence should be constrained to begin with GGAA, an optimal leader sequence for the T7 RNA polymerase:

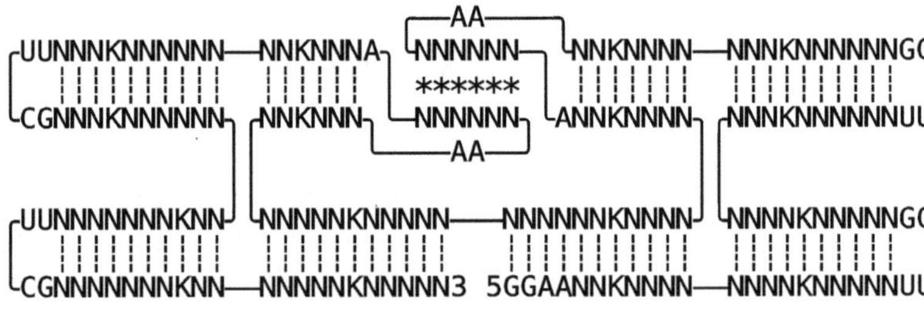

6. Introduce external kissing loops for tile polymerization. 120° KL will result in hexagonal lattices, and 180° KL in rectilinear lattices. Adjust the lengths of the edges to optimally orient the 120° KL to join the tiles in plane. An optimal number of base pairs between the crossovers of consecutive tiles is 22 bp [7].

```
120KLs

3N──────────┐                  ┌NNNNNNN─N5
:           │                  │         :
5N─NNNNNNN──┘                  └─────────N3

                                  ┌──AA─────┐
 ┌NNNNNNN─NNNNN───NNKNNNA┐ └NNNNNN──┐ └NNKNNNN──NNNKNNNNNN──────────┐
 │           │││││    ││││││  │    ******  │  │││││││     ││││││││││            │
 └──────────NNNNN┐ ┌NNKNNN┐ └─NNNNNN┐  └ANNKNNNN┐ ┌NNNKNNNNNN─NNNNNNN┘
                 │ │          └──AA───┘            │ │
┌NNNNNNN─NNNNNNNKNN┘ └NNNNNKNNNNN─────NNNNNNKNNNN┘ └NNNNN───────┐
│        ││││││││││     │││││││││││        │││││││││││     │││││        │
└─────────NNNNNNNKNN──NNNNNKNNNNN3  5GGAANNKNNNN──NNNNK─NNNNNNN┘
```

7. Manually introduce sequences into the programmable kissing loops to build lattices. Kissing loop sequences were obtained from the ones experimentally tested in [25, 26].

```
>2AE_120kl_A

                                   ┌──AA─────┐
 ┌UCGGACG─NNNNN───NNKNNNA┐ └NNNNNN──┐ └NNKNNNN──NNNKNNNNNN──────────┐
 │           │││││    ││││││  │    ******  │  │││││││     ││││││││││            │
 └──────────NNNNN┐ ┌NNKNNN┐ └─NNNNNN┐  └ANNKNNNN┐ ┌NNNKNNNNNN─GAUGGUU┘
                 │ │          └──AA───┘            │ │
┌UCUUGCG─NNNNNNNKNN┘ └NNNNNKNNNNN─────NNNNNNKNNNN┘ └NNNNN───────┐
│        ││││││││││     │││││││││││        │││││││││││     │││││        │
└─────────NNNNNNNKNN──NNNNNKNNNNN3  5GGAANNKNNNN──NNNNK─GUCCACU┘
```

```
>2AE_120kl_B

                                   ┌──AA─────┐
 ┌CGCAAGA─NNNNN───NNKNNNA┐ └NNNNNN──┐ └NNKNNNN──NNNKNNNNNN──────────┐
 │           │││││    ││││││  │    ******  │  │││││││     ││││││││││            │
 └──────────NNNNN┐ ┌NNKNNN┐ └─NNNNNN┐  └ANNKNNNN┐ ┌NNNKNNNNNN─GAGACGU┘
                 │ │          └──AA───┘            │ │
┌AGCACCA─NNNNNNNKNN┘ └NNNNNKNNNNN─────NNNNNNKNNNN┘ └NNNNN───────┐
│        ││││││││││     │││││││││││        │││││││││││     │││││        │
└─────────NNNNNNNKNN──NNNNNKNNNNN3  5GGAANNKNNNN──NNNNK─AACCAUC┘
```

8. Substitute the two adenines upstream of one of the hairpins that form the internal KL to generate the branched kissing loop motif (Fig. 2).

```
>2AE_120kl_bkl
```

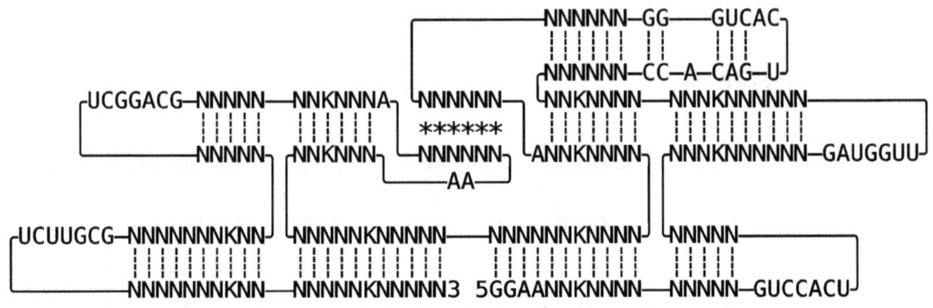

9. It is possible to use this bKL hairpin extension to introduce new functional motifs into the origami. Introduce the protein-binding F6 RNA aptamer as in D. Liu et al. [14] by modifying the terminal loop.

```
>2AE_120kl_bkl
```

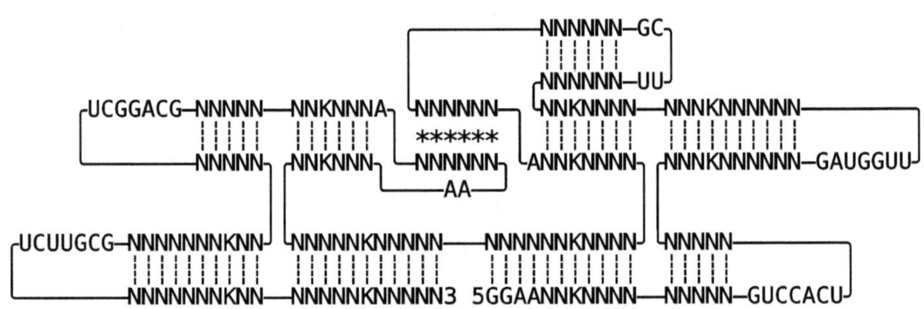

10. Open a terminal window and navigate to the working directory. It is a good idea to have separate directories for the different designs.

11. Name the text file as "patten.txt." Translate the sequence constrains specified on the blueprint onto a dot-bracket notation using the trace_pattern.pl script. Execute the following command:

    perl trace_pattern.pl pattern.txt > target.txt

    The output target.txt file should look like this:

```
2AE_120kl_bkl
((((((((((((((.......)))))((((((((((.......))))))))))((((((((((((((.(((...))))))))))[[[[[.))))))))))))))))))((((((((((((((((..]]]]].))))))(((((.......)))))
((((((((((.......))))))))))))))))))))))))
GGAANNKNNNNNNNNNNGUCCACUNNNNNNNNNKNNNNNNNGAUGGUUNNNNNNKNNNNNNNKNNNNNNNNCCACAGUCACUGGGNNNNNNNNNNNNANNKNNNNNNNNNNKNNNNNNNNNNNKNNNNNNNNKNNNAANNNNNNANNNKNNNNNNNGCAGGCUNNNNNN
NNKNNNNNNNGCGUUCUNNNNNNNNKNNNNNNNNKNNNNN
```

12. The Revolvr program reads the target and pattern files and repeatedly mutates a randomized sequence until it folds according to the required constrains under the ViennaRNA folding model [23, 27]. To execute it a predefined number of times and produce multiple candidate sequences, the user can execute the Batch_revolvr.pl script. To do so, execute the following command:

    perl batch_revolvr.pl

13. Choose a sequence that has a low ensemble diversity and convert it to DNA (we suggest the Biomodel transcription-translation tool (http://biomodel.uah.es/en/lab/cybertory/analysis/trans.htm)).

14. Introduce the T7 RNA polymerase promoter sequence upstream of the DNA sequence.

    T7 Promoter sequence:

    5′-TAATACGACTCACTATA-3′

    Example sequence 1 (T7 promoter in yellow, origami coding sequence in grey):

TAATACGACTCACTATA
GGAATTGCACCCGTGCGTCCACTGCACGAGGTCTTGCTGATGGTTAGCAAGGCCTGCTCGACC
ACTGCCCACAGTCACTGGGGCAGTGACGGCTAGTTGAGCGGTGTAATTCCTGGATTTGCGTGC
GCGGAAAGCCGTACCGTGCTGTTCGCAGGCTGGGCATCGAGCACCCGCGTTCTGGGTGTTTG
AACGCAGATCCA

15. Design primers for PCR amplification (*see* **Note 5**). Primers and gBlock gene fragments can be ordered from a DNA synthesis company (i.e., Integrated DNA Technologies or Twist Biosciences).

    It is possible to include an arbitrary sequence on the 5′ end of the DNA sequence to design primers that match the annealing temperature (Table 1). Alternatively, it is also possible to introduce a sequence constraint on the 3′ end of the origami before designing the sequence.

    Example sequence 2 (primer-binding sites in blue, T7 promoter in yellow, origami coding sequence in grey):

CATGTGTCTCAGGAGTGCCAGTAATACGACTCACTATAGGAATTGCACCCGTGCGTCCACT
GCACGAGGTCTTGCTGATGGTTAGCAAGGCCTGCTCGACCACTGCCCACAGTCACTGGGG
CAGTGACGGCTAGTTGAGCGGTGTAATTCCTGGATTTGCGTGCGCGGAAAGCCGTACCGT
GCTGTTCGCAGGCTGGGCATCGAGCACCCGCGTTCTGGGTGTTTGAACGCAGATCCA

3.1.2 DNA Template Production

1. Dissolve gBlock DNA to 100 nM and primers to 10 μM in nuclease-free H_2O or 1X Tris acetate-EDTA (TAE) buffer (*see* **Note** 7).

2. Prepare PCR mix for a 50 μL reaction: 10 μL 5X Q5 reaction buffer, 1 μL 10 mM dNTP mix, 2.5 μL of each 10 μM forward and reverse primers, 2 μL of 100 nM DNA template, 31.5 μL nuclease-free H_2O.

Table 1
Example primers for PCR amplification (*see* Note 6)

	Sequence (5′-3′)	Length (nt)	Tm (°C)
Forward	CATGTGTCTCAGGAGTGCCAG	21	68
Reverse	TGGATCTGCGTTCAAACACC	20	66

3. Add 0.5 μL of Q5 DNA polymerase, mix by pipetting, and place the samples into the thermocycler, use a standard PCR procedure such as follows:
   (a) Initial denaturation at 98 °C for 2 min
   (b) 30 cycles of:
      (i) Denaturation at 98 °C for 10 s.
      (ii) Annealing at 69 °C for 15 s (*see* **Note 6**).
      (iii) Polymerization at 72 °C for 20 s.
   (c) Final extension at 72 °C for 2 min.
   (d) Storage at 4 °C.
4. Purify using a standard PCR cleanup kit following the manufacturer's instructions (*see* **Note 8**).

3.1.3 RNA Transcription on Mica and AFM Imaging

1. Cleave mica and fix it to a metal disk. Use the aluminum block incubator to pre-heat to 37 °C and set aside.
2. Prepare the mix in an Eppendorf tube: 5 ng of DNA template with 1X transcription buffer, 0.5 mM of each NTP, and 1 mM DTT in a total volume of 50 μL (*see* **Note 9**).
3. Add the T7 RNA polymerase (0.1 U/50 μL).
4. Run reactions for 10 min.
5. Dilute reactions by a factor of 5X in AFM buffer.
6. Deposit 50 μL of the solution on the pre-heated mica and image immediately.

3.2 Design and Characterization of an RNA Origami-Based FRET Sensor System

In Jepsen et al. [10], a FRET sensor was developed using the 2-helix tile RNA origami tile as scaffold. This was done by incorporating, two different fluorogenic aptamers, Mango and Spinach [20, 21], with YO3-biotin and DFHBI-1T as respective cognate fluorophores that qualified as a FRET pair, at an optimal distance from each other at the end of the two helices of the RNA tile. Then, the capability to respond to a small ssRNA input was introduced through a switching mechanism based on the bKL motif. The sequence of this newly introduced hairpin can be changed to base pair with a desired single-stranded RNA or DNA input. This input will break apart the KL interaction by strand displacement, resulting in an increased mobility of the aptamers and therefore a decrease of the FRET signal (Fig. 3).

A similar system was used to detect S-adenosylmethionine (SAM) by introducing the SAM riboswitch [28] on the helix at which Mango was positioned (*see* Fig. 3c and **Note 10**). In this part of the chapter, we will guide users on how to incorporate the particular motifs used in Jepsen et al. to produce the different sensing devices on a text-based level.

3.2.1 Incorporation of Functional Motifs into the Origami

1. Add the fluorogenic aptamers to the end of the 2-helix tile:

```
Spinach aptamer

3-NNNNNGGAGUGUGAGAUG—AG—UU—GNNNN-5
       |||           ||    ||||
5-NNNNNCAA—GGACGGG—UC———CNNNN-3


Mango aptamer

┌CGU—GG—CAG—GG—AAG—CNNN—5
│                   ||||
└———GG—AGA—GG—AGA—GNNN—3
```

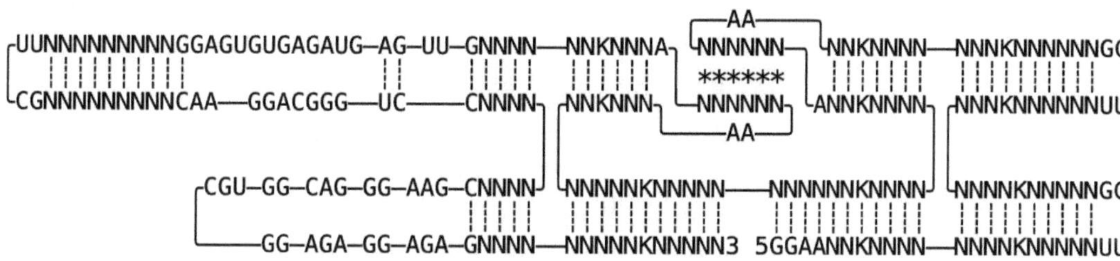

2. To build a reversible dynamic device that senses ssRNA inputs, introduce the branched kissing loop motif [14] (*see* **Note 11**).

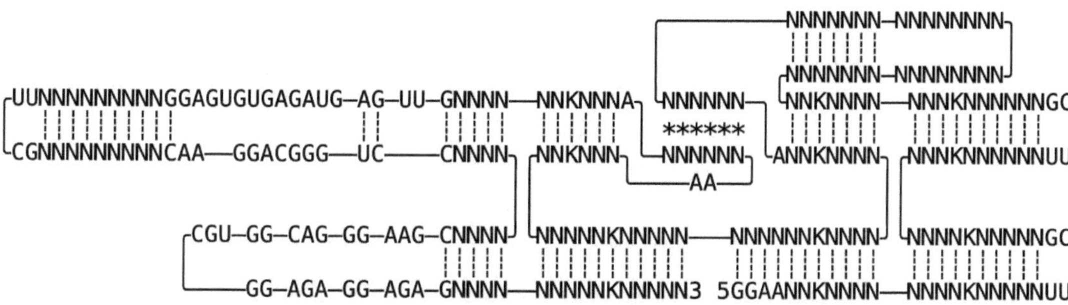

3. It is also possible to introduce the SAM riboswitch as a conformational change actuator that, upon target binding, will move the fluorogenic aptamers closer to each other.

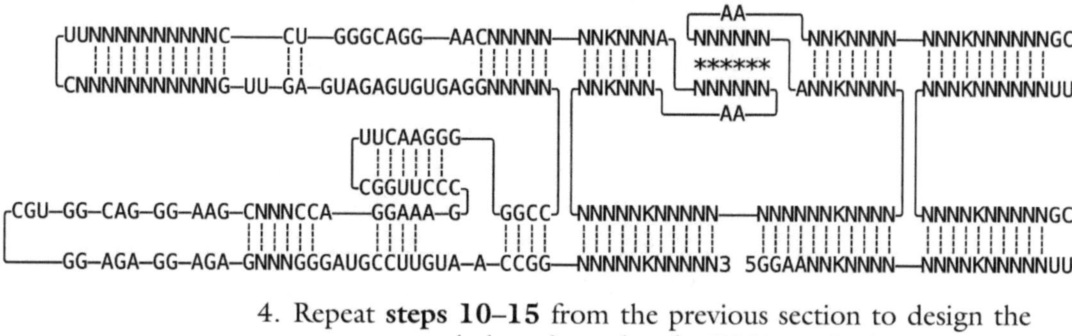

4. Repeat **steps 10–15** from the previous section to design the sequences and place the orders for DNA synthesis.

3.2.2 RNA Production and Purification

1. DNA template production as stated in the previous section (Subheading 3.1.2).
2. Prepare transcription by mixing DNA template, 1X transcription buffer 2, 1 mM DTT, 2.5 mM each NTP, T7 RNA polymerase.
3. Incubate reaction at 37 °C for 4 h (*see* **Note 12**).
4. Stop reaction by adding 1 U/100 μL of DNase I and incubating at 37 °C for 15 min.
5. Mix 1:1 with denaturing loading buffer and heat at 95 °C for 5 min.
6. Run at 15 W for 35 min.
7. Visualize gel bands using UV shadowing and cut the RNA bands out of the gel with a clean scalpel.
8. Elute overnight with nuclease-free H_2O.
9. Transfer the mix into the Freeze'n'Squeeze gel extraction spin columns to separate the liquid from the gel (*see* **Note 13**). Spin at 13,000× *g* for 3 min at room temperature.
10. Add 1/10 of the solution volume of 3 M NaOAc to a final concentration of 0.3 M NaOAc.
11. Add 3 times of the solution volume of 96% ethanol, vortex, and spin down.
12. Incubate sample on dry ice (−80 °C) for 15–60 min or in the freezer (−20 °C) for 1 h to overnight.
13. Centrifuge at 14,000 rpm for 30 min. This should be done at 4 °C if possible.
14. Carefully discard the supernatant and add 1.5 times the initial volume of 70% EtOH.
15. Centrifuge again at 14000 rpm for 10 min.
16. Discard the supernatant and dry the pellet by leaving the tube in the fume-hood for a few minutes.
17. Re-dissolve the pellet in nuclease-free H_2O and quantify the concentration at the NanoDrop spectrophotometer using absorption at 260 nm. Store at −20 °C.

3.2.3 FRET Measurements

1. Heat-anneal the RNA in the thermocycler by heating it up to 95 °C for 5 min and quickly cool it down by putting it in the freezer (−20 °C) for 3 min.
2. Add assembly buffer to a final concentration of 1X and leave the samples at 37 °C for 30 min.
3. Prepare 100 nM samples for the fluorometer measurements.
4. Add 1 μM of DFHBI-1T and 1.7 μM YO3-biotin.

5. DFHBI-1T should be excited at 450 nm, and its emission should be recorded at 503 nm; YO3-biotin should be excited at 580 nm, and its emission should be recorded at 620 nm.
6. Relative FRET calculations should be done as in Jepsen et al. [10], by using the equation: $\mathrm{FRET} = \frac{ID\Upsilon(exD, em\Upsilon)}{ID\Upsilon(exD, em\Upsilon)+ID\Upsilon(exD, emD)}$, where $ID\Upsilon(exD, em\Upsilon)$ is the emission at YO3-biotin wavelength after DFHBI-1T excitation and $ID\Upsilon(exD, emD)$ is the emission at DFHBI-1T wavelength after DFHBI-1T excitation. Both measurements should be performed with both fluorophores present.

4 Notes

1. This is a high-fidelity polymerase. It has a lower mutation rate and is therefore more expensive. Phusion polymerase is another high-fidelity alternative; however, it is perfectly possible to use a standard DNA polymerase such as Taq.
2. The font type should be "Menlo," and a font size 11 is recommended for visualization.
3. Holding the "option" key down on Mac or the "Alt" on PC allows rectangles of text characters to be copied and moved like modular blocks.
4. The crossover pattern introduced here corresponds to an antiparallel even crossover, i.e., an even number of half-turns (5.5 bp). It is also possible to make an origami with an odd number of helical half-turns between crossovers [7].
5. Alternatively, for high yield production or for in vivo expression, it is possible to include restriction sites on the gBlock sequence for cloning or to directly order the sequence on a plasmidic vector (Twist bioscience).
6. The New England Biolabs T_m Calculator tool (https://tmcalculator.neb.com/#!/main) is of great use for primer design and for melting temperature calculation.
7. Spin down tubes before dissolving the lyophilized powder, as it might fly out of the tube when opening it.
8. It is recommended to verify the PCR-amplified products via agarose gel electrophoresis.
9. Additionally, RNase Inhibitor (NEB) can be added to the reaction to a concentration of 1 U/μL to avoid degradation by RNases.
10. In the absence of the small molecule, the riboswitch is expected to have a moderately unstructured conformation, and upon

binding of SAM, the riboswitch acquires a more rigid conformation, allowing the fluorophores to be positioned at a distance to produce a FRET output (Fig. 3c).

11. The sequence of the branched kissing loop can be changed to have specificity to sense different RNA targets.
12. During this time, it is recommended to cast the polyacrylamide gels, as they might take up to an hour to polymerize.
13. If the volume of the gel slice is too big to fit into one column, use two or more.

Acknowledgment

This work was supported by the European Union's Horizon 2020 Research and Innovation Program, as part of the Interactive Training Network, DNA Robotics, under the Marie Sklodowska-Curie grant agreement n° 765703.

References

1. Winfree E et al (1998) Design and self-assembly of two-dimensional DNA crystals. Nature 394(6693):539–544
2. He Y et al (2008) Hierarchical self-assembly of DNA into symmetric supramolecular polyhedra. Nature 452:198
3. Ke Y et al (2012) Three-dimensional structures self-assembled from DNA bricks. Science 338(6111):1177–1183
4. Rothemund PWK (2006) Folding DNA to create nanoscale shapes and patterns. Nature 440(7082):297–302
5. Guo P (2010) The emerging field of RNA nanotechnology. Nat Nanotechnol 5:833
6. Chworos A et al (2004) Building programmable jigsaw Puzzles with RNA. Science 306(5704):2068–2072
7. Geary C, Rothemund PWK, Andersen ES (2014) A single-stranded architecture for cotranscriptional folding of RNA nanostructures. Science 345(6198):799–804
8. Batey RT, Rambo RP, Doudna JA (1999) Tertiary motifs in RNA structure and folding. Angew Chem Int Ed 38(16):2326–2343
9. Li M et al (2018) In vivo production of RNA nanostructures via programmed folding of single-stranded RNAs. Nat Commun 9(1):2196
10. Jepsen MDE et al (2018) Development of a genetically encodable FRET system using fluorescent RNA aptamers. Nat Commun 9(1):18
11. Jasinski D et al (2017) Advancement of the emerging field of RNA nanotechnology. ACS Nano 11(2):1142–1164
12. Krissanaprasit A et al (2019) Genetically encoded, functional single-strand RNA origami: anticoagulant. Adv Mater 31(21):1808262
13. Simmel FC, Yurke B, Singh HR (2019) Principles and applications of nucleic acid strand displacement reactions. Chem Rev 119(10):6326–6369
14. Liu D et al (2020) Branched kissing loops for the construction of diverse RNA homooligomeric nanostructures. Nat Chem 12(3):249–259
15. Shu D et al (2011) Thermodynamically stable RNA three-way junction for constructing multifunctional nanoparticles for delivery of therapeutics. Nat Nanotechnol 6:658
16. Geary CW, Andersen ES (2014) Design principles for single-stranded RNA origami structures. In: DNA computing and molecular programming. Springer International Publishing, Cham
17. Molinaro M, Tinoco I Jr (1995) Use of ultra stable UNCG tetraloop hairpins to fold RNA structures: thermodynamic and spectroscopic applications. Nucl Acids Res 23(15):3056–3063
18. Fu TJ, Seeman NC (1993) DNA double-crossover molecules. Biochemistry 32(13):3211–3220

19. Chang KY, Tinoco I Jr (1994) Characterization of a "kissing" hairpin complex derived from the human immunodeficiency virus genome. Proc Natl Acad Sci U S A 91(18):8705–8709
20. Paige JS, Wu KY, Jaffrey SR (2011) RNA mimics of green fluorescent protein. Science 333(6042):642–646
21. Dolgosheina EV et al (2014) RNA mango aptamer-fluorophore: a bright, high-affinity complex for RNA labeling and tracking. ACS Chem Biol 9(10):2412–2420
22. Montange RK, Batey RT (2006) Structure of the S-adenosylmethionine riboswitch regulatory mRNA element. Nature 441(7097):1172–1175
23. Geary C et al (2021) RNA origami design tools enable cotranscriptional folding of kilobase-sized nanoscaffolds. Nat Chem 13(6):549–558
24. Zadeh JN, Wolfe BR, Pierce NA (2011) Nucleic acid sequence design via efficient ensemble defect optimization. J Comput Chem 32(3):439–452
25. Severcan I et al (2010) A polyhedron made of tRNAs. Nat Chem 2(9):772–779
26. Grabow WW et al (2011) Self-assembling RNA nanorings based on RNAI/II inverse kissing complexes. Nano Lett 11(2):878–887
27. Lorenz R et al (2011) ViennaRNA package 2.0. Algorithms Mol Biol 6(1):26
28. Lu C et al (2008) Crystal structures of the SAM-III/SMK riboswitch reveal the SAM-dependent translation inhibition mechanism. Nat Struct Mol Biol 15(10): 1076–1083

Chapter 4

Reconfigurable Two-Dimensional DNA Molecular Arrays

Donglei Yang, Fan Xu, and Pengfei Wang

Abstract

In biology, molecular cascade signaling is an essential tool to mediate various pathways and downstream behaviors. Mimicking these molecular cascades plays an important role in synthetic biology. The use of DNA self-assembly represents an elegant way to build sophisticated molecular cascades. For instance, a DNA molecular array connected by a number of dynamic anti-junction units was able to realize prescribed, multistep, long-range cascaded transformation. The dynamic DNA molecular array is able to execute transformations with programmable initiation, propagation, and regulation. The transformation of the array can be initiated at selected units and then propagated, without addition of extra triggers, to neighboring units and eventually the entire array.

Key words DNA nanostructure, Reconfigurable 2D arrays, Cascaded transformation, Self-assembly

1 Introduction

Reconfigurable components responsive to external triggers are essential for designing reconfigurable DNA structures, which may be integrated into static DNA structures to realize dynamic motion [1]. Conventional methods to design reconfigurable DNA structures are summarized in Fig. 1. Hydrogen bond-mediated DNA strand base pairing may be the simplest method to induce dynamic motion (Fig. 1a). The switch between single- and double-stranded states in DNA may be induced via changing environmental parameters such as temperature, light exposure (e.g., UV versus visible light for azobenzene-modified DNA) [2], or ionic concentration [3] (e.g., metal ions, pH value). Another method involves enzymatic reactions to DNA strands (Fig. 1b), such as strand degradation, cleavage, or ligation [4, 5]. Strand displacement is probably the most commonly used strategy to make dynamic DNA structures, in which one strand is displaced from a double-stranded DNA, typically mediated by a toehold design (Fig. 1c) [6]. Base stacking between blunt ends of DNA helices are ion- and temperature-sensitive (Fig. 1d) and are therefore able to

Julián Valero (ed.), *DNA and RNA Origami: Methods and Protocols*, Methods in Molecular Biology, vol. 2639, https://doi.org/10.1007/978-1-0716-3028-0_4,

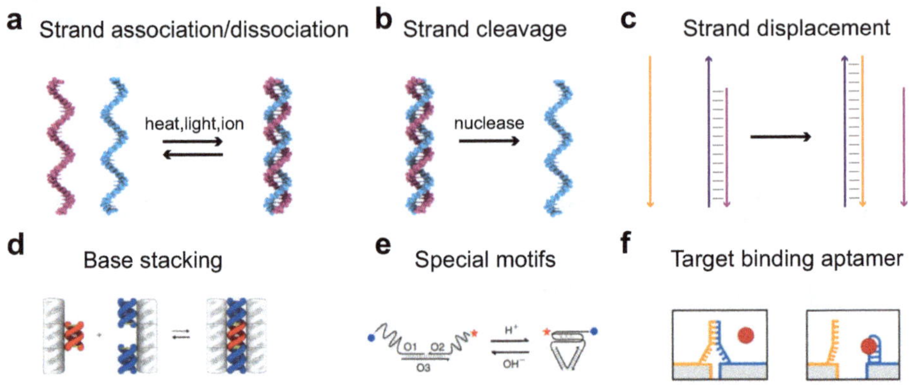

Fig. 1 Strategies for designing reconfigurable structures. (**a**) Association and dissociation between single-stranded DNA. (**b**) Nuclease-mediated DNA strand cleavage. (**c**) Toehold-mediated strand displacement. (**d**) Hydrophobic interaction-induced base stacking. (**e**) pH-sensitive structural motif. (**f**) Target-binding aptamers. (Figure adapted with permission from Ref. [1])

reconfigure upon stimulation [7]. Certain special structural motifs are reconfigurable as well, such as guanine- and cytosine-rich sequences that can fold into stable quadruplexes in the presence of specific metal ions and low pH (Fig. 1e) [8, 9]. Strand displacement may also happen if one of the strands has higher affinity to another target (e.g., proteins). DNA strands with such capability are designated as DNA aptamers (Fig. 1f) [10].

In contrast to the above conventional reconfigurable strategies, where merely partial components within a static structure are actually mobile, DNA molecular arrays built from anti-junction units are able to achieve global transformation via collective reconfiguration by every interconnected unit [11, 12]. As illustrated in Fig. 2, one anti-junction contains four DNA duplex domains of equal length and four dynamic nicking points (Fig. 2a,b), which can switch between two stable conformations, "red" and "green," driven by base stacking, through an unstable open conformation—"orange." An anti-junction is classified by the distance between two opposite dynamic nicking points (i.e., a 42-bp anti-junction). 2D DNA molecular relay array was built via self-assembly of the anti-junctions (Fig. 2c). A relay array can transform from one array conformation (e.g., all anti-junctions are in the red conformation) to another array conformation (e.g., all anti-junctions are in the green conformation). The array transformation follows specific pathways, depending on the array's geometry and binding locations of trigger strands. For instance, if the trigger strands were added to the units at a corner (Figs. 2d and 3a), the relay would undergo a step-by-step conversion from a red array conformation to a green array conformation via a diagonal track. In addition to regulating the transformation track, it may be also trapped at designated locations and then resumed by filling in a

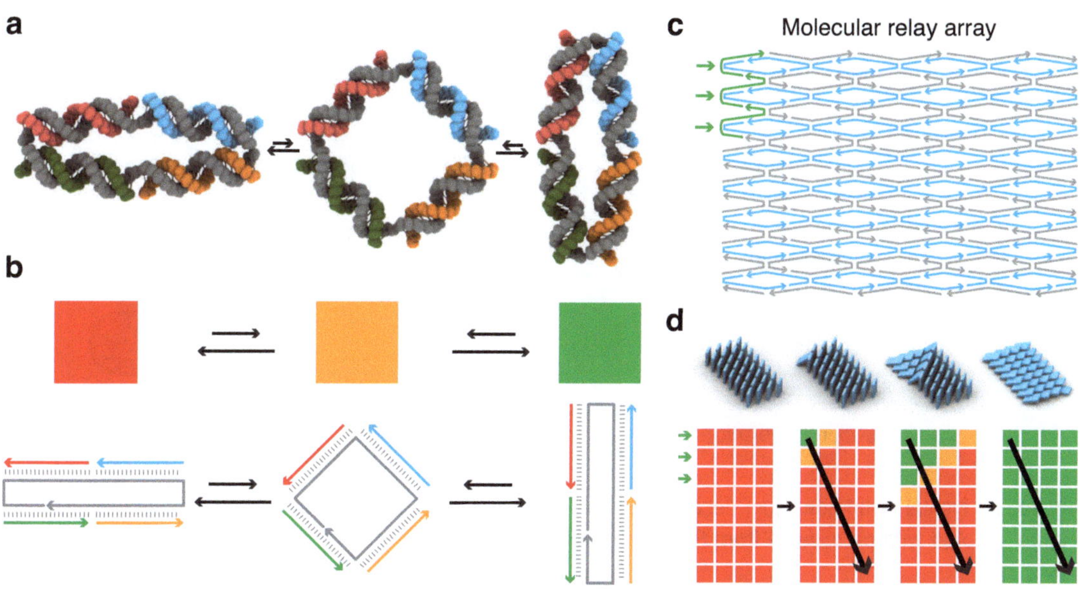

Fig. 2 Reconfiguration of molecular array mediated by dynamic DNA anti-junction units. (**a**) Anti-junction motif that can transform between two stable conformations. (**b**) Different diagrams for a DNA anti-junction: stable conformations "red" and "green" and unstable conformation "orange." (**c**) Strand diagram of an interconnected 2D DNA relay array with 4 units by 8 units. Three trigger strands (green) are added to three units in the upper left corner of the array to initiate the transformation. (**d**) The information of transformation propagates along prescribed pathways, causing the units to convert sequentially in this molecular array. (Figure adapted with permission from Ref. [11])

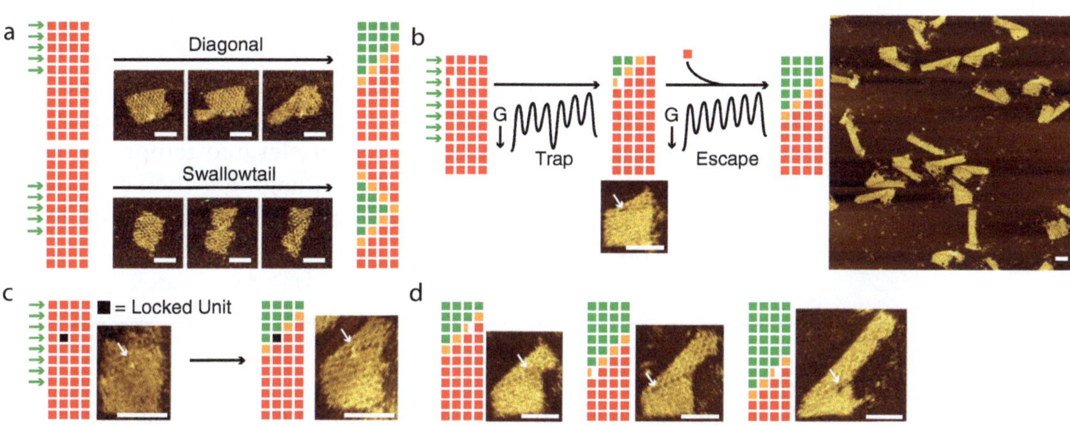

Fig. 3 Programming the reconfiguration pathway of DNA molecular arrays. (**a**) Control of the initiation of transformation via selection addition of green triggers. (**b**) The transformation pathways can be blocked and resumed by the removal and addition of units. (**c**) Blocking of transformation pathways via "lock" strands. (**d**) The transformation can be blocked at any designated location. (Figure adapted with permission from Ref. [11])

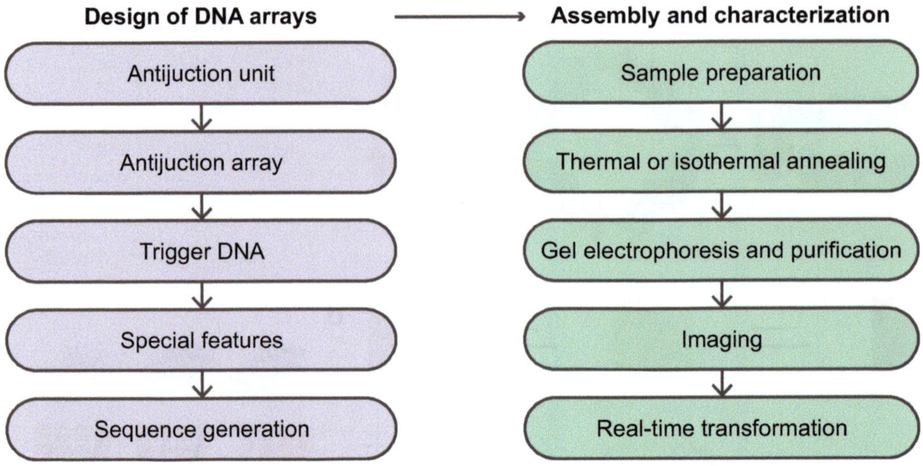

Fig. 4 A typical workflow for designing and characterizing reconfigurable DNA molecular arrays. (Figure adapted with permission from Ref. [12])

specific missing unit (Fig. 3b). Or the transformation may be locked permanently at designated pathway stages by locking the units via a molecular fastener or by omitting certain units that generate void traps (Fig. 3c). Basic design features, such as domain length and nicking-point positions, must be determined first while designing the DNA molecular relay arrays. The array is then connected by a number of interconnected anti-junction units. Special design features such as transformation pathway mediation may be then incorporated. DNA sequences will then be automatically generated by in silico design platform. Conventional thermal annealing protocol may be used for assembling DNA molecular relay arrays, which will then subject to gel electrophoresis and AFM/TEM imaging characterizations. Transformation may be achieved by adding molecular triggers at elevated temperatures to bake for a certain amount of time. AFM imaging is then conducted in situ to visualize real-time structural transformations of DNA molecular arrays (Fig. 4).

2 Materials

2.1 Reagents

Note that all chemical reagents might be potentially harmful:

1. Ethidium bromide.
2. Formamide.
3. Gel loading buffer.
4. LE agarose/gold agarose.
5. Polyethylene glycol 8000.
6. Nickel(II) chloride hexahydrate.

7. Single-stranded DNA staple strands.
8. Single-stranded M13 bacteriophage-derived scaffold p7560.
9. Uranyl formate (1 g).
10. 6× loading buffer.
11. 1 kb plus DNA ladder.

2.2 Equipment

1. AFM system.
2. AFM tips.
3. Aluminum sealing tape for 96-well plates.
4. Centrifuge.
5. Centrifuge filters.
6. DNA LoBind tube (0.5 mL).
7. Double-sided adhesive tape.
8. Erlenmeyer flask (250 mL).
9. Freeze'n Squeeze spin columns.
10. Freezer.
11. Gel chamber.
12. Gel imager.
13. Mica, 15-mm diameter.
14. Microwave.
15. Multichannel pipettes.
16. Multipette M4 pipette.
17. PCR machine.
18. PCR tubes.
19. Razor blade.
20. Reagent reservoir for multichannel pipettors. Round metal plates, 15-mm diameter.
21. UV transilluminator.
22. UV-visible spectrophotometer.
23. 1.5-mL tubes.
24. Software for DNA nanostructure design and sequence generation (cadnano, http://cadnano.org).

2.3 Reagent Setup

All buffer solutions should be prepared in deionized water. We suggest that fresh buffers and solutions are prepared regularly:

1. Gel buffer (0.5× TBE): 45 mM Tris, 45 mM boric acid, 1 mM EDTA, 12 mM $MgCl_2$. Store the buffer at room temperature.
2. 1 M $MgCl_2$.
3. 10× TE-Mg^{2+} buffer: 400 mM Tris, 10 mM EDTA, 120 mM $MgCl_2$. The buffer can be stored at room temperature.

4. 1× TE-Mg^{2+} buffer (dilution of 10× TE-Mg^{2+}): 40 mM Tris, 1 mM EDTA, 12 mM $MgCl_2$. Store the buffer at room temperature.
5. 15% (wt/vol) PEG solution: 15% (wt/vol) PEG 8000, 5 mM Tris base, 1 mM EDTA, 505 mM NaCl. Store the buffer at room temperature.
6. Staining solution: 1% (wt/vol) uranyl formate. Store the solution at room temperature in the dark. Centrifuge the solution at 12,000 g for 5 min at room temperature before use to remove the aggregates and impurities.

3 Methods

3.1 Folding of DNA Origami Structures

1. Staple DNA preparation (**steps 1–4**). Add dH_2O to each lyophilized oligonucleotide well to make the final concentration ~ 100 μM in 96-well plates.
2. Seal the plates and vortex to dissolve the DNA powder.
3. Spin down the plates at 1000 g for 5 min at room temperature. The dissolved oligonucleotides can be stored at −20 °C.
4. Prepare the core staple mixture for DNA nanostructures. Take the 96-well plates containing staple DNA from −20 °C storage, and leave at room temperature for 1–2 h to unfreeze the staple DNA.
5. Take a reagent reservoir for multichannel pipettors. Take 5 μL from each well of the 96-well plates, excluding the wells for trigger strands or modified strands, and put the liquid in the reservoir, using a multichannel pipette (*see* **Note 1**).
6. Mix the droplets and transfer the solution to a 1.5-mL DNA low-binding tube with a pipette.
7. Prepare other mixtures of trigger strands or modified strands.
8. Seal the plates with aluminum sealing tape for the 96-well plates.
9. Prepare folding samples in a PCR tube. The sample should have a total final volume of 40 μL and contain the following reagents (*see* **Notes 2** and **3**):

Component	Amount (μL)	Final concentration
SCAFFOLD P7560, 100 nM	4	10 nM
TE buffer, 10×	4	1×
$MgCl_2$, 100 mM	4	10 nM

(continued)

Component	Amount (μL)	Final concentration
Core staple mixture, 1 μM	4	100 nM per strand
Red or green trigger DNA, 2 μM each	2	100 nM
H_2O	22	
Total	40	

10. Assemble the DNA origami and DNA brick arrays by either ramp annealing (*see* **step 11**) or isothermal annealing (*see* **step 12**). Ramp annealing can be used for both DNA origami and DNA brick samples. We have tested isothermal annealing only on DNA brick samples, but we anticipate that it could also be used for DNA origami samples.
11. Ramp annealing of DNA arrays: Run the specific temperature program for the DNA nanostructures. Use the following program for 52-bp and 64-bp DNA brick arrays: The first temperature (95 °C) stays 5 minutes, the second ramp (from 85 to 25 °C) is kept at a constant speed of 20 min per °C, and finally hold at 4 °C.

    Use the following program for DNA origami arrays:

    The first temperature (95 °C) stays 5 minutes, the second ramp (from 85 to 25 °C) is kept at a constant speed of 10 min per °C, and finally hold at 4 °C.
12. Isothermal annealing assembly of DNA arrays (*see* **Note 4**):

    Fold the DNA array samples by incubating them at constant temperatures (e.g., 45–65 °C for every 1 °C with a heat shock at 95 °C for 5 min) to screen the optimal T_{fold}. Correctly folded samples will show clear bands in the agarose gel and well-defined geometry in the AFM images under the optimal T_{fold}.

3.2 Purification of DNA Nanostructures

Purify the assembled DNA structures using PEG precipitation (**steps 1–9**), agarose gel electrophoresis (**steps 10–19**), or ultracentrifugation (**steps 20–28**). Typically, gel electrophoresis is used in this protocol for AFM imaging and quality control. PEG precipitation or ultracentrifugation is also feasible for the purpose of high throughput or fast purification:

1. For PEG precipitation purification, first adjust the magnesium concentration of the DNA origami or DNA bricks to 20 mM with 1 M $MgCl_2$.
2. Add 1× TE and 20 mM $MgCl_2$ buffer to make the total volume 200 μL in a 1.5-mL or 2-mL Eppendorf tube.
3. Mix the DNA nanostructure solution with 200 μL of 15% (wt/vol) PEG solution.

4. Place the tube in a high-speed centrifuge and centrifuge for 17,000 g for 30 min at room temperature.
5. Remove the supernatant with a pipette and add 200 μL of 1× TE and 20 mM $MgCl_2$ to the pellet.
6. Insert the tube in a shaker at room temperature for 5 min to dissolve the pellet.
7. Repeat steps 2–5 two to three times.
8. Centrifuge at 17,000 g for 30 min at room temperature and remove the supernatant. Dissolve the pellet with 1× TE and 10 mM $MgCl_2$ to adjust the concentration of DNA nanostructures.
9. Measure the concentration of DNA nanostructures with a UV-visible spectrophotometer. The concentration should be ~5–10 nM.
10. The folding quality of DNA origami can be tested by agarose gel electrophoresis. The product bands can be separated from the short staples for further AFM or TEM imaging. For agarose gel electrophoresis purification (*see* **Notes 5**, **6**, and 7), follow the next steps:
11. For a 50-mL gel of 1% (wt/vol) agarose, weigh 0.5 g of agarose and add it to a 250-mL flask.
12. Add 50 mL of 0.5× TBE buffer with 12 mM $MgCl_2$ to the flask. Stir the flask to mix the agarose with buffer.
13. Put the flask in a microwave at high power for 1–3 min until the agarose is fully dissolved and the solution is completely clear.
14. Leave the bottle for 5–10 min to cool the solution before the next few steps.
15. Add 5 μL of ethidium bromide (10000×) and swirl it gently until the stain is evenly distributed.
16. Pour the gel solution into the casting tray and insert a gel comb.
17. Wait for 30–60 min until the gel has been solidified. Remove the gel comb and put the casting tray with gel into the gel box. Fill the gel box with 0.5× TBE containing 12 mM $MgCl_2$. Put the gel box in an ice-water bath to prevent heat damage.
18. Use M13-derived scaffold p7560 or 10 k DNA marker as a DNA ladder. Combine 15 μL of sample from **step 10** with 3 μL of loading buffer and load the samples into the gel wells.
19. Set the voltage at 50–70 V and run the gel for 1.5–2 h. Make sure there are bubbles near the electrode, indicating the flow of current through the gel. Stop running the gel, and use a UV transilluminator to visualize the bands. Adjust the focus and light intensity for good contrast.

20. For ultrafiltration purification (*see* **Note 8**), first insert a centrifugal filter into an affiliate tube.
21. Add 100 μL of the DNA array sample from **step 10** to the filter.
22. Add 300 μL of folding buffer to the filter. Seal with the cap.
23. Put the centrifugal filter into a centrifuge and spin for 3 min at 5000 g at room temperature.
24. Empty the tube that contains the folding buffer together with free DNA.
25. Repeat steps 23–25 three times.
26. Remove the filter set from the tube, flip it, and put it into a new tube.
27. Centrifuge the tube for 2 min at 2000 g at room temperature.
28. Pipette the 10–60 μL of DNA nanostructure solution into a DNA low-binding tube.

3.3 AFM Imaging of DNA Nanostructures

Atomic force microscope (AFM) is an analytical instrument used to characterize the surface morphology of materials, which is an important tool for nanoscale manipulation, imaging, and measurement of materials. When scanning the sample, the force distribution can be obtained by detecting the change of the signal, and the surface topography and surface roughness information can be obtained with nanometer resolution.

Therefore, it is widely used in DNA nanotechnology for imaging of DNA origami and DNA bricks. AFM is more useful for 2D DNA nanostructures such as rectangular DNA origami, while TEM is more suitable for measuring 3D structures. The samples need to be adsorbed to a mica surface through electrostatic interactions for AFM imaging to be performed:

1. Turn on the AFM system.
2. Prepare the freshly cleaved mica on the round metal surface with double-sided tape.
3. Place 2–3 μL of the purified sample (*see* Subheading 3.2) on the mica. The sample concentration should be ~1–5 nM. Dilute the sample if the concentration is too high.
4. Add 50–70 μL of TE buffer with 12 mM $MgCl_2$.
5. Wash the sample with TE buffer containing 12 mM $MgCl_2$ two to three times by pipetting up and down to remove any unbound sample or impurities.
6. (Optional) Add 2 μL of 100 mM $NiCl_2$ solution to the mica surface to increase the binding of DNA nanostructures to the mica surface.
7. Assemble an AFM tip on the liquid cell.

8. Transfer the sample disk to the AFM scanner and secure the liquid cell on top of the sample disk.
9. Set the scanner stage at a low position to prevent contact between the AFM tip and the mica surface.
10. Move the AFM tip to the sample surface.
11. Be careful when moving the tip close to the mica surface, as it is not a visible feature on the mica surface. Find the metal disk first and move the focus up a little, and then move the AFM tip close to the focus.
12. Maximize the SUM signal and adjust the VERT and HORZ values to zero.
13. Engage the AFM tip.
14. Adjust the parameters to obtain a good image (*see* **Note 9**).

3.4 TEM Imaging of DNA Nanostructures

TEM is a method of shooting electron beams onto very thin samples, where electrons collide with atoms in the sample to change their direction, resulting in scattering. DNA structure has low scattering capability, so it needs to be stained by heavy metal ions:

1. Deposit 3 μL of the purified DNA sample (*see* Subheading 3.2) onto a carbon-coated copper grid; incubate for 3 min, and then remove the solution from the grid by absorbing it with a piece of filter paper at the edge of the grid.
2. Add 3 μL of the staining solution (1% (wt/vol) uranyl formate) to the grid and incubate for 15 s, and then remove the solution using a piece of filter paper. Dry the grid.
3. Examine the grids immediately or store them in an EM grid case until examination. Image the grids using an electron microscopy operation at 100 kV. Scan at low magnification (10,000–12,000×) to get an overall view of sample composition, and then examine the finer details of the sample structures at higher magnification (30,000×).

3.5 Regulation of DNA Molecular Array Transformation

The regulation of DNA molecular array transformation after folding is performed with the 11 × 4 52-bp DNA origami array because the DNA brick arrays cannot transform after folding and the 32-bp DNA origami array has a higher transformation-energy barrier than the 52-bp design. One should first screen the transformation conditions with different temperatures (**steps 1**–**3**) and formamide concentrations (**steps 4**–**6**) and then perform real-time imaging with the optimized conditions (**steps** 7–**13**):

1. For the transformation of the DNA molecular array under high temperature, first add excess of trigger strands (~10–20 nM) into the purified DNA samples (*see* Subheading 3.2) (~5 nM).

2. Incubate the mixed samples at a constant temperature (from room temperature to 60 °C) for 5 min–12 h.
3. Carry out agarose gel electrophoresis (*see* **steps 10–19** from Subheading 3.2) or AFM imaging of the samples (Subheading 3.3).
4. For the transformation of the DNA molecular array in formamide, first add formamide at a concentration of 10–40% (vol/vol) to the purified DNA samples (*see* Subheading 3.2) (~5 nM) together with the trigger DNA (~10–20 nM).
5. Incubate the samples at room temperature for 30–60 min.
6. Carry out agarose gel electrophoresis (*see* **step 11B** of the main procedure) or AFM imaging of the samples.
7. To perform the real-time imaging of DNA relay array transformation in solution, mix the purified DNA samples (*see* Subheading 3.2; ~5 nM) with excess of trigger strands (generally ~10–20 nM) for 1 min.
8. Deposit 5uL sample onto the mica surface.
9. Deposit 80 μL of 1× TE 10 mM $MgCl_2$ buffer with the optimized concentration of formamide on the mica.
10. Incubate for 5 min at room temperature.
11. Measure the sample with AFM. Approach the mica surface at a relatively low force. Scan the samples continuously until no further transformation of DNA arrays is observed in the scan area.
12. To perform the real-time imaging using temperature-controlled AFM, first set the temperature at 60 °C (or the optimized temperature from Subheading 3.5, **steps 1–3**) via a resistive heating stage (temperature range, ambient temperature to 250 °C; resolution, 0.1 °C). A cooling water fluid circuit refrigerates the piezo-scanner.
13. Scan the DNA array samples until no further transformation of DNA arrays is observed in the scan area.

4 Notes

1. Each staple DNA should have the same molar concentration.
2. Magnesium concentration has been observed to have a major effect on the quality of DNA origami and DNA bricks. The optimal $MgCl_2$ concentration may differ depending on the structure of the DNA origami or DNA bricks and on the specific needs of a given DNA origami structure.

3. Make sure to use the correct concentration of $MgCl_2$ and high-purity magnesium chloride hexahydrate. EDTA is added to 1 mM final concentration in the folding buffer to chelate divalent ion impurities that can compete with magnesium during the folding process.
4. For 42-bp DNA brick array, the samples can be subjected to a one-step isothermal annealing over 12 h. The optimized isothermal condition is 53 °C with 1× TE buffer and 10 mM $MgCl_2$ for 18 h. DNA molecular array purification and quality control.
5. To separate the two conformations of DNA arrays, use 2% (wt/vol) or 2.5% (wt/vol) agarose gel.
6. If the gel bands of DNA samples are too weak, a possible reason is that sample volume is too low. Increasing the DNA sample concentration and volume can solve the problem.
7. If sample retarded in the gel wells, maybe the sample aggregated or the agarose gel concentration was too high. Adding poly(T) overhangs on the boundary strands of DNA molecular arrays or adjusting the gel concentration and sample loading volume can solve the problem.
8. If total volume after filtering is more than 60 μL, a possible reason is that the centrifugal speed is too low or the centrifugal filters are too small. Increasing the speed of centrifugation or using 100 k centrifugal filters can solve the problem.
9. If the DNA nanostructures cannot be found, it might be due to a low DNA origami concentration. Increasing sample loading can solve the problem and check whether the samples are stored probably.

References

1. Zhang Y, Pan V, Li X, Yang X, Li H, Wang P, Ke Y (2019) Dynamic DNA structures. Small 15(26):e1900228. https://doi.org/10.1002/smll.201900228
2. Kuzyk A, Yang Y, Duan X, Stoll S, Govorov AO, Sugiyama H, Endo M, Liu N (2016) A light-driven three-dimensional plasmonic nanosystem that translates molecular motion into reversible chiroptical function. Nat Commun 7:10591. https://doi.org/10.1038/ncomms10591
3. Day HA, Pavlou P, Waller ZA (2014) i-Motif DNA: structure, stability and targeting with ligands. Bioorg Med Chem 22(16): 4407–4418. https://doi.org/10.1016/j.bmc.2014.05.047
4. Chen Y, Wang M, Mao C (2004) An autonomous DNA nanomotor powered by a DNA enzyme. Angew Chem Int Ed Engl 43(27): 3554–3557. https://doi.org/10.1002/anie.200453779
5. Yin P, Yan H, Daniell XG, Turberfield AJ, Reif JH (2004) A unidirectional DNA walker that moves autonomously along a track. Angew Chem Int Ed Engl 43(37):4906–4911. https://doi.org/10.1002/anie.200460522
6. Yurke B, Turberfield AJ, Mills AP Jr, Simmel FC, Neumann JL (2000) A DNA-fuelled molecular machine made of DNA. Nature 406(6796):605–608. https://doi.org/10.1038/35020524
7. Gerling T, Wagenbauer KF, Neuner AM, Dietz H (2015) Dynamic DNA devices and assemblies formed by shape-complementary, non–base pairing 3D components. Science

347(6229):1446–1452. https://doi.org/10.1126/science.aaa5372

8. Henderson E, Hardin CC, Walk SK, Tinoco I Jr, Blackburn EH (1987) Telomeric DNA oligonucleotides form novel intramolecular structures containing guanine-guanine base pairs. Cell 51(6):899–908
9. Zeraati M, Langley DB, Schofield P, Moye AL, Rouet R, Hughes WE, Bryan TM, Dinger ME, Christ D (2018) I-motif DNA structures are formed in the nuclei of human cells. Nat Chem 10(6):631–637. https://doi.org/10.1038/s41557-018-0046-3
10. Shangguan D, Li Y, Tang Z, Cao ZC, Chen HW, Mallikaratchy P, Sefah K, Yang CJ, Tan W (2006) Aptamers evolved from live cells as effective molecular probes for cancer study. Proc Natl Acad Sci U S A 103(32): 11838–11843. https://doi.org/10.1073/pnas.0602615103
11. Song J, Li Z, Wang P, Meyer T, Mao C, Ke Y (2017) Reconfiguration of DNA molecular arrays driven by information relay. Science 357(6349). https://doi.org/10.1126/science.aan3377
12. Wang D, Song J, Wang P, Pan V, Zhang Y, Cui D, Ke Y (2018) Design and operation of reconfigurable two-dimensional DNA molecular arrays. Nat Protoc 13(10):2312–2329. https://doi.org/10.1038/s41596-018-0039-0

Chapter 5

Two-Dimensional DNA Origami Lattices Assembled on Lipid Bilayer Membranes

Yuki Suzuki, Hiroshi Sugiyama, and Masayuki Endo

Abstract

Molecular self-assembly has attracted much attention as a method to create novel supramolecular architectures. The scaffolded DNA origami method has enabled the construction of almost arbitrarily shaped DNA nanostructures, which can be further used as components of higher-order architectures. Here, we describe a method to construct and visualize two-dimensional (2D) lattices self-assembled from DNA origami tiles on lipid bilayer membranes. The weak adsorption of DNA origami tiles onto the mica-supported lipid bilayer allows their lateral diffusion along the surface, facilitating interactions among the tiles to assemble and form large 2D lattices. Depending on the design (i.e., shape, size, and interactions with each other) of DNA origami tiles, a variety of 2D lattices made of DNA are constructed.

Key words Self-assembly, DNA nanotechnology, DNA nanostructures, DNA origami, Lipid bilayer membranes, Supported lipid bilayer, Atomic force microscopy

1 Introduction

DNA molecules form canonical double helices through their sequence-complementary base pairing. Using this unique property, i.e., an information-coding programmable polymer, various nanostructures are self-assembled from DNA strands (oligonucleotide strands) with artificially designed sequences [1, 2]. Among the methods available for constructing DNA nanostructures, the DNA origami method is now used routinely to obtain DNA nanostructures with user-defined shapes.

In the DNA origami method, a long single-stranded DNA (typically M13mp18 ssDNA, 7249 bases), called scaffold strand, is annealed to many short single-stranded DNAs (staple strands) whose sequences are designed to form the desired 2D/3D shape [3–5]. Each staple strand required to produce a certain DNA origami structure has an individual base sequence and therefore forms a double helix at a prescribed position in the origami structure. Thus, each position of the constructed DNA origami becomes

Julián Valero (ed.), *DNA and RNA Origami: Methods and Protocols*, Methods in Molecular Biology, vol. 2639, https://doi.org/10.1007/978-1-0716-3028-0_5, © Springer Science+Business Media, LLC, part of Springer Nature 2023

addressable through the sequence information of the staple strand. This addressability is a remarkable feature of DNA origami. Various molecules can be placed at "prescribed positions" by 3′/5′-end extensions or by chemical modification of the staple strand.

One of the limitations of DNA origami technology is that the size of the constructed DNA nanostructure is limited to the length of the scaffold strand. However, if the DNA origamis are used as building blocks and further assembled into higher-order structures, the advantages described above can be utilized over much larger areas. Among a variety of approaches [6–11], surface-assisted self-assembly is a promising way to obtain one- or two-dimensionally ordered DNA origami arrays [12–14]. The usable surfaces for this approach are not only solid surfaces but also soft surfaces [15–17]. In this chapter, we describe a method to construct micrometer-sized DNA origami lattices on the surface of an artificial lipid bilayer membrane (Fig. 1).

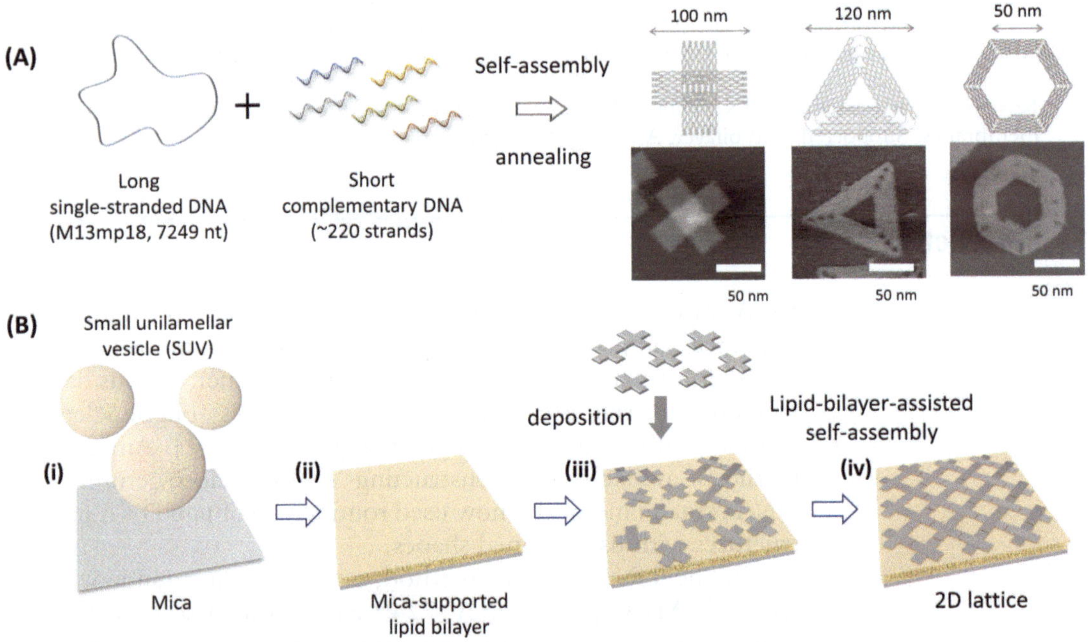

Fig. 1 (**a**) Preparation of DNA origami. DNA origami is assembled by annealing long single-stranded DNA with short complementary DNA strands. Designed DNA origami structures and their AFM images are shown on the right. (**b**) Schematic illustration of the process for DNA origami assembly on mica-supported lipid bilayers. (i) Small unilamellar vesicles are placed onto the mica surface. (ii) Lipid bilayer covered mica surface. (iii) Deposition of DNA origami onto the mica-supported lipid bilayer. (iv) DNA origami assembly and lattice formation on the mica-supported lipid bilayer

2 Materials

2.1 DNA Origami Structures

1. Scaffold single-stranded DNA: M13mp18 (Tilibit Nanosystems, München, Germany).
2. Staple DNA strands (Eurofins, Japan).
3. Thermal cycler.
4. Sephacryl S-300 (GE Healthcare, UK).
5. Gel filtration column (Bio-Rad Laboratories, CA, USA).
6. Folding buffer: 5 mM Tris–HCl (pH 7.5), 10 mM $MgCl_2$, 1 mM EDTA.
7. Electrophoresis apparatus.
8. Electrophoresis buffer: 0.5× TBE (Tris/borate/EDTA buffer), 5 mM $MgCl_2$.
9. Agarose gel: 1.0% agarose, 0.5× TBE, 5 mM $MgCl_2$.
10. SYBR Gold (BioDynamics Laboratory Inc., Japan).
11. Gel imager.
12. Milli-Q water.

2.2 Mica-Supported Lipid Bilayers

1. 1,2-Dioleoyl-*sn*-glycero-3-phosphocholine (DOPC, Avanti Polar Lipids, AL, USA).
2. Chloroform.
3. Water bath sonicator.
4. Tris-buffer: 20 mM Tris–HCl (pH 7.5), 10 mM $MgCl_2$, 1 mM EDTA.
5. Kimwipes.
6. Plastic dishes.

2.3 Lipid-Bilayer-Assisted Self-Assembly of 2D DNA Origami Lattices

1. Mica disks with a diameter of 1.5 mm (Furuuchi Chemical Corporation, Tokyo, Japan).
2. Mica-supported lipid bilayer prepared in 3.2.
3. Solution of 10 nM DNA origami structures prepared in Sect. 3.1.
4. Imaging buffer: 20 mM Tris–HCl (pH 7.6), 10 mM $MgCl_2$, 1 mM EDTA.
5. High-speed atomic force microscopy (AFM) (Nano Live Vision, RIBM, Japan).
6. Cantilever (BL-AC10EGS, Olympus, Japan).

3 Methods

3.1 DNA Origami Structures

The DNA origami shape and the sequence of staple strands for the desired shape are designed using caDNAno software [18]. DNA origami nanostructures used in this chapter are constructed through thermal annealing of a mixture containing a scaffold strand and staple strands (Fig. 1a). Folding of DNA origami structures is confirmed by agarose gel electrophoresis and by direct imaging with AFM.

3.1.1 Preparation of DNA Origami Structures

1. Mix 10 nM M13mp18 single-stranded DNA, 40 nM staple strands, and 2 μL of 10× folding buffer in a final volume of 20 μL (*see* **Note 1**).
2. Anneal the reaction mixture by reducing the temperature from 85 to 65 °C at a rate of −1.0 °C/min and then from 65 to 15 °C at a rate of −0.5 °C/min.
3. After annealing, purify the DNA origami solutions using a Sephacryl S-300 gel filtration column (*see* **Note 2**).

3.1.2 Agarose Gel Electrophoresis

1. Dilute the sample by mixing 1 μL of the purified DNA origami solution, 1 μL of 10× folding buffer, and Milli-Q water in a final volume of 10 μL.
2. Add 2 μL of 6× loading buffer to the diluted solution, and load the sample on the 1.0% agarose gel containing 0.5× TBE and 5 mM $MgCl_2$.
3. Electrophorese in electrophoresis buffer at 4 °C at constant voltage for an appropriate length of time.
4. Stain the gel with 1 × SYBR Gold for 10 min.
5. Rinse the stained gel with Milli-Q water for 10 min (*see* **Note 3**).
6. Observe the gel with an appropriate gel imager.

3.1.3 AFM Imaging

1. Cleave the mica disk which glued onto a glass stage to obtain a fresh surface.
2. Deposit a drop (~2 μL) of 1 nM DNA origami solution onto the freshly cleaved mica surface (*see* **Note 4**).
3. After incubation for 1 min at room temperature (25 °C), rinse the surface with imaging buffer (~10 μL).
4. Place the cantilever on the cantilever holder (*see* **Note 5**).
5. Fill the liquid cell with ~120 μL of imaging buffer.
6. Align the laser focusing position so that the intensity of the laser light reflected back from the cantilever is maximized.
7. Align the photodetector position so that the reflected laser makes a spot at the center of the photodetector.

8. Mount the sample prepared in **step 3** on the scanner stage.
9. Mount the scanner over the liquid cell in which the cantilever is immersed in imaging buffer.
10. Find the resonant frequency of the cantilever using an FFT analyzer.
11. Excite the cantilever at the resonant frequency by applying sinusoidal AC voltage.
12. Execute this approach until the software automatically stops the motor.
13. Adjust the set point voltage to ~75–95% of the free oscillation amplitude.
14. Gradually decrease the set point voltage until the sample is clearly imaged (*see* **Note 6**).

3.2 Lipid-Bilayer-Assisted Self-Assembly of DNA Origami Lattices

Surfaces of mica-supported lipid bilayers (mica-SLBs) are used for lipid-bilayer-assisted self-assembly of 2D DNA origami lattices (Fig. 1b) [17, 19, 20]. Mica-SLBs are prepared by the vesicle-fusion method [21, 22]. In the following sections, DNA origami tiles are electrostatically adsorbed onto the mica-supported zwitterionic lipid bilayer via divalent cations [23]. The large 2D lattices which are obtained are directly visualized with AFM (Fig. 2). Depending on the design of the DNA origami tiles, different types of 2D lattices are constructed (Fig. 3).

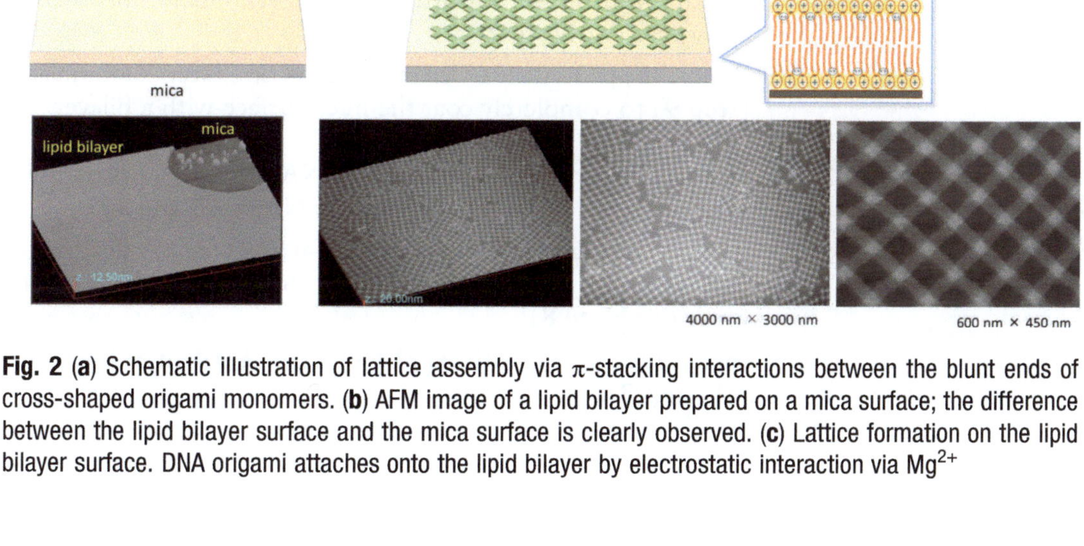

Fig. 2 (**a**) Schematic illustration of lattice assembly via π-stacking interactions between the blunt ends of cross-shaped origami monomers. (**b**) AFM image of a lipid bilayer prepared on a mica surface; the difference between the lipid bilayer surface and the mica surface is clearly observed. (**c**) Lattice formation on the lipid bilayer surface. DNA origami attaches onto the lipid bilayer by electrostatic interaction via Mg^{2+}

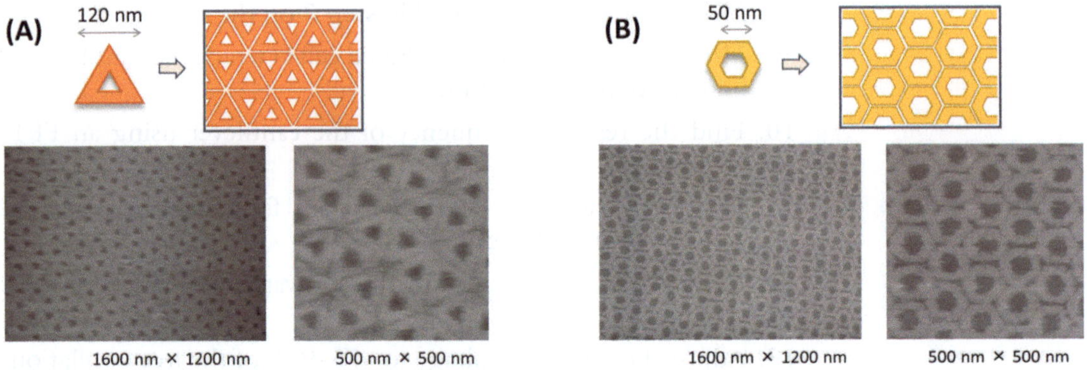

Fig. 3 2D lattices formed by close packing of DNA origami tiles. (**a**) Schematic illustration and AFM images of the close-packed triangular DNA origami tiles. (**b**) Schematic illustration and AFM images of the close-packed hexagonal DNA origami tiles. (Reproduced from Ref. [17])

3.2.1 Preparation of Mica-SLBs

1. Place 20 μL of DOPC solution (10 mg/mL) dissolved in chloroform into a round-bottom glass tube.
2. Evaporate the chloroform and dry the lipid solution under the flow of N_2 gas.
3. Further dry the lipid film under vacuum for over 3 h.
4. Add 100 μL of ultrapure water to the lipid film.
5. Vortex the glass tube to dissolve the lipid film.
6. Sonicate the glass tube to obtain small unilamellar vesicles (SUVs) (*see* **Note 7**).
7. Deposit 2 μL of SUV solution followed by 1 μL of 20 mM Tris–HCl (pH 7.6), 1 mM EDTA, and 10 mM $MgCl_2$ buffer onto freshly cleaved mica disks with a diameter of 1.5 mm.
8. Incubate the sample for 30 min at 25 °C in a sealed plastic dish containing a piece of Kimwipe moistened with ultrapure water (*see* **Note 8**).
9. Gently rinse the surface with buffer to eliminate unadsorbed liposomes (*see* **Note 9**).
10. Repeat the procedure twice from deposition (**step 7**) to rinsing (**step 9**) to completely coat the mica surface with a bilayer.

3.2.2 Lipid-Bilayer-Assisted Self-Assembly of DNA Origami Lattices

1. Deposit a drop (2 μL) of 10 nM DNA origami solution onto the preformed mica-SLB surface (*see* **Notes 10** and **11**).
2. Incubate the sample for over 60 min at 25 °C in a sealed plastic dish containing a piece of Kimwipe moistened with ultrapure water (*see* **Note 8**).
3. Image the surface with AFM (*see* **steps 4–14** in Subheading 3.1, **step 3**).

4 Notes

1. Scaffold strand is mixed with staple strands in a 1:4 ratio. This ratio is adjusted depending on the desired DNA origami structure.
2. The purification step should be repeated 2–3 times to completely remove the excess staple strands.
3. This rinsing can be omitted. However, bands on the gel are more clearly observed by reducing the background dye signal by rinsing with water.
4. The DNA origami solution is diluted with imaging buffer (or folding buffer) to an appropriate concentration which allows well-dispersed adsorption of DNA origami structures on the mica surface.
5. For AFM imaging, small cantilevers are used. Small cantilevers (9 μm long, 2 μm wide, and 130 nm thick; BL-AC10DS, Olympus, Tokyo, Japan) made of silicon nitride with a spring constant ~0.1 N/m and a resonant frequency of ~300–600 kHz in water and ~ 1500 kHz in air are commercially available from Olympus. These cantilevers have bird beak-like tips; however, the apex of the tips is not sharp enough to obtain high-resolution images of DNA nanostructures in solution. Therefore, we use custom-made cantilevers with electron-beam deposited tips at the top of the bird beak-like tips (BL-AC10EGS, Olympus, Tokyo, Japan).
6. The samples are imaged in imaging buffer with a scanning rate of 0.2–0.5 frame/s.
7. Sonicate the sample until the hydrated lipid solution becomes transparent.
8. The sample should be incubated under wet conditions to prevent drying of the lipid bilayer on the mica surface.
9. Too much rinsing will cause detachment of lipid bilayers from the mica surface.
10. Deposit the DNA origami solution immediately after the rinsing described in **step 10** in Subheading 3.2, **step 1**.
11. The optimal amount (concentration and volume) of DNA origami deposited onto the mica-SLB will change depending on origami size, origami shape, and interactions among the origami structures.

Acknowledgments

This work was supported by the Japan Society for the Promotion of Science (JSPS) Grant-in-Aid for Scientific Research (KAKENHI; grant numbers 18 K19831 and 19H04201 to Y.S., 16H06356 to H.S., and 18KK0139 to M.E.). Financial support from the Uehara Memorial Foundation and the Nakatani Foundation to M.E. are also acknowledged.

References

1. Seeman NC (1999) DNA engineering and its application to nanotechnology. Trends Biotechnol 17:437–443
2. Seeman NC (2003) DNA in a material world. Nature 421:427–431
3. Rothemund PW (2006) Folding DNA to create nanoscale shapes and patterns. Nature 440: 297–302
4. Douglas SM, Dietz H, Liedl T, Hogberg B, Graf F, Shih WM (2009) Self-assembly of DNA into nanoscale three-dimensional shapes. Nature 459:414–418
5. Dietz H, Douglas SM, Shih WM (2009) Folding DNA into twisted and curved nanoscale shapes. Science 325:725–730
6. Liu W, Zhong H, Wang R, Seeman NC (2011) Crystalline two-dimensional DNA-origami arrays. Angew Chem Int Ed Engl 50:264–267
7. Rajendran A, Endo M, Katsuda Y, Hidaka K, Sugiyama H (2011) Programmed two-dimensional self-assembly of multiple DNA origami jigsaw pieces. ACS Nano 5:665–671
8. Woo S, Rothemund PW (2011) Programmable molecular recognition based on the geometry of DNA nanostructures. Nat Chem 3:620–627
9. Zhao Z, Liu Y, Yan H (2011) Organizing DNA origami tiles into larger structures using preformed scaffold frames. Nano Lett 11:2997–3002
10. Gerling T, Wagenbauer KF, Neuner AM, Dietz H (2015) Dynamic DNA devices and assemblies formed by shape-complementary, non--base pairing 3D components. Science 347: 1446–1452
11. Tikhomirov G, Petersen P, Qian L (2017) Programmable disorder in random DNA tilings. Nat Nanotechnol 12:251–259
12. Sun X, Hyeon Ko S, Zhang C, Ribbe AE, Mao C (2009) Surface-mediated DNA self-assembly. J Am Chem Soc 131:13248–13249
13. Aghebat Rafat A, Pirzer T, Scheible MB, Kostina A, Simmel FC (2014) Surface-assisted large-scale ordering of DNA origami tiles. Angew Chem Int Ed Engl 53:7665–7668
14. Woo S, Rothemund PW (2014) Self-assembly of two-dimensional DNA origami lattices using cation-controlled surface diffusion. Nat Commun 5:4889
15. Johnson-Buck A, Jiang S, Yan H, Walter NG (2014) DNA-cholesterol barges as programmable membrane-exploring agents. ACS Nano 8:5641–5649
16. Kocabey S, Kempter S, List J, Xing Y, Bae W, Schiffels D, Shih WM, Simmel FC, Liedl T (2015) Membrane-assisted growth of DNA origami nanostructure arrays. ACS Nano 9: 3530–3539
17. Suzuki Y, Endo M, Sugiyama H (2015) Lipid-bilayer-assisted two-dimensional self-assembly of DNA origami nanostructures. Nat Commun 6:8052
18. Douglas SM, Marblestone AH, Teerapittayanon S, Vazquez A, Church GM, Shih WM (2009) Rapid prototyping of 3D DNA-origami shapes with caDNAno. Nucleic Acids Res 37:5001–5006
19. Sato Y, Endo M, Morita M, Takinoue M, Sugiyama H, Murata S, Nomura SM, Suzuki Y (2018) Environment-dependent self-assembly of DNA origami lattices on phase-separated lipid membranes. *Adv Mater Interfaces* 5:5
20. Suzuki Y, Sugiyama H, Endo M (2018) Complexing DNA origami frameworks through sequential self-assembly based on directed docking. Angew Chem Int Ed Engl 57:7061–7065
21. Mingeot-Leclercq MP, Deleu M, Brasseur R, Dufrene YF (2008) Atomic force microscopy of supported lipid bilayers. Nat Protoc 3:1654–1659
22. Uchihashi T, Kodera N, Ando T (2012) Guide to video recording of structure dynamics and dynamic processes of proteins by high-speed atomic force microscopy. Nat Protoc 7:1193–1206
23. Mengistu DH, Bohinc K, May S (2009) Binding of DNA to zwitterionic lipid layers mediated by divalent cations. J Phys Chem B 113:12277–12282

Part II

Molecular Dynamics and Simulations of DNA Origami

Chapter 6

The oxDNA Coarse-Grained Model as a Tool to Simulate DNA Origami

Jonathan P. K. Doye, Hannah Fowler, Domen Prešern, Joakim Bohlin, Lorenzo Rovigatti, Flavio Romano, Petr Šulc, Chak Kui Wong, Ard A. Louis, John S. Schreck, Megan C. Engel, Michael Matthies, Erik Benson, Erik Poppleton, and Benedict E. K. Snodin

Abstract

This chapter introduces how to run molecular dynamics simulations for DNA origami using the oxDNA coarse-grained model.

Key words DNA origami, Molecular simulation, Coarse-grained models

1 Introduction

DNA origami provides an attractive approach for designing structures and devices on the nanoscale. Particular benefits include the relative ease of the design and assembly processes, the addressability of the resulting structures, and the fine structural control that is achievable. These features are some of the reasons that the field of DNA origami has seen spectacular growth since its inception in 2006 [1], and ever more complex (e.g., in terms of size [2], function [3], and programmable motions [4]) origami designs are being realized.

Being able to model the properties of DNA origami has the potential to contribute significantly to the future of this field. Potential benefits include (i) a more detailed view of origami structure than is typically available from experiment, (ii) a realistic picture of the effect of thermal fluctuations on origami structure and behavior (as opposed to the more static viewpoint inherent to design programs), (iii) the ability to pre-screen the properties of putative origamis prior to experimental realization, and (iv) the ability to identify the physical causes of observed behaviors and

Julián Valero (ed.), *DNA and RNA Origami: Methods and Protocols*, Methods in Molecular Biology, vol. 2639,
https://doi.org/10.1007/978-1-0716-3028-0_6,

thus to contribute to a rational design process. Furthermore, access to these types of insight is becoming more important as the desired functional complexity of origami increases and the design process becomes more challenging.

Modeling can be performed at a variety of levels of resolution with inevitable trade-offs between the detail available and the time scales required for computation. At the coarser end are models like CanDo [5, 6] and mrDNA [7] in which origamis are represented as a series of mechanical elements (e.g., duplexes, single strands, junctions) with known properties, where the ease of use, robustness, and short computation times have led to widespread usage. At the other end are atomistic simulations where all atoms of the origami and local solution environment are represented [8]. In the middle are models like oxDNA [9–11], the focus of the current chapter. oxDNA is a nucleotide-level model of DNA where the interactions illustrated in Fig. 1 have been fitted to capture well the structure and mechanics of duplex DNA (e.g., bend and twist persistence lengths) and single-stranded DNA (e.g., the force-extension curve) and the thermodynamics of hybridization. These features make oxDNA very generally applicable and allow it to realistically describe the many origami properties that are outworkings of these basic biophysical properties of DNA. In terms of structure, for example, it is able to describe the local splaying out

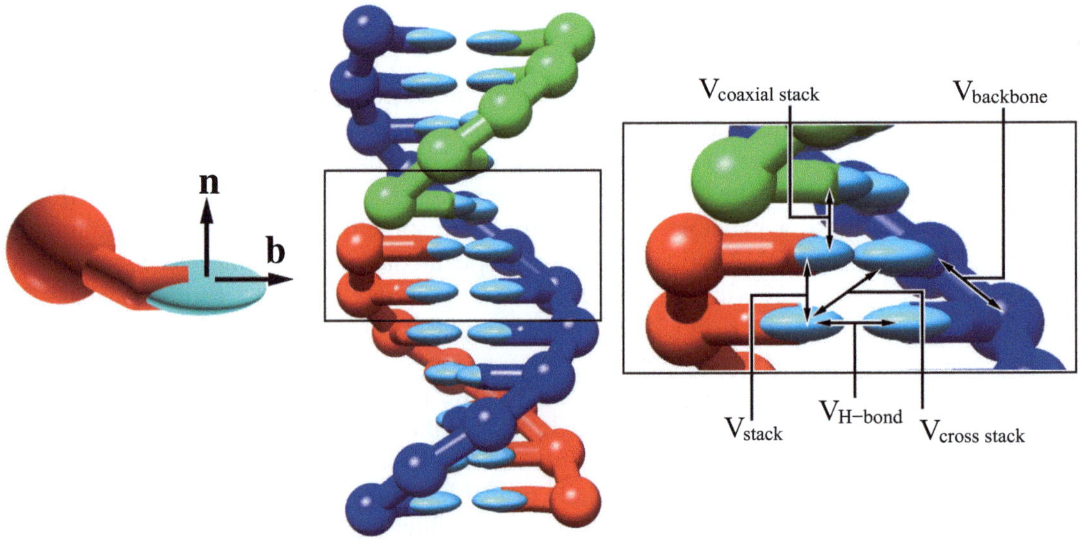

Fig. 1 The oxDNA model. Each nucleotide is a rigid body with sites corresponding to the centers of the different interactions. The position and orientation of a nucleotide are defined by **r** the position of the notional center of mass, **b** a "base" vector collinear with the stacking and hydrogen-bonding sites, and **n** a vector normal to the notional plane of the base. The basic interactions in the model are the (FENE) backbone potential connecting backbone sites, a hydrogen-bonding potential between complementary nucleotides, (coaxial) stacking interactions between bases that are (non)adjacent along the chain, electrostatic repulsion between backbone sites, and cross-stacking interactions between bases that are diagonally opposite in the duplex

of helices at four-way junctions in origami [12], the overall global structure of the most accurately determined origami structure to within the experimental resolution [12], the response of origamis to internal stresses that lead to an elastic response (e.g., twisting [11, 13] and bending [14]) or to coupling to internal degrees of freedom (e.g., breaking of base pairs, unstacking at nicks and junctions), and the properties of single strands as linkers in flexible origamis [4, 15, 16] or as bearers of tension [14, 17]. In terms of mechanics, it can describe the elastic properties of origamis, the yielding of origamis under tension [18], and even the chiral helicoidal fluctuations of twisted DNA nanotubes [13]. It can also be used to describe the thermodynamics and dynamics of hybridization processes associated with DNA origami, e.g., self-assembly [19] and the actuation of origami devices [20].

However, as with any coarse-grained models, there are features that cannot be fully described. Here, we wish to provide some caveats about the model, along with the possible implications for origami modeling. Firstly, isolated four-way junctions in their stacked form in oxDNA have a left-handed twist angle between the two helices [12], whereas experiments indicate that the preferred form is right-handed. For origami modeling this is unlikely to be an issue as most such junctions are constrained to adopt an antiparallel configuration. Secondly, the extensional modulus in oxDNA is significantly higher than for actual DNA [9]. In the parameterization of the model, it did not prove possible to generate a model that could simultaneously be a good fit of the bend, twist, and extensional moduli, and a choice was made to prioritize accurate modeling of bend and twist [9]. This shortcoming may affect mechanical responses of origami that couple to the stretching of individual helices.

Thirdly, the electrostatic interactions included in the model are of a relatively simple Debye-Hückel form that has an explicit dependence on the ionic strength of the solution. This has been fitted to reproduce the [Na^+] dependence of the hybridization thermodynamics for [Na^+] > 0.1 M. Such a simple description clearly cannot capture ion-specific effects. For example, $MgCl_2$ has an ionic strength that is just 3 times that of the ionic strength of NaCl, but has a much greater effect than this on both duplex and origami stability. Following Rothemund's original protocol [1], DNA origamis are often assembled at [$MgCl_2$] = 12.5 mM, whereas very high NaCl concentrations are required for origami assembly [21]. This difference is partly due to the particular stabilization of the stacked form of four-way junctions by Mg^{2+}. Although in oxDNA the transition from the open to the stacked form of four-way junctions occurs at a too low Na^+ concentration, this is probably helpful for modeling of origamis, as the junctions consequently behave more like in the typical Mg^{2+} conditions used in experiments. We recommend using [Na^+] = 1 M in oxDNA as representative of those conditions.

Fourthly, due to the relative lack of high-quality physical chemistry data, the parameterization of the coaxial stacking interaction in oxDNA should be viewed with some caution. Currently, the interaction has no sequence dependence, and its magnitude represents a compromise between experimental data that could not be simultaneously reproduced [11]; further, little is known about the orientation dependence of this interaction. This shortcoming may affect the modeling of blunt-ended stacking of helices that is being increasingly used to mediate multi-origami assembly [2, 22].

Currently, the oxDNA model is available as its original standalone simulation program, which can be used for both Monte Carlo and molecular dynamics, and also through LAMMPS [23], a widely used molecular dynamics package. Parallelization is much more straightforward for molecular dynamics than for Monte Carlo, and for this reason molecular dynamics is the favored approach for simulating large structures such as DNA origami.

Currently, there are two main versions of oxDNA, the original [9] and a second version [11] (sometimes called "oxDNA2" and specified by **interaction_type = DNA2** in the input file). For origami simulations one should always use oxDNA2, as its properties (e.g., DNA pitch, twist at junctions and nicks) have been fine-tuned to match experimental data on DNA origami [11]. Other additional features of oxDNA2 include major-minor grooving and electrostatics. The parameter sets for the models also come in sequence-averaged and sequence-dependent [10] varieties. In the sequence-averaged model, the interaction strengths are independent of the identity of the base (although of course base pairing can still only occur between Watson-Crick pairs). In the sequence-dependent model, the interaction strengths have been tuned to reproduce the sequence dependence of the thermodynamics of hybridization, but sequence-dependent structure and mechanics have not been explicitly incorporated. Generally, we use the sequence-averaged model to study the general behavior of the DNA and the sequence-dependent model when comparing to a specific experimental system or when we are specifically interested in the sequence dependence of the behavior. Note that there is also an RNA equivalent of the oxDNA model, which is called oxRNA [24] and has been parameterized in a similar manner, i.e., with a focus on reproducing the thermodynamics of RNA, and which is particularly useful for RNA nanotechnology.

2 Materials

2.1 Software

1. **oxDNA.** oxDNA is used to refer both to the coarse-grained model of that name and its dedicated simulation code. The oxDNA simulation code can be downloaded from https://github.com/lorenzo-rovigatti/oxDNA. Documentation

(somewhat in need of updating) is currently at https://lorenzo-rovigatti.github.io/oxDNA/. Installation is known to work on Linux and Mac OS X.

2. **tacoxDNA** is a web server (http://tacoxdna.sissa.it/) that provides a user-friendly interface to interconvert different DNA file formats (including those used by the most popular DNA design tools) in order to facilitate simulations with oxDNA. The stand-alone Python scripts are also available. A full description of tacoxDNA's capabilities is available in Ref. [25].
3. **oxView** is a browser-based viewer that allows oxDNA configurations and trajectories to be visualized and manipulated. It is also integrated with a package of Python analysis tools that provides for many common simulation analysis needs. See https://github.com/sulcgroup/oxdna-viewer and https://github.com/sulcgroup/oxdna_analysis_tools. A full description of the capabilities of oxView and its associated analysis tools is available in Ref. [26].
4. **cogli1** is a convenient lightweight viewer that can directly read in oxDNA configurations. It can be downloaded from https://sourceforge.net/projects/cogli1/. Calling cogli1 without any options lists the options available and the keyboard and mouse bindings. Note that, although it has been made publicly available, it is primarily a research tool for the developers and their collaborators.

2.2 Files

Files for the examples considered in this chapter can be obtained as a zip file both from the publisher's website and from the Oxford University Research Archive (https://ora.ox.ac.uk/objects/uuid:7a11527-3c1a-4c0f-af89-774b01f43abd). The input files for oxDNA, named input_min, input_relax, and input_sim, contain the simulation parameters (number of time steps, salt concentration, temperature, etc.) as well as the paths to the input and output for the simulation. They can be read and edited in any text editor.

3 Methods

Here we will illustrate how to simulate DNA origami using oxDNA for three examples. The first example is an asymmetric "pointer" origami block whose structure has been determined to high accuracy by cryoEM [27]. The second is a six-helix bundle that is designed to have two turns of left-handed twist [28]. The third is a "switch" that has two arms and can adopt open and closed states [22]. In the open state, which we study here, the two arms, which are connected by short single-stranded linkages, can rotate relatively freely with respect to each other. In the closed state, the two arms interlock and are held together by blunt-ended stacking between helices. The files for these examples are in the directories/folders "pointer," "2xLH," and "switch." In these case studies, we will perform the simulations using the native oxDNA simulation code rather than the implementation in LAMMPS, because only the former can be currently run on a GPU. Once GPU support is available for oxDNA in LAMMPS, this code could provide an appropriate alternative (note that tacoxDNA provides the means to interconvert between the formats required for the oxDNA simulation code and for LAMMPS).

For convenience, the oxDNA simulation code uses its own internal unit system (often referred to as simulation units). The values for parameters in the input files typically have to be given in these units. Similarly, input and output configurations are also in simulation units. The conversion factors are given in *see* **Note 1**.

Note that, although the following is written as commands entered via the command line, in practice, one will generally want to run the relaxation and simulation stages on a cluster and to submit the jobs via a queuing system rather than interactively from the command line.

3.1 Conversion to oxDNA Format

This chapter will not cover how to design an origami, but rather we assume that there is a design file available for the origami of interest that has been produced by one of the relevant computer-aided design programs:

1. First, take the design file and convert it into an initial configuration in the oxDNA format. tacoxDNA can be used to achieve this conversion, either using the interface on the website or the accompanying suite of Python scripts. Currently, it can convert from cadnano [29], Tiamat [30], CanDo [6], and vHelix [31] formats. Also, some of the more recent design tools also allow one to directly output into oxDNA format, e.g., vHelix, Adenita [32], and magicDNA. tacoxDNA can also convert an all-atom **.pdb** structure to oxDNA format; this is particularly useful for conversions from the suite of design programs for wireframe structures developed in the Bathe group [33–35] (*see* **Note 2**). These tools allow oxDNA configurations to be generated for all the most popular DNA nanotechnology design programs. The conversion tools generally output two files: an oxDNA configuration file (**.oxDNA**, **.dat**, or **.conf**) specifying the positions and orientations of the nucleotides and an oxDNA topology file (**.top**) specifying the sequence and which nucleotides are covalently bonded to which along the DNA backbone (*see* **Note 3**).
2. Once converted, add the configuration and topology files to the relevant directory (*see* **Note 4**).

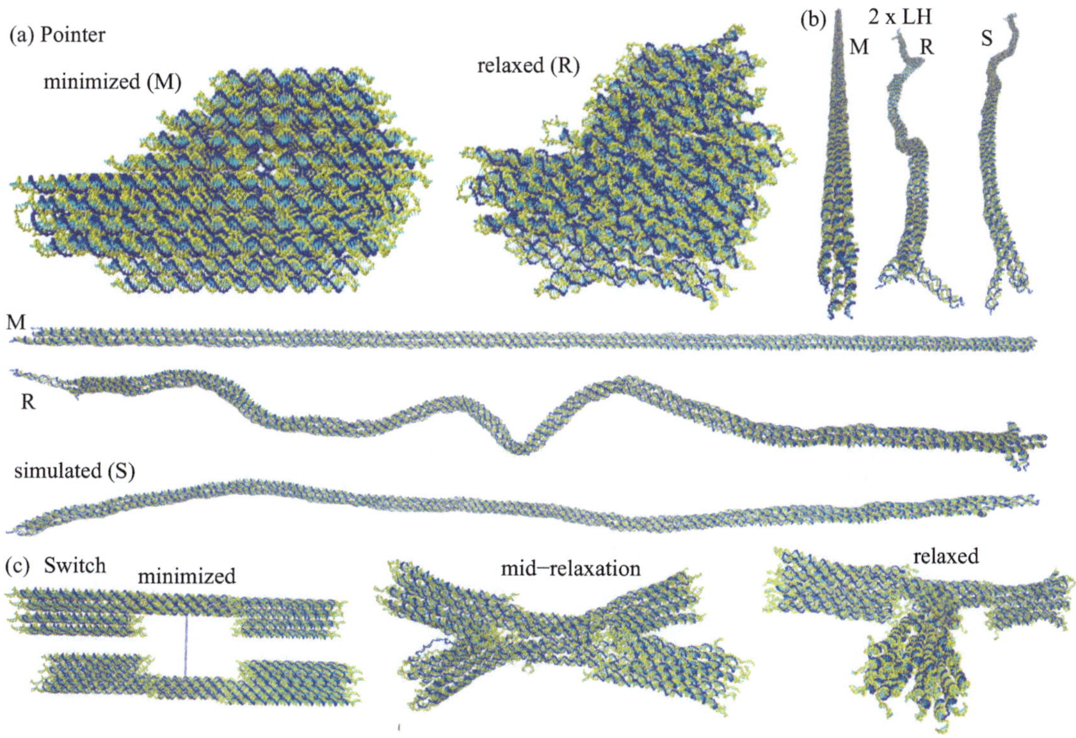

Fig. 2 Images of the three origami systems: (**a**) pointer, (**b**) 2xLH, and (**c**) switch after minimization (M), relaxation (R), and simulation (S)

3. To view the converted geometries (initially converted geometries are illustrated in Fig. 2), cogli1 can be used. A typical command to launch cogli1 and load the oxDNA configuration is

   cogli1 -m -v -t xxxx.top xxxx.conf

4. To view configurations in oxView, drag and drop the configuration and topology files into the browser window running oxView (*see* **Note 5**).

3.2 Relaxation of Initial Geometry

The configurations produced by the above conversion are typically not an appropriate starting point for a standard molecular dynamics simulation run, as there are usually some nucleotides whose excluded volumes overlap somewhat and maybe some backbone bonds that are too long. These give rise to extremely large forces that would cause the molecular dynamics simulation to fail. Therefore, it is first necessary to "relax" the structure to remove the above issues. We typically do this relaxation in two stages (*see* **Notes 6** and **7** and Fig. 3):

1. The first stage involves running a minimization algorithm for a few thousand steps (this is specified by setting **sim_type = min** in the input file) (*see* **Notes 8** and **9**). Run the minimization by entering the command

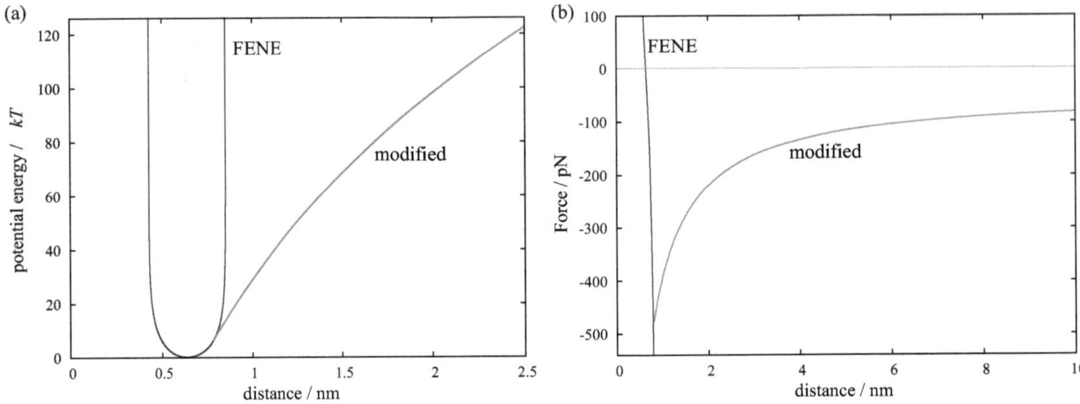

Fig. 3 (**a**) The FENE backbone potential and its modified form at long range for parameter values $F_{max} = 10$; $A = 1$ (values in simulation units; equivalent to 486 pN and 48.6 pN). (**b**) The forces due to these potentials

oxDNA input_min

from the relevant directory. The above assumes that the oxDNA executable is in one's path. If not, the full location of the executable should be given.

2. The standard set of files created by oxDNA provide the energy, trajectory, and final configuration. In the supplied input files, we have chosen them to be named **energy_min.dat, trajectory_min.dat**, and **last_conf_min.dat**. Visualize the configurations in the trajectory file by using cogli1 or oxView. Typically, the changes in structure will be very small and barely noticeable.

3. The second stage involves a fairly standard molecular dynamics simulation but with the modified backbone potential. This can be run on a GPU (*see* **Note 10**), and its aim is to allow relaxation that requires larger-scale motions (*see* **Note 11**). To run the relaxation simulation, simply enter (*see* **Notes 12–17** and Fig. 4)

oxDNA input_relax

3.3 Origami Simulation

Once one has a sufficiently relaxed origami configuration (*see* **Note 18**), it is then just a matter of running a molecular dynamics simulation. Typically for origami we will run the simulation on a GPU for it to occur in a reasonable timeframe:

1. For each origami (example input files for this stage are provided), start the simulations by running the command

oxDNA input_sim

2. Choose the appropriate timeframe to run the simulation, highly dependent on the structure and accuracy of the study (*see* **Note 19**).

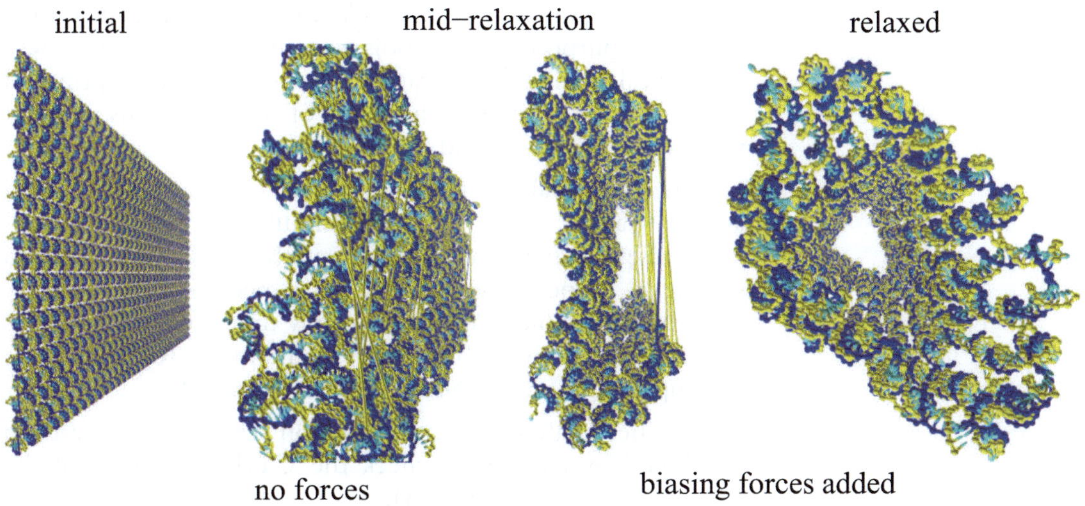

Fig. 4 Relaxation of an origami tube that was designed in cadnano as a flat sheet but with crossovers between the top and bottom helices. If no external biasing forces are added, the long bonds will cause the sheet to buckle nonuniformly, resulting in topological entanglements on relaxation. However, if forces are added that pull the top and bottom helices to the right, the sheet deforms into a C-like configuration. Consequently, none of the long bonds pass through the rest of the structure, and relaxation to a tube configuration proceeds smoothly. Note that there are two isomers for this system depending on which surface is on the outside. The other isomer can be obtained by applying the forces in the opposite direction

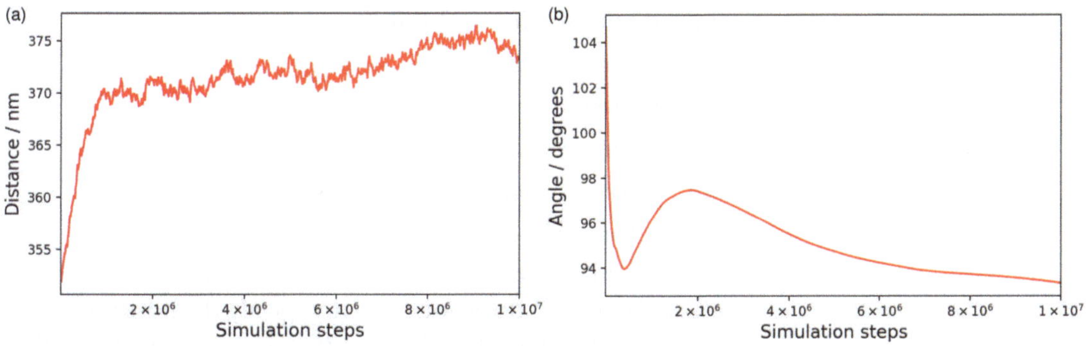

Fig. 5 Origami properties during the simulations: (**a**) "end-to-end" distance (measured between points slightly in from each end where the helices do not splay out) of the six-helix bundle, (**b**) twist angle (measured as the angle between vectors defined by ten parallel helices in each arm) between the arms of the switch

3.4 Analysis of a Simulation Trajectory

The general philosophy of the oxDNA code is to use the simulation code to produce trajectory files that can then be post-processed, rather than hard-coding lots of analysis options into the simulation code. Typically, this analysis has been done with bespoke Python scripts, such as that used to compute the twist angle of the switch in Fig. 5:

1. For analyzing simulations of DNA origami [25], use the variety of general-purpose analysis tools associated with oxView. These include calculating the average structure, performing principal component analysis, and clustering of configurations in a trajectory (*see* **Notes 20** and **21**).
2. Use the "observables" that the oxDNA code is equipped with and that can be output during the simulation. They are particularly useful when the frequency with which one needs to sample a property would otherwise necessitate excessively large trajectory files. Although mainly undocumented, details of some of the observables can be found in the **README** file accompanying the oxDNA source (*see* **Note 22**).
3. For further analysis and other Python scripts that have been developed by past users, check the UTILS directory of the oxDNA source (*see* **Note 23**).

4 Notes

1. The conversion factors for the oxDNA code's simulation units are:

   1 unit of length = 0.8518 nm

   1 unit of energy = thermal energy at 3000 K = 4.142×10^{-20} J

   1 unit of temperature = 3000 K

   1 unit of force = 48.63 pN

   1 unit of mass = 5.24×10^{-25} kg

   1 unit of time = 3.03 ps

   1 unit of force constant = 57.09 pN/nm

   1 unit of torque = 41.423 pN nM
2. Although these programs can also output in CanDo format, the CanDo format does not explicitly specify the position of the nucleotides in single-stranded sections. If one uses the CanDo to oxDNA converter, the positions this general algorithm generates for these nucleotides may not always be the most appropriate (e.g., topological entanglements may result). By contrast, the Bathe group design programs, when outputting to **.pdb**, have specific algorithms to generate sensible positions for the nucleotides in the single-stranded sections at the vertices of the wireframe structures.
3. In our three examples, the origami designs have been produced by cadnano [29], and the cadnano **.json** files are in the relevant directory of the files. When using tacoxDNA, the cadnano lattice used has to be specified. It is a square lattice for the pointer, and a hexagonal lattice for the six-helix bundle and the switch.

4. By default, the cadnano to oxDNA converter assigns a random sequence to the scaffold. As most origami properties have little dependence on sequence, this is expected to be unproblematic for the vast majority of applications. If there are particular reasons a specific sequence is needed (perhaps only in certain sections), the following discussion may be helpful: https://sourceforge.net/p/oxdna/discussion/general/thread/aa60af259b/.
5. Note that the tacoxDNA web server provides a direct link to view the converted cadnano files using oxView. In addition, oxView allows movies of trajectories to be easily created.
6. Note that for simple origamis that have a realistic starting configuration, the second stage is not always necessary. It is required for the three examples.
7. In both stages, we use a modified backbone potential. The standard oxDNA backbone potential is a FENE potential. This potential diverges beyond a certain distance (approximately 0.873 nm). The purpose of the modified potential is both to remove this divergence and to ensure that the forces resulting from stretched bonds do not damage (e.g., by breaking base pairs) the origami structure. The modified potential has the form

$$V_{\text{mod}} = V_{\text{FENE}} \text{ for } r \leq r_{\text{max}}$$

$$V_{\text{mod}} = A\, r + B \log(r) + C \quad \text{for } r > r_{\text{max}}$$

where V_{FENE} is the original backbone potential and r_{max} is the distance at which the force due the FENE potential is equal to F_{max} (the value of F_{max} is set in the input file using **max_backbone_force**). A corresponds to the limiting value of the force at large r and is set using the variable **max_backbone_force_far** in the input file; B and C are chosen so that the modified potential is continuous and differentiable at r_{max}. The original and modified potentials are illustrated in Fig. 3 along with the resulting force. The modified force increases relatively gently as r increases with the force tending to A at sufficiently large r.
8. An alternative to the use of the minimization algorithm is to run a short Monte Carlo simulation (see, e.g., the relevant input file on the tacoxDNA server).
9. This is usually sufficient to remove all particle overlaps and those overstretched bonds that are too long by only a relatively small amount (on the order of a few nanometers). It can also help nucleotides that are designed to be base-paired to correctly orient themselves to fully base-pair. The minimization runs on a single CPU core, but even for a full-size origami, it should take no more than a few minutes.

10. To run oxDNA on a GPU, the source code needs to be compiled using the flag -DCUDA = 1, and an appropriate GPU has to be available on your machine/cluster. Note also that there are two GPU parallelization approaches implemented in the oxDNA code. In the original, the force calculation is parallelized over particles, whereas in the second "edge-based" approach (specified by **use_edge = 1** in the input file), the parallelization is over interacting pairs of particles [36]. We recommend the edge-based approach as it is generally more computationally efficient.
11. For a design file that provides a pretty accurate representation of the three-dimensional structure of the origami (e.g., the pointer), the first stage is virtually sufficient, and this stage can be relatively short. The two other case studies both require larger-scale motions. The initially converted geometry of the six-helix bundle is untwisted, but has significant internal stresses (due to the "deletions" in the design [37]) that are partially resolved by the global twisting of the structure. The switch has two sections that are able to freely rotate with respect to each other and are connected by short single-stranded sections. These are significantly extended in the starting structure, and the two blocks are in an atypical parallel arrangement (Fig. 2).
12. In the relaxation MD run, it is helpful to use a tightly coupled thermostat to more quickly remove the energy liberated by relaxation to a lower-energy configuration. In particular, we use the thermostat due to Bussi et al. [38] (specified by **thermostat = bussi** in the input file with the algorithm parameter **bussi_tau** controlling the tightness of the coupling). By contrast, in standard MD runs, we typically use the Andersen-like thermostat of Ref. [39] (specified by **thermostat = brownian**), with parameters designed to lead to efficient diffusion of strands.
13. In these examples, the lengths of the runs are different, reflecting the differing extents of the motions required to achieve relaxation. For the pointer we only use 10^4 steps (where we use an integration time step of 0.005 (in simulation units)). The six-helix bundle requires on the order of 10^5 steps. The switch requires on the order of 10^6 steps, as the relaxation is hindered by the initially parallel arrangement of the two arms. The long bonds pull the two arms together until they come into contact with each other. The further relaxation of the long bonds is hindered until the arms diffuse sufficiently far from a parallel arrangement.
14. Pictures of the three examples at different stages of the relaxation are illustrated in Fig. 2, and movies of the relaxation trajectories are available at the oxDNA YouTube channel. The

following videos are available on the oxDNA YouTube channel:

Switch relaxation (side view): https://www.youtube.com/watch?v=XpjnjYIa2N8

Switch relaxation (top view): https://www.youtube.com/watch?v=4SPMRBIA_Rs

2xLH relaxation (side view): https://www.youtube.com/watch?v=cm2Gzrx1hVk

2xLH relaxation (end view): https://www.youtube.com/watch?v=w07rySiZ40w

2xLH simulation: https://www.youtube.com/watch?v=sxW_Fz46z4Y

Tube (Fig. 4) relaxation: https://www.youtube.com/watch?v=sAjCGKe_iwA

Tube relaxation (close-up): https://www.youtube.com/watch?v=pyO5zK5ndJY

In the videos of the relaxation of the six-helix bundle, the rotation of the ends as the origami adopts its overall twisted geometry is very apparent. For this origami, relief of the twist stress during the relaxation also leads to local bending that is not representative of the equilibrated state.

15. For a general example, visualizing the trajectory can help to show how close a system is to completing the relaxation, particularly for cases involving long bonds.

16. One alternative to the above scheme that has recently become available is to use the multi-resolution DNA (mrDNA) tool of Maffeo and Aksimentiev [7]. This allows modeling of DNA origami at coarser levels of detail than oxDNA. One potential advantage of this tool for relaxation is that the compute time required for relaxation may be significantly shorter when large-scale motion is required due to the lower resolution of the model. The final configurations can be output to oxDNA format and then used as starting points for a standard oxDNA molecular dynamics simulation (note that mrDNA can automatically perform an oxDNA minimization and relaxation on the configuration).

17. The conversion and relaxation of the three examples above should all be straightforward. However, this is not always the case. The first potential problem is in the conversion of the cadnano file. Although the underlying Python script can handle most cadnano files, it occasionally fails when less common features are present. If this occurs, we recommend trying one of the other tools that can load cadnano files and output in oxDNA format, e.g., mrDNA, vHelix, or Adenita.

The two most common problems in the relaxation stage are that (i) part of the DNA origami becomes irreversibly damaged due to the breaking of base pairs as a result of the internal stresses that are present in the initial structure and (ii) there are topological entanglements due to the layout of the origami structure in the design file.

To overcome the first problem, one can introduce artificial "mutual traps" between those base pairs that are liable to break. (These traps are simply a harmonic potential in the distance between the relevant nucleotides.)

Mutual traps are one example of a number of types of external forces that can be applied using the native oxDNA code. To activate this feature, one must add **external_-forces = true** in the input file and also supply an external force file (specified by the key **external_forces_file** = <**file**>). For mutual traps the format of this file is the following for each pair of nucleotides *i* and *j* interacting via a mutual trap:

```
{
type = mutual_trap
particle = <i>
ref_particle = <j>
stiff = 1.
r0 = 1.2
}
{
type = mutual_trap
particle = <j>
ref_particle = <i>
stiff = 1.
r0 = 1.2
}
```

The above provides sensible values for the bond stiffness and equilibrium internucleotide separation for the harmonic potential.

This scheme would require the identification of the relevant nucleotides, and so a simpler general solution is to apply mutual traps between all pairs of nucleotides that are designed to be base-paired. A file specifying these mutual traps can be generated by one of the Python scripts associated with oxView [26] and by the Tiamat converter in tacoxDNA. The latter approach also has the potential advantage that it allows one to change the parameters of the modified backbone potential to increase the forces applied and so enables sections connected by overstretched bonds to be brought together more quickly without having to worry about base pairs being broken.

Topological entanglements can result from the layout of the origami in the design file. This is particularly an issue for origamis designed with cadnano that involve multiple blocks or that have a structure that is not compatible with the hexagonal and square lattices available in cadnano. For example, one could represent a tube in cadnano as a flat sheet, but with crossovers between the top and bottom helices (Fig. 4). If one tries to relax such a structure in oxDNA, initially the bonds between the top and bottom helices will pass through the rest of the origami. The forces associated with these over-stretched bonds will cause the sheet to buckle. If one is lucky, the sheet might coherently buckle into a C-shape, removing potential entanglements, and relaxation into a tube could be successful. However, it is much more likely that some parts of the sheet will buckle more into an "S-shape," so that these long bonds still pass through the sheet (Fig. 4). In this case, relaxation will lead to a malformed structure which cannot escape from this topologically entangled state because the excluded volume in the oxDNA mode prevents nucleotides passing through each other (there is no excluded volume associated with a connection between nucleotides, so stretched bonds can pass through each other).

The best way to avoid such topological problems is of course to not have them in the first place, so if one is designing an origami in cadnano that one will want to model with oxDNA, we recommend trying to organize the layout of the design so topological problems are avoided if possible, e.g., by displacing parts of the structure so that extended bonds do not pass through the rest of the origami. However, if one is working with an existing cadnano file, there are a number of different approaches that have been used by different oxDNA users with new tools recently becoming available that makes this easier.

The first set of approaches involves manipulation of the structure prior to relaxation. Probably the most convenient way is to use a program that allows a real space representation of the structure to be manipulated on screen. For example, oxView has the functionality to select origami blocks and then to translate and rotate them. These manipulation features can also be sometimes used to aid relaxation even when there is no topological entanglement of strands. For example, if we rotate one arm of the switch by 90°, when the two arms get pulled together, they will no longer collide, and relaxation occurs more rapidly (order of 10^5 steps). Somewhat similarly, Adenita allows origami elements to be manipulated. In addition, oxView has a rigid-body dynamics mode that can aid the generation of sensible initial geometries. Also, manipulations of an origami structure can be performed in mrDNA via Python scripting.

An alternative way to manipulate the origami structure may be to directly edit the cadnano **.json** file prior to conversion using, for example, the online json editor available at https://jsoneditoronline.org/, although this of course requires a certain level of understanding of the cadnano format.

The second set of approaches directs the relaxation away from topological traps. The facility in the oxDNA code to add bespoke forces to different parts of the origami is particularly useful in this regard.

Adding the snippet below to an external forces file will cause a constant external force to be applied to the center of mass of a particular nucleotide *i* in the *z*-direction:

```
{
type = string
particle = <i>
F0 = 1.0
rate = 0.0
dir = 0.0, 0.0, 1.0
}
```

The options set the magnitude of the force (in simulation units) and its direction. For example, for the case of the flat sheet above, if one adds forces pulling the top and bottom helices out of the plane in the same direction, this will encourage the sheet to deform into a C-shape and relax correctly (Fig. 4).

In cases with potential topological entanglements, we recommend visualization of the relaxation trajectory as spotting the entanglements in a relaxed configuration is not always straightforward.

18. Hopefully, the relaxation runs for the three examples are sufficiently long that the end configuration will always successfully run in the standard MD run. However, as the relaxation is a stochastic process, this cannot be fully guaranteed. If the simulation does happen to fail initially, it is simply a matter of further relaxing the end configuration.

19. An important question is for how long to run the simulation. This will of course depend on what one is trying to achieve; this could range from getting a quick feel for what an origami "looks like" to comprehensively sampling the configuration space of the origami. In the latter case, the answer to this question depends on two main factors: the time required to equilibrate the system (i.e., achieve a representative starting configuration for sampling) and the time to appropriately sample the configuration space.

Even though the starting configuration needs to have been sufficiently relaxed to be a starting point for MD, it may not necessarily yet be representative of the equilibrium ensemble at the temperature of interest. Therefore, an important part of any simulation is an equilibration period that allows that to be achieved. When calculating average equilibrium properties of an origami, the data from the equilibration period should of course not be included. Typically, to determine when equilibration has occurred, one looks at the behavior of relevant properties of the system as a function of time. One obvious starting point is the energy. If we take the case of the six-helix bundle, the starting energy is actually not atypical of the ensemble. However, there are other slower degrees of freedom. For example, Fig. 5a shows the end-to-end distance. As well as the expected twisting, the release of internal stress during the relaxation leads to considerable bending, causing the end-to-end distance in the resulting configuration to be significantly shorter than in an equilibrated configuration, and it takes at least 10^6 steps to reach a more realistic value.

Somewhat similarly, an appropriate sampling time depends on the time scales associated with fluctuations in the slowest degrees of freedom (and also whether the property of interest depends on these degrees of freedom). The pointer is relatively rigid and should sample its configurational ensemble fairly rapidly. For the six-helix bundle, although again relatively rigid, if one wants to characterize the long-wavelength elastic thermal fluctuations accurately, one needs to use very long runs (see Ref. [13] for the unusual chiral fluctuations of this system). For the switch, the diffusive nature of the relative motion of the arms and the large length scales involved mean that fully sampling the open state will also require very long runs. As can be seen from Fig. 5, the 10^7 step simulation runs in the examples for these two cases are insufficient to accurately calculate the probability distributions for the end-to-end distance or the inter-arm angle of the switch. In the switch, there is an even longer time scale associated with its opening and closing.

One important point is that the absolute time scales associated with coarse-grained simulations, such as those using oxDNA, should be interpreted with caution. Typically, time scales are reported using the time unit defined by the units of the basic oxDNA interactions. However, coarse graining reduces the time scale separation between microscopic motions and diffusion times; thus, the effective time may be considerably larger, and it may be more appropriate to report times based on a mapping of diffusion times [19].

20. OxView also allows the results of analyses to be overlaid on structure, for example, to identify flexible sections or points in a structure where base pairs are more likely to break due to internal stresses.
21. Access to examples may also be a useful source to see how particular tasks can be achieved. A number of examples come with the oxDNA source, some of which are documented at https://dna.physics.ox.ac.uk/index.php/Category:Examples, and all the analysis scripts associated with oxView include example simulations. We also encourage users, when publishing results produced using oxDNA, to deposit relevant data, including input files and scripts used to process data. Links to these data deposits are included on the publications page of the oxDNA website: https://dna.physics.ox.ac.uk/index.php/Publications.
22. Additional bespoke observables can be created through the plugin infrastructure of the code.
23. Note, however, that they are not maintained and are often undocumented, the exception to the latter being the scripts documented here: https://oxdna-utils.readthedocs.io/.

Acknowledgments

We are grateful for support from the EPSRC Centre for Doctoral training, Theory and Modelling in Chemical Sciences, under grant EP/L015722/1.

References

1. Rothemund PWK (2006) Folding DNA to create nanoscale shapes and patterns. Nature 440:297–302
2. Wagenbauer KF, Sigl C, Dietz H (2017) Gigadalton-scale shape-programmable DNA assemblies. Nature 552:78–83
3. Ramezani H, Dietz H (2020) Building machines with DNA molecules. Nat Rev Genet 21:5–26
4. Zhou L, Marras AE, Huang C-M, Castro CE, Su HJ (2018) Paper origami-inspired design and actuation of DNA nanomachines with complex motions. Small 14:1802580
5. Castro CE, Kilchherr F, Kim D-N, Shiao EL, Wauer T, Wortmann P, Bathe M, Dietz H (2011) A primer to scaffolded DNA origami. Nat Methods 8:221–229
6. Kim D-N, Kilchherr F, Dietz H, Bathe M (2012) Quantitative prediction of 3D solution shape flexibility of nucleic acid nanostructures. Nucleic Acids Res 40:2862–2868
7. Maffeo C, Aksimentiev A (2020) MrDNA: a multi-resolution model for predicting the structure and dynamics of DNA systems. Nucleic Acids Res 48, Advance Article
8. Yoo J, Aksimentiev A (2013) In situ structure and dynamics of DNA origami determined through molecular dynamics simulations. Proc Natl Acad Sci U S A 110:20099–20104
9. Ouldridge TE, Louis AA, Doye JPK (2011) Structural, mechanical and thermodynamic properties of a coarse-grained DNA model. J Chem Phys 134:085101
10. Šulc P, Romano F, Ouldridge TE, Rovigatti L, Doye JPK, Louis AA (2012) Introducing sequence-dependent interactions into a coarse-grained DNA model. J Chem Phys 137:135101

11. Snodin BEK, Randisi F, Mosayebi M, Šulc P, Schreck JS, Romano F, Ouldridge TE, Tsukanov R, Nir E, Louis AA, Doye JPK (2015) Introducing improved structural properties and salt dependence into a coarse-grained model of DNA. J Chem Phys 142: 234901
12. Snodin BEK, Schreck JS, Romano F, Louis AA, Doye JPK (2019) Coarse-grained modelling of the structural properties of DNA origami. Nucleic Acids Res 47:1585–1597
13. Tortora MMC, Mishra G, Prešern D, Doye JPK (2020) Chiral shape fluctuations and the origin of chirality in cholesteric phases of DNA origamis. Sci. Adv. 6:eaaw8331
14. Shi Z, Castro CE, Arya G (2017) Conformational dynamics of mechanically compliant DNA nanostructures from coarse-grained molecular dynamics simulations. ACS Nano 11:4617–4630
15. Sharma R, Schreck JS, Romano F, Louis AA, Doye JPK (2017) Characterizing the motion of jointed DNA nanostructures using a coarse-grained model. ACS Nano 11:12426–12435
16. Huang CM, Kucinic A, Le J, Castro CE, Su H-J (2019) Uncertainty quantification of a DNA origami mechanism using a coarse-grained model and kinematic variance analysis. Nanoscale 11:1647–1660
17. Engel MC, Romano F, Louis AA, Doye JPK (2020) Measuring internal forces in single-stranded DNA: application to a DNA force clamp. J Chem Theory Comput. Submitted
18. Engel MC, Smith DM, Jobst MA, Sajfutdinow M, Liedl T, Romano F, Rovigatti L, Louis AA, Doye JPK (2018) Force-induced unravelling of DNA origami. ACS Nano 12:6734–6747
19. Snodin BEK, Romano F, Rovigatti L, Ouldridge TE, Louis AA, Doye JPK (2016) Direct simulation of the self-assembly of a small DNA origami. ACS Nano 10:1724–1737
20. Shi Z, Arya G (2020) Free-energy landscapes of salt-actuated reconfigurable DNA nanodevices. Nucleic Acids Res 48:548–560
21. Martin TG, Dietz H (2012) Magnesium-free self-assembly of multi-layer DNA objects. Nat Commun 3:1103
22. Gerling T, Wagenbauer KF, Neuner AM, Dietz H (2015) Dynamic DNA devices and assemblies formed by shape complementary, non-base pairing 3D components. Science 347:1446–1452
23. Henrich O, Gutierrez-Fosado YA, Curk T, Ouldridge TE (2018) Coarse-grained simulation of DNA using LAMMPS. Eur Phys J E 41:57
24. Šulc P, Romano F, Ouldridge TE, Doye JPK, Louis AA (2014) A nucleotide-level coarse-grained model of RNA. J Chem Phys 140: 235102
25. Suma A, Poppleton E, Matthies M, Šulc P, Romano F, Louis AA, Doye JPK, Micheletti C, Rovigatti L (2019) TacoxDNA: a user-friendly web server for simulations of complex DNA structures, from single strands to origami. J Comput Chem 40:2586–2595
26. Poppleton E, Bohlin J, Matthies M, Sharma S, Zhang F, Šulc P (2020) Design, optimization, and analysis of large DNA and RNA nanostructures through interactive visualization, editing, and molecular simulation. Nucleic Acids Res. Submitted. bioRXiv 2020.01.24.917419
27. Bai X-C, Martin TG, Scheres SHW, Dietz H (2012) Cryo-EM structure of a 3D DNA-origami object. Proc Natl Acad Sci U S A 109:20012–20017
28. Siavashpouri M, Wachauf CH, Zakhary MJ, Praetorius F, Dietz H, Dogic Z (2017) Molecular engineering of chiral colloidal liquid crystals using DNA origami. Nat Mater 16:849–856
29. Douglas SM, Marblestone AH, Teerapittayanon S, Vazquez A, Church GM, Shih WM (2009) Rapid prototyping of 3D DNA-origami shapes with caDNAno. Nucleic Acids Res 37:5001–5006
30. Williams S, Lund K, Lin C, Wonka P, Lindsay S, Yan H (2009) Tiamat: a three-dimensional editing tool for complex DNA structures. In: Lecture notes in computer science, vol 5347. Springer, Berlin/Heidelberg, pp 90–101
31. Benson E, Mohammed A, Gardell J, Masich S, Czeizler E, Orponen P, Högberg B (2015) DNA rendering of polyhedral meshes at the nanoscale. Nature 523:441–444
32. de Llano E, Miao H, Ahmadi Y, Wilson AJ, Beeby M, Viola I, Barisic I (2020) Adenita: interactive 3D modelling and visualization of DNA nanostructures. bioRxiv
33. Veneziano R, Ratanalert S, Zhang F, Yan H, Chiu W, Bathe M (2016) Designer nanoscale DNA assemblies programmed from the top down. Science 352:1534
34. Jun H, Zhang F, Shepherd T, Ratanalert S, Qi X, Yan H, Bathe M (2019) Autonomously designed free-form 2D DNA origami. Sci Adv 5:eaav0655
35. Jun H, Shepherd TR, Zhang K, Bricker WP, Li S, Chiu W, Bathe M (2019) Automated sequence design of 3D polyhedral wireframe DNA origami with honeycomb edges. ACS Nano 13:2083–2093

36. Rovigatti L, Šulc P, Reguly IZ, Romano F (2015) A comparison between parallelization approaches in molecular dynamics simulations on GPUs. J Comput Chem 36:1–8
37. Dietz H, Douglas SM, Shih WM (2009) Folding DNA into twisted and curved nanoscale shapes. Science 325:725–730
38. Bussi G, Donadio D, Parrinello M (2007) Canonical sampling through velocity-rescaling. J Chem Phys 126:014101
39. Russo J, Tartaglia P, Sciortino F (2009) Reversible gels of patchy particles: role of the valence. J Chem Phys 131:014504

Chapter 7

All-Atom Molecular Dynamics Simulations of Membrane-Spanning DNA Origami Nanopores

Himanshu Joshi, Chen-Yu Li, and Aleksei Aksimentiev

Abstract

Building on the recent technological advances, all-atom molecular dynamics (MD) simulations have become an indispensable tool to study the molecular behavior at nanoscale. Molecular simulations have been used to characterize the structure, dynamics, and mechanical and electrical properties of DNA origami objects. In this chapter we describe a method to build all-atom model of lipid-spanning DNA origami nanopores and perform molecular dynamics simulations in explicit electrolyte solutions.

Key words DNA origami nanopores, Lipid bilayer membrane, Molecular dynamics simulation, Ionic current

1 Introduction

State-of-the-art DNA nanotechnology has enabled an efficient route to construct and control synthetic nanoscale systems which perform specific functions [1]. Introduced in 2006, DNA origami method [2] has filled fresh aspirations to the field of DNA nanotechnology. Computer-aided design of staple strands has simplified the protocols for synthesizing DNA origami constructs [3]. In 2012, Simmel lab at the Technical University of Munich demonstrated the insertion of a cholesterol-anchored DNA origami barrel in a lipid vesicle [4]. Simultaneously, the Howorka group at the university college of London also characterized a series of lipid-spanning DNA nanopores using various hydrophobic modifications including ethyl-phosphorothioate [5], streptavidin, porphyrin [6], and added functionalities [7]. Membrane-spanning DNA nanostructures have opened new avenues in the area of synthetic membrane channels for various applications [8] such as transmembrane molecular transport [9], DNA translocation [10], mimicking the natural enzymes [11], etc. The highly programmable functionalities of the DNA backbone provide DNA nanopores an advantage over the biological protein nanopores. One of the

Julián Valero (ed.), *DNA and RNA Origami: Methods and Protocols*, Methods in Molecular Biology, vol. 2639,
https://doi.org/10.1007/978-1-0716-3028-0_7,

important features of the DNA origami nanopore is that their diameter can be controlled externally by sequence design. Keyser group at the Cambridge University has synthesized DNA channels with variable shape and size [12–14].

With the recent advancement in computer architecture and numerical algorithms, computational methods, in particular coarse-grained and all-atom molecular dynamics (MD) simulations, can now provide accurate microscopic account of the structure and dynamics of self-assembled DNA nanosystems [15, 16]. Previously, all-atom MD simulations were successfully used to characterize the structural [17–19], mechanical [20, 21], and electrical [22] properties of DNA nanostructures. MD simulation of DNA nanostructures using coarse-grained oxDNA model [23] successfully reproduced the experimental observations such as cryo-EM structures [24]. Due to their contrasting interaction, the assembly of DNA and lipid is not very common in nature. Molecular simulations can be particularly helpful in understanding the molecular mechanism and interaction governing the self-assembly of DNA nanopore in lipid bilayer membranes [25].

Our group has pursued the application-oriented computational exploration of membrane-tethered DNA nanosystems using all-atom and coarse-grained molecular dynamics simulations. Our 2015 study characterized the mechanism of ionic conductance, mechanical gating, and electro-osmotic transport in membrane-embedded DNA nanopores using the all-atom MD method [26]. In the subsequent studies, based on the results of all-atom MD simulation trajectories, we demonstrated a novel mechanism of toroidal pore formation in lipid bilayer membrane by DNA nanopores [12, 13, 27]. In one of the recent studies, all-atom and coarse-grained MD simulation revealed that a DNA origami nanopore in a lipid bilayer membrane catalyzes the spontaneous transport of lipids from one leaflet to other with a rate which is at least three orders of magnitude higher than in similar biological enzymes [11]. The fluorescence microscopy experiments confirmed the predication of microscopic MD simulations. In this chapter, we summarize the general protocols used to perform all-atom MD simulations of membrane-embedded DNA origami nanopore in lipid bilayer membranes.

2 Materials

2.1 Software and Online Servers

1. **caDNAno**. Developed by Douglas et al., caDNAnano [3] is the most widely used computer program to design the DNA origami nanostructures. The latest version of caDNAno can be downloaded from http://cadnano.org. There are several other interactive and command-based interfaces available to design

the DNA nanostructure including Nanoengineer-1, DAEDALUS [28], NAB [29], and Tiamat [30] which can also be useful to create DNA nanopore depending on the design of the nanopore. More details about this can be found in our previous tutorial [31].

2. **caDNAno to pdb converter**. The output "json" file of caDNAno design can be uploaded into the ENRG-MD server http://bionano.physics.illinois.edu/origami-structure and converted to the PDB file format representing all atoms in the structure. Alternatively, one can also use mrDNA package [32] for json to all-atom PDB conversion; please refer to the website for the details about mrDNA: https://gitlab.engr.illinois.edu/tbgl/tools/mrdna.
3. **MarvinSketch.** Developed by ChemAxon, MarvinSketch is a useful and open-source software to draw the chemical structure of nonstandard residues and export their 3D all-atom model. It can be downloaded from https://chemaxon.com/products/marvin.
4. **VMD.** VMD [33] is a molecular visualization program to display and animate atomistic structures. It can be downloaded from http://www.ks.uiuc.edu/Research/vmd. VMD is supported by majority of operating systems (Windows, UNIX, Mac OS) and also provides a tcl-based programming interface which is very helpful for the analysis of the simulation trajectories. For more instruction, please refer to VMD use guide [34] and VMD tutorials [35].
5. **CGenFF.** CGenFF [36] web server can be accessed at https://cgenff.umaryland.edu/. It is a utility to create the CHARMM-compatible topology and parameters of small organic molecules to perform all-atom MD simulations.
6. **CHARMM-GUI.** CHARMM-GUI [37] is a web server to interactively build the membrane systems. The preferred lipid-type membrane system can be assembled and downloaded from http://www.charmm-gui.org.
7. **NAMD.** NAMD is a highly parallel molecular dynamics code which also supports CUDA-based acceleration. NAMD is compatible with Linux/UNIX, Mac OS X, or Windows operating systems and runs on laptops as well. However, for performing the all-atom MD simulation of membrane-spanning DNA nanopore, it is recommended to use supercomputers with parallel programming environment. For more details on NAMD, please refer to the NAMD user guide [38] and NAMD tutorials [39].
8. **CHARMM topology and parameter files**. CHARMM topology files are required to create the all-atom structures. Also, we use CHARMM force field parameters [40] with latest CUFIX

[41] to describe the inter- and intramolecular interaction in the system. CHARMM topology and parameters can be downloaded from https://www.charmm.org/charmm/. The latest non-bonded (CUFIX) corrections to ion-DNA and ion-lipid interaction parameters can be downloaded from http://bionano.physics.illinois.edu/CUFIX.

9. **PERL.** Perl is a general-purpose scripting language available across all the major operating systems. In this chapter, we will use a file created from perl script to enforce hexahydrate water structure around Mg^{2+} ions.

2.2 Required Files

1. **DNPinMembraneTutorial package**. The files used in this chapter (scripts and other support files) are available at http://bionano.physics.illinois.edu/sites/default/files/dnpinmembranetutorial.zip
2. **origamiTutorial package**. Some files used in our guide to simulate DNA origami nanostructure as described in our previous chapter [42] will also be used in this chapter. These files can be found on our website at http://bionano.physics.illinois.edu/sites/default/files/origamitutorial.tar.gz

3 Methods

In this section, we describe the steps involved in building an all-atom model of DNA origami nanopore in lipid bilayer membranes, performing MD simulations in explicit electrolyte solution and calculating the ionic current in MD simulation. The whole section is arranged in the following manner: in Subheading 3.1 we will describe how to build and assemble atomistic models of various components of the system including DNA origami nanopore, hydrophobic lipid anchor, lipid bilayer membrane, water, and ions; in Subheading 3.2 we will describe the methodology to run equilibrium MD simulations; in Subheading 3.3 we will describe the method of applying the electric field in MD simulation using NAMD. Finally, in Subheading 3.4 we will be discuss how we calculate ionic currents from the all-atom MD simulation trajectories.

3.1 Assembling All-Atom Model of DNA Origami Nanopore in Lipid Bilayer Membrane

1. The first step in building the system is to create the caDNAnano design of DNA origami nanopore. We chose a square lattice and draw the design of a four-helix DNA nanopore in caDNAno as per the design used in experiments [11] (Fig. 1).
2. Save the output "json" file of the caDNAnano design used in this chapter which can be find in the step1 subdirectory of the DNPinMembraneTutorial folder.

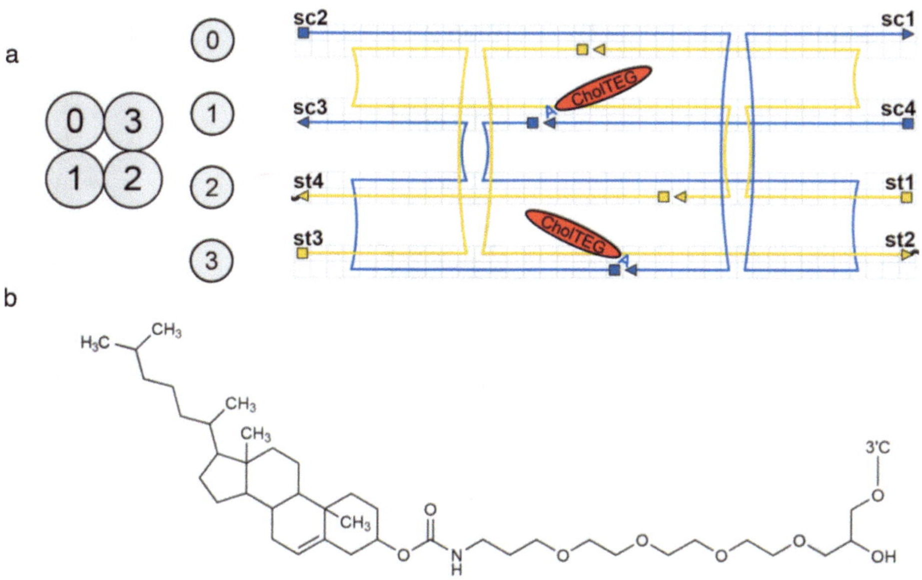

Fig. 1 Design of DNA origami nanopore with cholesterol anchor for membrane tethering. (**a**) caDNAno design of a DNA nanostructure. The sequence of the structure is given in the reference article [11]. The "CholTEG" tag indicates a cholesterol group chemically linked to a nucleotide. (**b**) Chemical structure of a cholesterol group containing a TEG linker (CholTEG). The "3′C" tag indicates the 3′ carbon of the modified nucleotide

3. Next, upload the json file, Fig. 2a, into the ENRG-MD web server http://bionano.physics.illinois.edu/origami-structure, and follow the instruction provided over the webpage.

4. Download and unzip the output file from the ENRG-MD web server, in the subdirectory step2 of the DNPinMembraneTutorial folder. The output folder contains the necessary files to run NAMD simulations. These files are the CHARMM format structure file (.psf), coordinate (.pdb), the extrabonds file to implement the restraints of elastic network in the origami structure (.exb), and NAMD configuration file (.namd). The downloads will also contain a folder with CHARMM parameter files. A detailed description on how to assemble and simulate DNA origami is provided in our practical guile for DNA origami simulations [42]. Figure 2b shows the all-atom pdb structure of DNA origami nanopore.

5. Introduce cholesterol groups in the DNA origami (*see* **Note 1**): After obtaining the all-atom topology and coordinate of DNA origami nanopore (**steps 3–4**), the next step is to covalently connect the hydrophobic lipid anchors (cholesterol group with tetraethylene glycol (chol-TEG) linker in our case) to the O3′ atom of the DNA backbone as schematically shown in Fig. 1a. Before connecting the lipid anchors, equilibrate the DNA origami nanopore in vacuum using ENRG-MD restraints as discussed in our chapter on DNA origami simulation protocols [42].

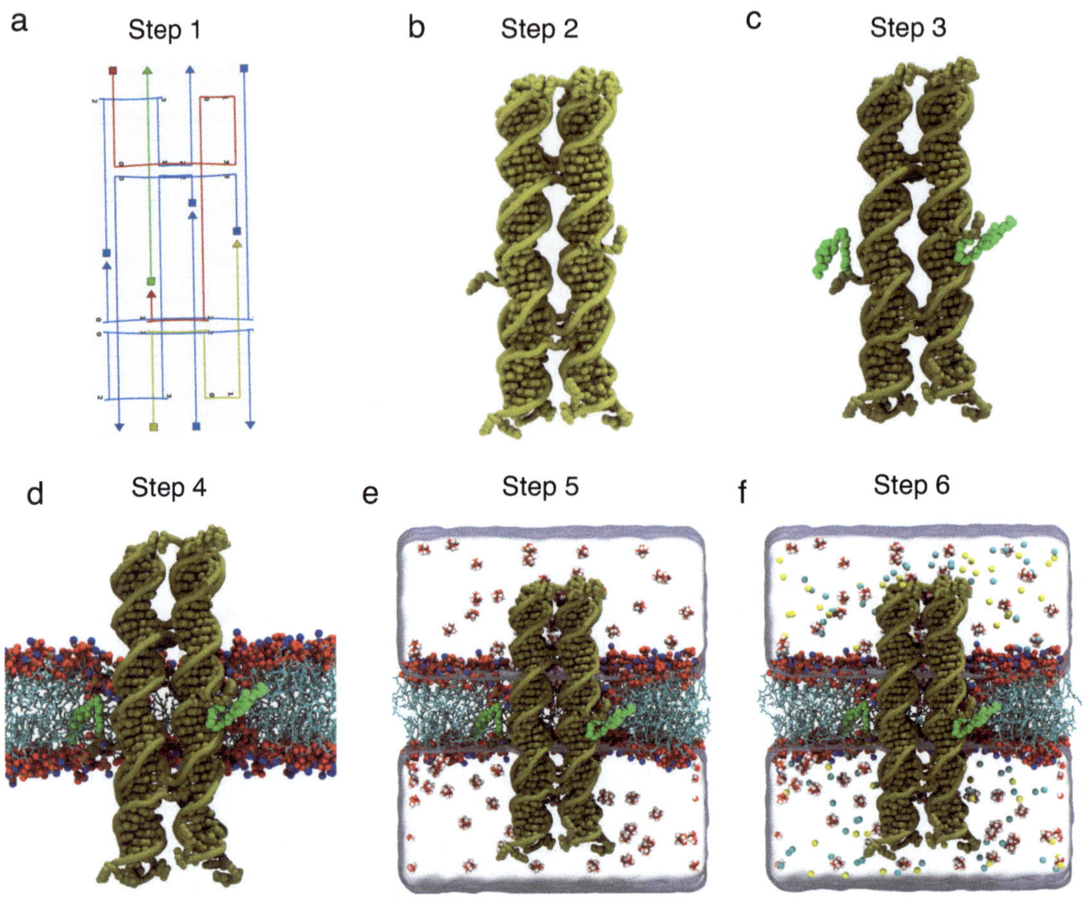

Fig. 2 Steps involved in assembling the all-atom MD simulation system of a membrane-spanning DNA origami nanopore. (**a**) The design of a DNA origami nanopore as visualized in caDNAnano. (**b**) An all-atom model of the DNA nanopore obtained by converting the json file to atomistic representation using ENRG-MD web server; atoms of DNA are shown using tan spheres. (**c**) All-atom representation of cholTEG covalently conjugated to the DNA nanopore; the atoms of chol-TEG are shown using green spheres. (**d**) The DNA nanopore embedded into a patch of a lipid bilayer membrane. Nitrogen, phosphorus, and oxygen atoms of the lipid headgroups are highlighted in blue, tan, and red spheres respectively, and the rest of the lipid is shown in cyan licorice representation. (**e**) Membrane-embedded DNA origami nanopore with added magnesium hexahydrate ions and solvated in a box of water. (**f**) Fully assembled DNA origami nanopore embedded in a DPhPE lipid bilayer membrane and solvated in aqueous solution of K+ (yellow) and Cl− ions (cyan)

6. Create a pdb (*cholTEG.pdb*) of chol-TEG using the MarvinSketch and save it. For the purpose of providing additional flexibility to the anchor, an unpaired DNA nucleotide (adenine) can be added to the chol-TEG molecule. The file of the modeled chol-TEG molecule can be also found in the step3 subdirectory of the DNPinMembraneTutorial folder.
7. Upload the pdb structure of chol-TEG into the CGenFF webserver [36], and obtain the topology and parameter file (*cholteg.str*) for the cholesterol anchors.

8. Using VMD, strategically connect the chol-TEG with O3′ atom of DNA such that chol-TEG extends diagonally opposite directions away from the DNA nanopore (Fig. 2c).
9. Create the necessary "patches" to make covalent connections between DNA and chol-TEG using psfgen plugin of VMD. A complete tcl script, *gen_psf.tcl*, to create the topology of chol-TEG-conjugated DNA origami pore can be found in subdirectory step3 of the DNPinMembraneTutorial folder.
10. The next step is to insert the chol-TEG-conjugated DNA origami nanopore into the diphytanoyl phosphatidylethanolamine (DPhPE) lipid bilayer membrane. The all-atom model of an equilibrated patch of the DPhPE lipid bilayer membrane can be obtained from the CHARMM-GUI website. The dimensions of the membrane patch should be large enough to avoid the undesirable interaction between the periodic images of the DNA origami nanopores. If only a smaller patch of equilibrated lipid is available, "writemol" commands from the psfgen plugin of VMD can be utilized to create a larger patch by adding multiple periodic images of the original patch.
11. Use tcl commands in VMD to embed the atomistic structure of DNA origami nanopore in the middle of the square patch (12.6 nm × 12.6 nm) of pre-equilibrated lipid bilayer membrane. After embedding the DNA nanostructure into the lipid membrane, lipid molecules located either within 3 Å of the nanostructure or inside the nanostructure are removed (Fig. 2d). The psf and pdb files of the membrane-embedded DNA nanopore are kept in the step4 subfolder of the DNPinMembraneTutorial folder.
12. Mg^{2+} ions are crucial for the stability of the DNA origami nanostructures (*see* **Note 2**). Using a perl script available in subfolder step5_6, place 151 of magnesium hexahydrate ($MGHH^{2+}$) ions to compensate the electrical charge of DNA.
13. Following that, solvate the structure in box of water using Solvate plugin of VMD (Fig. 2e).
14. Finally, add K^+ and Cl^- ions into the system using Autoionize plugins of VMD to acquire 1 M KCl electrolyte concentration. A script, *solvate_ionize.tcl*, used to solvate and ionize the system is given in the subfolder step5_6 of the DNPinMembraneTutorial folder. Thus, obtained topology (psf) and coordinate (pdb) files will be used to run the simulations. The final assembled system measured 12.5 × 12.5 × 17 nm^3 and contained 235, 646 atoms. It is advisable to load the structure and coordinate files into VMD and visualize the individual components carefully. Figure 2f shows the fully assembled all-atom model of the system.

3.2 Equilibrating the Structure of DNA Origami Nanopores in Lipid Bilayer Membrane (See Note 3)

1. After creating the psf and pdb files of solvated structure of a DNA origami nanopore embedded in a lipid bilayer membrane, prepare the NAMD configuration files to run the equilibrium MD simulations.
2. To remove possible clashes between the DNA, lipid, and solvent in the system, we first minimize the system for 1200 steps using the conjugate gradient method.
3. Subsequently, to achieve the correct density of the simulation system, equilibrate the system using constant number of atoms (N), pressure ($P = 1$ atms), and temperature ($T = 295$ K), i.e., the NPT ensemble. To maintain the pressure and temperature in the system, we use Nosé-Hoover Langevin piston [43, 44] and Langevin thermostat [45, 46], respectively.
4. Use anisotropic pressure coupling in these simulations; the ratio of system's dimensions in the membrane plane (x-y plane) is kept constant using the "useConstantRatio" keyword in NAMD program. The system's dimension along the bilayer normal (Z axis) is allowed to adjust independently.
5. Initially, equilibrate the system for 205 ns having all non-hydrogen atoms of the DNA nanostructure harmonically restrained to their initial coordinates using a spring constant of 1 kcal mol^{-1} $Å^{-2}$ which allowed the lipid and water to adopt equilibrium configurations around DNA nanopore.
6. Following that, decrease the spring constants of the restraints to 0.5 and then to 0.1 kcal mol^{-1} $Å^{-2}$; the system is equilibrated at each spring constant value for 4.8 ns.
7. Next, replace spatial restraints by a network of harmonic restraints that maintain distances between atomic pairs at their initial values; such elastic restraints exclude hydrogen atoms, phosphate groups, atoms in the same nucleotide, and pairs separated by more than 8 Å. These types of restraints are implemented using the "extraBonds" utility of the NAMD program; the required extrabonds file for the origami structure is obtained from the ENRG-MD server.
8. Simulate the system under such network of elastic restraints for 14.4 ns; the spring constants of the restraints are decreased from 0.5 to 0.1 and then to 0.01 kcal mol^{-1} $Å^{-2}$ in 4.8 ns steps. All MD simulations are performed using the program NAMD [50], periodic boundary conditions, the CHARMM36 parameter set for water [51], ions [52], nucleic acids [53], and lipid bilayer [54]. Use the latest CUFIX parameters for ion-DNA, ion-ion, and DNA-lipid interactions [41, 52].
9. Invoke a 2-2-6 fs multiple time-stepping method in NAMD to integrate the equation of motion.

10. Employ SETTLE algorithm [56] to keep water molecules rigid and RATTLE algorithm [57] to keep all other covalent bonds involving hydrogen atoms rigid. An 8-10-12 Å cutoff scheme is adopted to compute the van der Waals and short-range electrostatic forces. Long-range electrostatic interactions are computed using the particle mesh Ewald (PME) method [58] over a 1.2-Å resolution grid [59].
11. Save the coordinates of the system every 2.4 picoseconds.
12. Finally, perform the 2.2 μs production simulation of the DNA origami nanopore embedded in a lipid bilayer membrane recording the system's coordinates every 240 ps. The original production simulation of the system was performed on Anton 2 supercomputer using simulation parameters equivalent to those described above, except that temperature and pressure were maintained using the Nosé-Hoover thermostat [60, 61] and the Martyna-Tobias-Klein barostat [46]. Figure 3a, b shows the instantaneous snapshot of the system at the beginning, 0 μs, and after 2.2 μs-long production MD simulations, respectively. The DNA origami nanostructure overall maintained its structure during the simulation. Also, we observed significant scrambling of the lipids from one leaflet to the other [11].

3.3 Electric Field Simulations

1. In order to calculate the ionic current through the DNA origami nanopores, perform MD simulations under transmembrane potential differences of ±30 mV, ± 0.100 V, ± 0.150 V, and ± 0.250 V along the bilayer normal. The voltage values are similar to those used in typical experiments to measure the ionic conductance across a membrane-spanning DNA origami nanopore [4, 6, 7]. Figure 4a shows a cutaway view of the system set up to measure the ionic currents in all-atom MD simulations.
2. Use the conformation obtained at the end of the production simulation and apply constant electric field along the bilayer normal (z axis). All simulations are performed using a constant number of atoms (N), volume (average system size), and temperature ($T = 295$ K) ensemble, i.e., the NVT ensemble. The dimensions of the system in all three directions are kept constant to the average system dimensions from the last 5 ns of the production MD simulation. The desired voltage difference across the system is maintained using an externally applied electric field, E, given by $E = -V/L$, where L is the length of the simulation box along the direction of applied field (z axis). In order to obtain the desired value of the electric field in the unit of kcal/mol/Å/e, which is used in the NAMD configuration file, a factor of 23.0609 needs to be multiplied in the above expression (given that the voltage and length are provided in volts and Å).

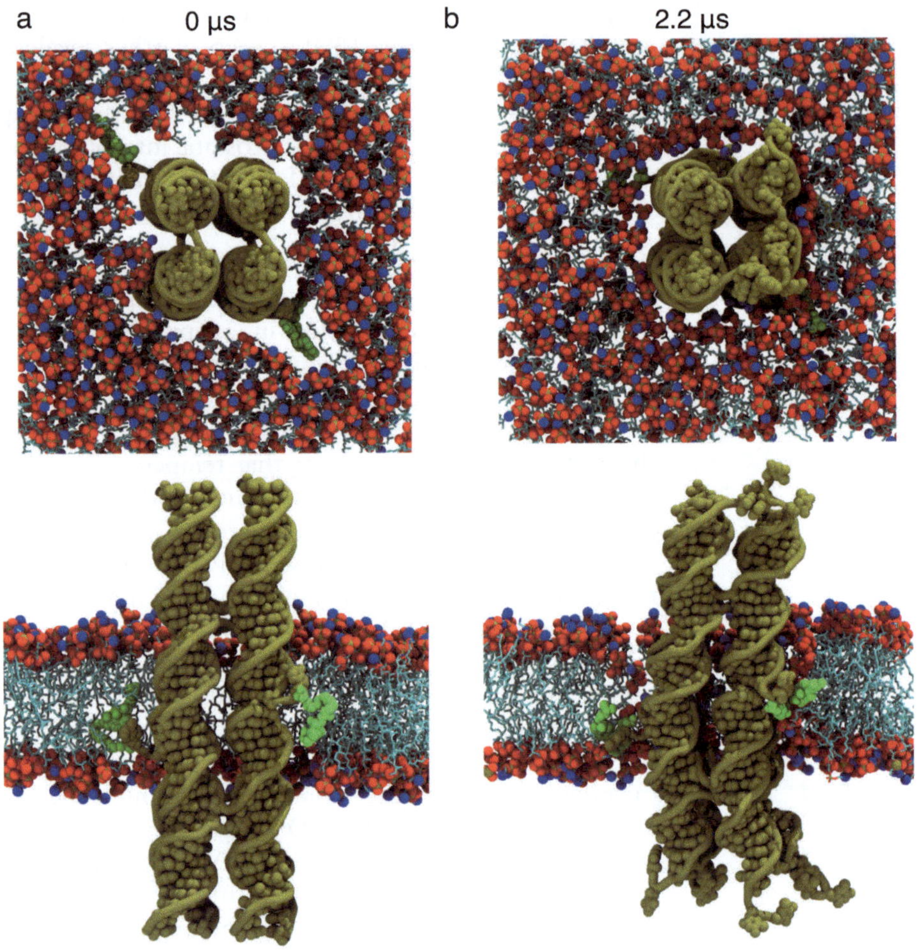

Fig. 3 Instantaneous snapshots of the simulated system. Microscopic configuration of the simulated system (**a**) at the beginning of the simulation and (**b**) at the end of the 2.2 μs equilibration MD simulation. Top panel shows the top view of the system, whereas the bottom panel illustrates a cutaway view. DNA base pairs are shown using tan spheres, whereas the backbone of DNA is shown using a tubular representation. Lipid bilayer membrane is shown using a cyan licorice representation, whereas the nitrogen, phosphorus, and oxygen atoms of the lipid headgroups are highlighted using blue, tan, and red spheres, respectively. CholTEG is shown using green spheres; the electrolyte solution is not shown for clarity

3. Run simulation at each bias for approximately 100 ns, and save the coordinate of the system every 2.4 ps.

3.4 Calculation of Ionic Current and Lipid Scrambling

1. Conceptually, we measure the transmembrane ionic current (I) by counting the effective number of ions translocated through the DNA origami nanopore:

$$I = \frac{Nq}{t}$$

where the N is the number of ions permeated across the membrane and q is charge of ion and t is the simulation time.

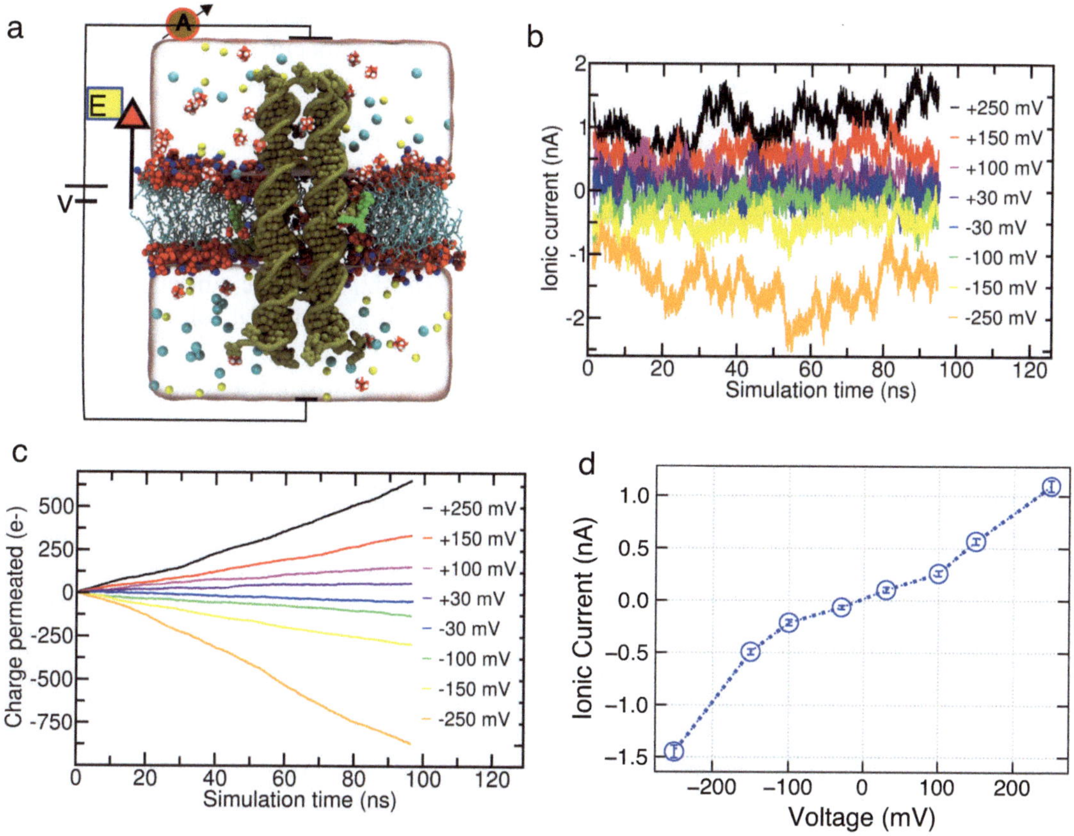

Fig. 4 Simulated ionic current through a DNA origami nanopore. (**a**) A cutaway view of the simulation system with applied voltage bias. (**b**) Total ionic current through the DNA origami nanopore at ±30 mV, ±100 mV, ±150 mV, and ± 250 mV as a function of simulation time. (**c**) Total charge permeated across the membrane through the DNA origami nanopore at different applied voltage biases as a function of the simulation time. (**d**) The average ionic current obtained as the block average of the instantaneous ionic current. The error bars show the standard error of the mean ionic current

2. In a steady state, the current through any cross section of the system must be the same in the direction of the applied electric field. Center all the frames of the simulation trajectory around DNA origami nanopore.
3. To reduce the thermal noise originating from the stochastic displacements of ions in the bulk solution, compute the ionic current in the region within $L/2 \leq z \leq L/2$ using the following equation as explained in our previous work [62]:

$$I\left(t + \frac{\Delta t}{2}\right) = \frac{1}{\Delta t\, L} \sum_{i}^{N} q_i[(\zeta_i(t + \Delta t) - \zeta_i(t))],$$

where $L = 30$ Å (the region where the DNA nanopore is inside the membrane), Δt is the output frequency of the simulation trajectory, N is the number of ions present in the system, q_i is the charge of the respective ion, and ζ_i is the z coordinate of the respective ion at that instance:

$$if abs(\zeta_i) \leq L/2; \zeta_i = z_i(t)$$

$$if abs(\zeta_i) > L/2; \zeta_i = L/2$$

$$if\ abs(\zeta_i) < -L/2; \zeta_i = -L/2$$

A tcl script (*currentTraj.tcl*) available under the analysis subfolder of the DNPinMembraneTutorial folder utilizes this equation to calculate the ionic current.

4. Sample the ionic current values every 2.4 ps, and then take running average over 1000 windows for difference biases, Fig. 4b.
5. By integrating the instantaneous ionic current values over the whole interval of the simulation time, obtain the total charge permeated across the membrane in the direction of applied field (Fig. 4c). In MD simulation, we can count the number of ions translocated across the membrane which must be similar to the integrated current values as plotted in Fig. 4c. Comparing the number of ions translocated to the integrated ionic current, one can cross-check if the script used to calculate the ionic currents is working correctly. Also, the absolute number of effective ion translocations can be used to obtain an independent assessment of the relative error (one over the square root of the number of translocated ions) of the average current value.
6. Finally, average the ionic current for the entire 100 ns MD simulations using the blocks of 70 ps (Fig. 4d). The I-V response of the DNA nanopore appears to be nonlinear at the higher voltage biases (>100 mV).

4 Notes

1. Inserting a negatively charged barrel of DNA into lipid bilayer membrane can disrupt the structure of the membrane. In order to overcome this instability in membrane-embedded DNA nanostructures, several modifications to the DNA backbone using hydrophobic moieties such as cholesterol [4, 7, 11–14], ethyl groups [5], porphyrins [6, 60], and biotin-streptavidin [9], etc. have been proposed and realized experimentally. Using the coarse-grained MD simulations, we have shown that cholesterol anchors can account for the free energy of inserting a DNA barrel in a lipid bilayer membrane [12].

2. Previously, we have shown that instead of bare Mg^{2+} ions, a magnesium hexahydrate ($MGHH^{2+}$) complex model reproduces the simulated DNA-DNA forces in better agreement with experiments [61–63].

3. During the initial equilibration, DNA origami simulation can crash due to various reasons. It is recommended to monitor the energy, temperature, pressure, density, etc. in the log file during the course of the simulation. In UNIX-based operating system (Linux, Ubuntu, Fedora, Red Hat, etc.), *awk*, *sed*, *grep*, etc. commands can be useful to print and plot these quantities from NAMD simulation log files. For example, following terminal command will extract simulation time step, kinetic energy, potential energy, temperature, and pressure from the NAMD log file to a new file "output.dat":

   awk '/^ENERGY:/{print $2, $11, $14}' npt1.log >output.dat

   When the simulation crashes unexpectedly, the log file generally writes the reason of the crash. One of the common causes which can lead to the unstable simulation during the initial stages is the piercing of bonds through the aromatic rings in DNA or the linker molecule. The best way to avoid these errors is to altogether remove the clashes between DNA and lipids at the first place. Alternatively, we can load the molecule into VMD and move the coordinates of the atoms or residues interactively (by using keys 5 and 6) and remove the clashes.

Acknowledgments

We gratefully acknowledge support from the National Institutes of Health (P41-GM104601), National Science Foundation (DMR-1827346), and supercomputer time provided through XSEDE Allocation Grant MCA05S028 and the Blue Waters petascale supercomputer system (UIUC).

References

1. Seeman NC (2016) Structural DNA nanotechnology. Cambridge University Press
2. Rothemund PWK (2006) Folding DNA to create nanoscale shapes and patterns. Nature 440(7082):297–302. https://doi.org/10.1038/nature04586
3. Douglas SM, Marblestone AH, Teerapittayanon S, Vazquez A, Church GM, Shih WM (2009) Rapid prototyping of 3D DNA-origami shapes with caDNAno. Nucleic Acids Res 37(15):5001–5006
4. Langecker M, Arnaut V, Martin TG, List J, Renner S, Mayer M, Dietz H, Simmel FC (2012) Synthetic lipid membrane channels formed by designed DNA nanostructures. Science 338(6109):932–936. https://doi.org/10.1126/science.1225624
5. Burns JR, Stulz E, Howorka S (2013) Self-assembled DNA nanopores that span lipid bilayers. Nano Lett 13(6):2351–2356
6. Burns JR, Göpfrich K, Wood JW, Thacker VV, Stulz E, Keyser UF, Howorka S (2013) Lipid-

bilayer-spanning DNA Nanopores with a bifunctional porphyrin anchor. Angew Chem Int Ed 52(46):12069–12072

7. Burns JR, Seifert A, Fertig N, Howorka S (2016) A biomimetic DNA-based channel for the ligand-controlled transport of charged molecular cargo across a biological membrane. Nat Nanotechnol 11(2):152–156. https://doi.org/10.1038/nnano.2015.279
8. Czogalla A, Franquelim HG, Schwille P (2016) DNA nanostructures on membranes as tools for synthetic biology. Biophys J 110(8): 1698–1707
9. Krishnan S, Ziegler D, Arnaut V, Martin TG, Kapsner K, Henneberg K, Bausch AR, Dietz H, Simmel FC (2016) Molecular transport through large-diameter DNA nanopores. Nat Commun 7:12787. https://doi.org/10.1038/ncomms12787
10. Hernández-Ainsa S, Bell NA, Thacker VV, Gopfrich K, Misiunas K, Fuentes-Perez ME, Moreno-Herrero F, Keyser UF (2013) DNA origami nanopores for controlling DNA translocation. ACS Nano 7(7):6024–6030
11. Ohmann A, Li C-Y, Maffeo C, Al Nahas K, Baumann KN, Göpfrich K, Yoo J, Keyser UF, Aksimentiev A (2018) A synthetic enzyme built from DNA flips 107 lipids per second in biological membranes. Nat Commun 9(1): 2426. https://doi.org/10.1038/s41467-018-04821-5
12. Göpfrich K, Li C-Y, Ricci M, Bhamidimarri SP, Yoo J, Gyenes B, Ohmann A, Winterhalter M, Aksimentiev A, Keyser UF (2016) Large-conductance transmembrane Porin made from DNA origami. ACS Nano 10(9): 8207–8214
13. Göpfrich K, Li C-Y, Mames I, Bhamidimarri SP, Ricci M, Yoo J, Mames A, Ohmann A, Winterhalter M, Stulz E, Aksimentiev A, Keyser UF (2016) Ion channels made from a single membrane-spanning DNA duplex. Nano Lett 16(7):4665–4669. https://doi.org/10.1021/acs.nanolett.6b02039
14. Göpfrich K, Zettl T, Meijering AE, Hernández-Ainsa S, Kocabey S, Liedl T, Keyser UF (2015) DNA-tile structures induce ionic currents through lipid membranes. Nano Lett 15(5):3134–3138
15. Cheatham TE III, Case DA (2013) Twenty-five years of nucleic acid simulations. Biopolymers 99(12):969–977. https://doi.org/10.1002/bip.22331
16. Shaw DE, Grossman J, Bank JA, Batson B, Butts JA, Chao JC, Deneroff MM, Dror RO, Even A, Fenton CH (2014) Anton 2: raising the bar for performance and programmability in a special-purpose molecular dynamics supercomputer. In: Proceedings of the international conference for high performance computing, networking, storage and analysis. IEEE Press, pp 41–53
17. Yoo J, Aksimentiev A (2013) In situ structure and dynamics of DNA origami determined through molecular dynamics simulations. Proc Natl Acad Sci 110(50):20099–20104. https://doi.org/10.1073/pnas.1316521110
18. Maingi V, Lelimousin M, Howorka S, Sansom MSP (2015) Gating-like motions and wall porosity in a DNA nanopore scaffold revealed by molecular simulations. ACS Nano 9(11): 11209–11217. https://doi.org/10.1021/acsnano.5b06357
19. Maffeo C, Yoo J, Aksimentiev A (2016) De novo reconstruction of DNA origami structures through atomistic molecular dynamics simulation. Nucleic Acids Res 44(7): 3013–3019. https://doi.org/10.1093/nar/gkw155
20. Joshi H, Dwaraknath A, Maiti P (2015) Structure, stability and elasticity of DNA nanotubes. PCCP 17(2):1424–1434
21. Joshi H, Kaushik A, Seeman NC, Maiti PK (2016) Nanoscale structure and elasticity of pillared DNA nanotubes. ACS Nano 10(8): 7780–7791. https://doi.org/10.1021/acsnano.6b03360
22. Li C-Y, Hemmig EA, Kong J, Yoo J, Hernández-Ainsa S, Keyser UF, Aksimentiev A (2015) Ionic conductivity, structural deformation, and programmable anisotropy of DNA origami in electric field. ACS Nano 9(2):1420–1433
23. Doye JP, Ouldridge TE, Louis AA, Romano F, Šulc P, Matek C, Snodin BE, Rovigatti L, Schreck JS, Harrison RM (2013) Coarse-graining DNA for simulations of DNA nanotechnology. Phys Chem Chem Phys 15(47): 20395–20414
24. Schreck JS, Romano F, Zimmer MH, Louis AA, Doye JP (2016) Characterizing DNA star-tile-based nanostructures using a coarse-grained model. ACS Nano 10(4):4236–4247
25. Maingi V, Burns JR, Uusitalo JJ, Howorka S, Marrink SJ, Sansom MS (2017) Stability and dynamics of membrane-spanning DNA nanopores. Nat Commun 8:14784
26. Yoo J, Aksimentiev A (2015) Molecular dynamics of membrane-spanning DNA channels: conductance mechanism, electro-osmotic transport, and mechanical gating. J Phys Chem Lett 6(23):4680–4687
27. Joshi H, Maiti PK (2018) Structure and electrical properties of DNA nanotubes embedded

in lipid bilayer membranes. Nucleic Acid Res 46(5):2234–2242

28. Veneziano R, Ratanalert S, Zhang K, Zhang F, Yan H, Chiu W, Bathe M (2016) Designer nanoscale DNA assemblies programmed from the top down. Science 352(6293):1534–1534
29. Macke TJ (1998) Case DA Modeling unusual nucleic acid structures. In: ACS symposium series. ACS Publications, pp 379–393
30. Williams S, Lund K, Lin C, Wonka P, Lindsay S, Yan H (2008) Tiamat: a three-dimensional editing tool for complex DNA structures. In: International workshop on DNA-based computers. Springer, pp 90–101
31. Joshi H, Maffeo C, Aksimentiev A (2018) Molecular dynamics simulations of self-assembled DNA nanostructures. http://www.ks.uiuc.edu/Training/Workshop/Urbana2018c/tutorials/universal-all-atomTutorial.pdf
32. Maffeo C, Aksimentiev A (2020) MrDNA: a multi-resolution model for predicting the structure and dynamics of DNA systems. *Nucleic Acids Res 48*(9):5135–5146. https://doi.org/10.1093/nar/gkaa200
33. Humphrey W, Dalke A, Schulten K (1996) VMD: visual molecular dynamics. J Mol Graph 14(1):33–38. https://doi.org/10.1016/0263-7855(96)00018-5
34. VMD User Guide http://wwwksuiucedu/Research/vmd/current/ug/ughtml
35. VMD Tutorial. http://wwwksuiucedu/Training/Tutorials/vmd/tutorialhtml/indexhtml
36. Vanommeslaeghe K, Hatcher E, Acharya C, Kundu S, Zhong S, Shim J, Darian E, Guvench O, Lopes P, Vorobyov I (2010) CHARMM general force field: a force field for drug-like molecules compatible with the CHARMM all-atom additive biological force fields. J Comput Chem 31(4):671–690
37. Lee J, Cheng X, Swails JM, Yeom MS, Eastman PK, Lemkul JA, Wei S, Buckner J, Jeong JC, Qi Y (2015) CHARMM-GUI input generator for NAMD, GROMACS, AMBER, OpenMM, and CHARMM/OpenMM simulations using the CHARMM36 additive force field. J Chem Theory Comput 12(1):405–413
38. NAMD User Guide. http://www.ksuiucedu/Research/namd/current/ug/
39. NAMD Tutorial. http://www.ksuiucedu/Training/Tutorials/namd/namd-tutorial-unix-html/indexhtml
40. MacKerell AD Jr, Bashford D, Bellott M, Dunbrack RL Jr, Evanseck JD, Field MJ, Fischer S, Gao J, Guo H, Ha S (1998) All-atom empirical potential for molecular modeling and dynamics studies of proteins. J Phys Chem B 102(18):3586–3616
41. Yoo J, Aksimentiev A (2018) New tricks for old dogs: improving the accuracy of biomolecular force fields by pair-specific corrections to non-bonded interactions. Phys Chem Chem Phys 20(13):8432–8449
42. Yoo J, Li C-Y, Slone SM, Maffeo C, Aksimentiev A (2018) A practical guide to molecular dynamics simulations of DNA origami systems. In: Zuccheri G (ed) DNA nanotechnology: methods and protocols. Springer, New York, New York, NY, pp 209–229. https://doi.org/10.1007/978-1-4939-8582-1_15
43. Martyna GJ, Tobias DJ, Klein ML (1994) Constant pressure molecular dynamics algorithms. J Chem Phys 101(5):4177–4189
44. Feller SE, Zhang Y, Pastor RW, Brooks BR (1995) Constant pressure molecular dynamics simulation: the Langevin piston method. J Chem Phys 103(11):4613–4621
45. Brünger AT (1992) X-PLOR: version 3.1: a system for x-ray crystallography and NMR. Yale University Press
46. Sindhikara DJ, Kim S, Voter AF, Roitberg AE (2009) Bad seeds sprout perilous dynamics: stochastic thermostat induced trajectory synchronization in biomolecules. J Chem Theory Comput 5(6):1624–1631
47. Phillips JC, Braun R, Wang W, Gumbart J, Tajkhorshid E, Villa E, Chipot C, Skeel RD, Kale L, Schulten K (2005) Scalable molecular dynamics with NAMD. J Comput Chem 26(16):1781–1802
48. Jorgensen WL, Chandrasekhar J, Madura JD, Impey RW, Klein ML (1983) Comparison of simple potential function for simulating liquid water. J Chem Phys 79(2):926–935. https://doi.org/10.1063/1.445869
49. Beglov D, Roux B (1994) Finite representation of an infinite bulk system: solvent boundary potential for computer simulations. J Chem Phys 100(12):9050–9063
50. Hart K, Foloppe N, Baker CM, Denning EJ, Nilsson L, MacKerell AD Jr (2011) Optimization of the CHARMM additive force field for DNA: improved treatment of the BI/BII conformational equilibrium. J Chem Theory Comput 8(1):348–362
51. Klauda JB, Venable RM, Freites JA, O'Connor JW, Tobias DJ, Mondragon-Ramirez C, Vorobyov I, MacKerell AD Jr, Pastor RW (2010) Update of the CHARMM all-atom additive force field for lipids: validation on six lipid types. J Phys Chem B 114(23):7830–7843
52. Yoo J, Aksimentiev A (2015) Improved parameterization of amine–carboxylate and amine–

phosphate interactions for molecular dynamics simulations using the CHARMM and AMBER force fields. J Chem Theory Comput 12(1): 430–443

53. Miyamoto S, Kollman PA (1992) Settle: an analytical version of the SHAKE and RATTLE algorithm for rigid water models. J Comput Chem 13(8):952–962

54. Andersen HC (1983) Rattle: a "velocity" version of the shake algorithm for molecular dynamics calculations. J Comput Phys 52(1): 24–34

55. Batcho PF, Case DA, Schlick T (2001) Optimized particle-mesh Ewald/multiple-time step integration for molecular dynamics simulations. J Chem Phys 115(9):4003–4018

56. Skeel RD, Hardy DJ, Phillips JC (2007) Correcting mesh-based force calculations to conserve both energy and momentum in molecular dynamics simulations. J Comput Phys 225(1):1

57. Nosé S (1984) A unified formulation of the constant temperature molecular dynamics methods. J Chem Phys 81(1):511–519

58. Hoover WG (1985) Canonical dynamics: equilibrium phase-space distributions. Phys Rev A 31(3):1695

59. Aksimentiev A, Schulten K (2005) Imaging α-hemolysin with molecular dynamics: ionic conductance, osmotic permeability, and the electrostatic potential map. Biophys J 88(6): 3745–3761

60. Seifert A, Göpfrich K, Burns JR, Fertig N, Keyser UF, Howorka S (2014) Bilayer-spanning DNA nanopores with voltage-switching between open and closed state. ACS Nano 9(2):1117–1126

61. Yoo J, Aksimentiev A (2012) Improved parametrization of Li+, Na+, K+, and Mg2+ ions for all-atom molecular dynamics simulations of nucleic acid systems. J Phys Chem Lett 3(1): 45–50

62. Yoo J, Aksimentiev A (2012) Competitive binding of cations to duplex DNA revealed through molecular dynamics simulations. J Phys Chem B 116(43):12946–12954

63. Yoo J, Aksimentiev A (2016) The structure and intermolecular forces of DNA condensates. Nucleic Acids Res 44(5):2036–2046

Part III

Single-Molecule Characterization of DNA Origami

Chapter 8

Single-Molecule Imaging of Enzymatic Reactions on DNA Origami

An Yan, Lele Sun, and Di Li

Abstract

The interaction between enzymes is very important for understanding of enzyme functions and shed light on enzymatic reaction mechanisms. With the development of DNA nanotechnology, DNA origami has become a powerful tool for analyzing the dynamic behavior of enzyme molecules. In this protocol, a method for imaging and analysis of single-molecule cascade enzyme reactions on DNA origami raft by total internal reflection fluorescence microscopy (TIRFM) is described. Through trajectory analysis and calculation, the diffusion of downstream enzymes in enzymatic reaction and chemotaxis of enzymatic reactions were elucidated at the single molecular level.

Key words DNA origami, Single molecular imaging, Enzyme cascades, Protein modification, TIRFM

1 Introduction

Single-molecule fluorescence technology has been widely used to address biological problems that cannot be solved by classical molecular biology methods. With the aid of single-molecule tools, the biomolecular structure and intermolecular interaction can be revealed at single-molecule level, which is of great significance for further understanding receptor recognition, biological processes, and structure-function relationship [1–3]. Total internal reflection fluorescence microscopy (TIRFM) has received extensive attention as a powerful single-molecule fluorescence tool. TIRFM is an effective technology to image the location and interaction of biomolecules at liquid/solid interface [4–6]. Abundant dynamic information of catalytic reaction, molecular structure, and receptor binding can be obtained by TIRFM [7–9]. Moreover, TIRFM has been widely used for the real-time imaging of the single molecular behavior at or near the cell surface.

DNA origami provides researchers the ability to design multi-dimensional nanostructures [10, 11]. The spatial localization of molecules and intermolecular distance could be precisely controlled

Julián Valero (ed.), *DNA and RNA Origami: Methods and Protocols*, Methods in Molecular Biology, vol. 2639,
https://doi.org/10.1007/978-1-0716-3028-0_8,

at nanoscale with DNA origami [12, 13]. Various DNA origami have been used for the nanopatterning of proteins, DNA strands, and other biomolecules [14, 15]. Therefore, DNA origami enables to regulate the position of different molecules of interest at molecular level [16, 17]. Therefore, DNA origami has been a promising material in single molecular imaging.

Herein, we combine TIRFM with the positioning ability of DNA origami to study the chemotaxis of enzymes at single molecular level. We construct a cascade enzymatic reaction on DNA origami. The relative locations and moving trail of GOx and catalase were monitored by TIRFM. The data is analyzed by ImageJ software and calculated by a program wrote with MATLAB.

2 Materials

2.1 Chemicals

1. DOPC (1,2-dioleoyl-sn-glycero-3-phosphocholine).
2. DMPC (1,2-dimyristoyl-sn-glycero-3-phosphocholine).
3. NBD-PE (1,2-dimyristoyl-sn-glycero-3-phosphoethanolamine-N-(7-nitro-2-1,3-benzoxadiazol-4-yl)).
4. Biotin-PE (1,2-dioleoyl-sn-glycero-3-phosphoethanolamine-N-(cap biotinyl)).
5. Catalase (EC 1.11.1.6).
6. Glucose oxidase (GOx, EC 1.1.3.4).
7. Horseradish peroxidase (HRP, EC 1.11.1.7).
8. Streptavidin-HRP.
9. SPDP (N-succinimidyl 3-(2-pyridyldithio)-propionate).
10. NHS-biotin (+)-biotin N-hydroxysuccinimide ester.
11. Glucose.
12. ABTS (2,2′-azino-bis(3-ethylbenzothiazoline-6-sulfonic acid)).
13. NHS-alexa647 (Alexa Fluor 647 NHS Ester).
14. NHS-atto488 (ATTO 488 NHS Ester).
15. Amplex Red.
16. Polycarbonate (PC) membrane with a pore diameter of 100 nm.
17. Single-stranded oligonucleotides (Takara Bio; Dalian, China).
18. M13mp18 single-stranded DNA (New England Biolabs Ltd.; Beijing, China).

2.2 Buffer Solutions

All solutions are prepared with ultrapure water (18 MΩ cm) from a Millipore system (Milli-Q Synthesis A10) and analytical grade reagents (*see* **Note 1**):

1. Cleaning buffer: 50 mM sodium HEPES, pH 7.5.
2. SPDP stock solution: 20 mM of SPDP dissolved in DMSO.
3. 50× TAE stock solution: 2 M Tris base, 1 M acetic acid, 0.1 M EDTA.
4. 10× TAE-Mg^{2+} stock solution: 400 mM Tris base, 200 mM acetic acid, 20 mM EDTA, 125 mM magnesium acetate. The pH is adjusted to 8.0.
5. 1× TAE-Mg^{2+} buffer is diluted from 10× TAE-Mg^{2+} buffer with final concentration of 40 mM Tris, 20 mM acetic acid, 2 mM EDTA, and 12.5 mM magnesium acetate.
6. 1× PBS buffer: 135 mM NaCl, 4.7 mM KCl, 10 mM Na_2HPO_4, and 2 mM NaH_2PO_4. The pH value is adjusted to 7.4 or 8.5 using NaOH and hydrochloric acid.
7. 1 M KOH.
8. 0.5% hydrofluoric acid (HF).
9. Mixed solution: 10 nM HRP, 10 mM glucose, 100 μM Amplex Red.

3 Methods

3.1 DNA-Enzyme Conjugation and Characterization

As shown in Fig. 1, SPDP is used to conjugate thiol-modified DNA strands and fluorophore to catalase and GOx. NHS-alexa647 and NHS-atto488 are, respectively, used for label catalase and GOx. The detailed conjunction steps are described as the following:

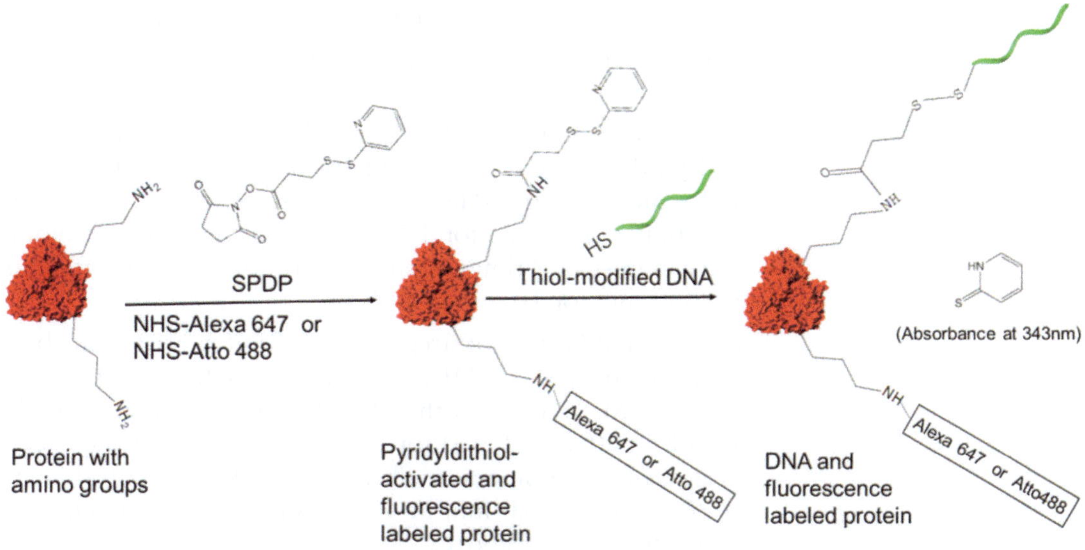

Fig. 1 Schematic for the modification of enzymes with DNA and fluorophore

1. Before modifying enzymes with DNA or fluorophores, small impurities and primary amine contaminates are removed with cleaning buffer by using 30 kDa cutoff filter. The concentration of enzymes is determined by UV absorbance at 405 nm for catalase (extinction coefficient, 420,000 M^{-1} cm^{-1}) and 450 nm for GOx (extinction coefficient, 270,000 M^{-1} cm^{-1}) (*see* **Note 2**).
2. For 50 μL of 40 μM catalase solution, add 2 μL SPDP stock solution and tenfold excess NHS-alexa647. The reaction is performed in 1× PBS buffer (pH 8.5) for 1.5 h under room temperature. For 50 μL of 40 μM GOx solution, add 2 μL SPDP stock solution and tenfold excess NHS-atto488 in 1× PBS buffer (pH 8.5) at 37 °C for 1.5 h at room temperature (*see* **Note 3**).
3. Remove excess of SPDP and free fluorophore by 3 times ultrafiltration with 30 kD cutoff filters at 3000 rpm for 10 min at 4 °C. The modification efficiency of alexa647or atto488 dyes is evaluated by monitoring the absorbance increase at 651 nm (extinction coefficient, 270,000 M^{-1} cm^{-1}) or 501 nm (extinction coefficient, 90,000 M^{-1} cm^{-1}) in UV-vis spectrum (*see* **Note 4**).
4. Conjugate the thiol-modified DNA to alexa647-modified catalase or atto488-modified GOx by adding threefold excess DNA into enzymes solution and incubating the reaction mixture in 1× PBS buffer (pH 7.4) for 8 h under room temperature (*see* **Note 5**).
5. Evaluate the ligation efficiency of DNA by monitoring the absorbance increase at 343 nm (extinction coefficient, 8080 M^{-1} cm^{-1}) caused by the free of pyridine-2-thione. Finally, remove the excess DNA strands by 3 times ultrafiltration with 30 kD cutoff filters at 3000 rpm for 10 min in 4 °C (*see* **Note 6**).
6. Conjugate biotin to catalase by the following steps. For 50 μL of 40 μM catalase solution, add 20-fold excess of NHS-biotin and tenfold excess NHS-alexa647. The reaction runs in 1× PBS buffer (pH 8.5) for 1.5 h. Remove the excess of biotin and alexa647 by 3 times ultrafiltration with 30 kD cutoff filters at 3000 rpm for 10 min in 4 °C.
7. For the modification of streptavidin-HRP (SA-HRP), incubate 20 μL of 60 μM SA-HRP (extinction coefficient, 403 nm/ 101,000 M^{-1} cm^{-1}) with 20-fold excess NHS-alexa647 in 1× PBS buffer (pH 8.5) for 1.5 h. Then, remove the excess of alexa647 by 3 times ultrafiltration with 100 kD cutoff filters at 3000 rpm for 10 min in 4 °C. Evaluate the coupling efficiency as described above.

8. To compare the activity of unmodified and modified GOx, mix 10 μL 60 mM glucose, 10 μL 6 mM ABTS, 20 μL 300 nM HRP, and 20 μL 3 nM GOx or GOx-DNA-atto488 in PBS buffer (final volume 60 μL). Then, measure the absorbance of $ABTS^+$ at 420 nm and analyze the data.
9. Measure the activity of catalase or catalase-DNA-alexa647 by monitoring the absorbance of H_2O_2 at 240 nm. 40 μL 0.04% H_2O_2 and 40 μL 1 nM catalase or catalase-DNA-alexa647 were mixed in PBS buffer (final volume 80 μL). For all experiments, the reactive kinetic was monitored within 10 min.

3.2 Assembly and Characterization of GOx on DNA Origami

In this protocol, DNA sequences of a previously published DNA origami triangle are used to assemble GOx [18]:

1. All preparation and assembly processes are performed in 1× TAE-Mg^{2+} buffer. Briefly, 2 nM single-stranded M13mp18 DNA is incubated with a tenfold molar excess of staple stands and a 400-fold molar excess of Cy3-labeled strands.
2. The mixture is annealed using a PCR thermocycler from 95 °C to 4 °C with the temperature gradient shown in Table 1 (*see* **Notes** 7 and **8**).
3. Excess strands are removed by 2 times ultrafiltration in 1× TAE-Mg^{2+} buffer and one time in 1× PBS (pH 7.4) with 100 kD cutoff filters at 3000 rpm for 10 min in 4 °C.
4. Next, GOx-DNA-atto488 is mixed with the triangle DNA origami with a molar ratio of 120:1 in 1× PBS buffer (pH 7.4). After 2-h incubation under room temperature, excess enzymes are removed by 3 times ultrafiltration in 1× PBS (pH 7.4) with 100 kD cutoff filters at 3000 rpm for 10 min in 4 °C. The sample was stored at 4 °C (*see* **Note 9**).

Table 1
The temperature gradient program for assembling triangle DNA origami

Temperature gradient	Gradient
95 °C	30 s
86–71 °C	1 min/step
70–60 °C	10 min/step
59–30 °C	15 min/step
29–26 °C	10 min/step
25 °C	25 min
4 °C	Hold

5. Two μL of GOx-modified triangle DNA origami is deposited onto freshly cleaved mica. After 2 min of adsorption, rinse the sample with Milli-Q water and dry up with N_2. Then, the sample is imaged by atomic force microscopy (AFM) (Multimode 8 AFM) using peak-force mode (Fig. 1b).

3.3 Preparation of Liposome and Supported Lipid Bilayer (SLB)

Solid-supported phospholipid bilayer (SLB) is prepared by the fusion of small unilamellar vesicles (SUVs) on cover slides:

1. According to previously reported method, dissolve 5 mg DOPC and 5 mg DMPC in chloroform. Then, pour the DOPC and DMPC solution in a 25 mL round-bottomed flask. In addition, NBD-PE or biotin-PE at a molar ratio of 1% are introduced for further experiments.
2. Subsequently, chloroform is evaporated and dried by blowing with N_2. The dried mixture is resuspended in 1 mL deionized (DI) water directly added to the round-bottom flask. After 10 min of sonication, a 5 mg/mL phospholipid suspension is obtained.
3. Next, the suspension is extruded more than 35 times through polycarbonate (PC) membrane with a pore diameter of 100 nm. Finally, the resulting SUV solution is stored at 4 °C and can be used in 1 week.
4. Clean the cover glass by successively sonicating for 15 min in chloroform, acetone, ethanol, and 1 M KOH. Then, wash carefully three times with DI water. Immerse immediately the clean cover glass in 0.5% HF for 30 s, then wash with DI water three times, and dry with N_2.
5. Construct a sample chamber by stacking a square chip fence with 0.5 cm edge length onto the cover glass. Then, add 50 μL of SUV solution into the sample chamber. After incubation for 30 min at 25 °C, remove the excess of unfused SUVs by washing with DI water for 15–20 times (*see* **Note 9**). Store the prepared SLB at 4 °C (*see* **Note 10**).

3.4 GOx and Catalase Anchored on Phospholipid Bilayer

1. For the location of enzymes on phospholipid bilayer via DNA hybridization, drip 50 μL of 200 nM cholesterol-modified ssDNA on the prepared SLB. After 1 h of incubation under room temperature, wash the excess of DNA strands with 1× PBS buffer for 15–20 times.
2. Then, add 50 μL of 100 pM catalase-DNA-alexa647 in 1× PBS buffer onto the SLB containing ssDNA. After 10 min incubation under room temperature, remove the excess of catalase-DNA-alexa647 conjugates by washing with 1× PBS buffer for 15–20 times (*see* **Note 11**).

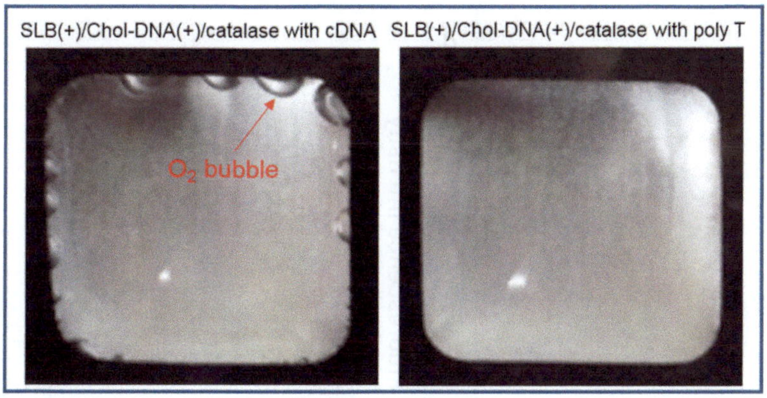

Fig. 2 Schematic for the activity study of catalase-DNA-alexa647 on SLB via different enzyme modification DNA

3. Add 50 μL of 0.2 pM GOx-DNA-atto488-triangle DNA origami onto the surface. After 10 min of incubation under room temperature, remove the excess of DNA origami by washing with 1× PBS buffer for 15–20 times.
4. Determine the enzymatic activity of catalase and GOx on SLB by the following steps. Incubate 50 μL of 400 pM of catalase-alexa647-cDNA or catalase-alexa647-polyT on the DNA-SLB for 20 min. Then, add 50 μL 20 mM H_2O_2 into the sample chamber. As shown in Fig. 2, the O_2 bubble formed only in the left chamber.
5. Add 50 μL of 200 nM cholesterol-modified DNA or 50 μL PBS with the SLB to 50 μL of 100 pM GOx-triangle DNA origami. After 20 min incubation, add 50 μL mixed solution with 10 nM HRP, 10 mM glucose, and 100 μM Amplex Red into the sample chamber. The specific tethering of GOx on the SLB is shown in Fig. 3.
6. For the localization of enzymes on the phospholipid bilayer via biotin-streptavidin connection, biotin-PE is used. Biotin-SLB is prepared by the above method, and then 50 μL of 0.1 mg/mL streptavidin in PBS buffer is added onto the surface. After 10 min of incubation at room temperature, remove the excess of streptavidin by washing with 1× PBS buffer for 15–20 times (*see* **Note 12**).
7. Add 50 μL of 100 pM catalase-biotin-alexa647 onto the sample chamber. After 10 min of incubation under room temperature, remove the excess of enzymes by washing with 1× PBS buffer for 15–20 times.
8. For the localization of SA-HRP-alexa647 on phospholipid bilayer, the 3′-biotin-modified DNA (biotin-DNA) is used. Firstly, mix SA-HRP-alexa647 with fourfold excess of biotin-

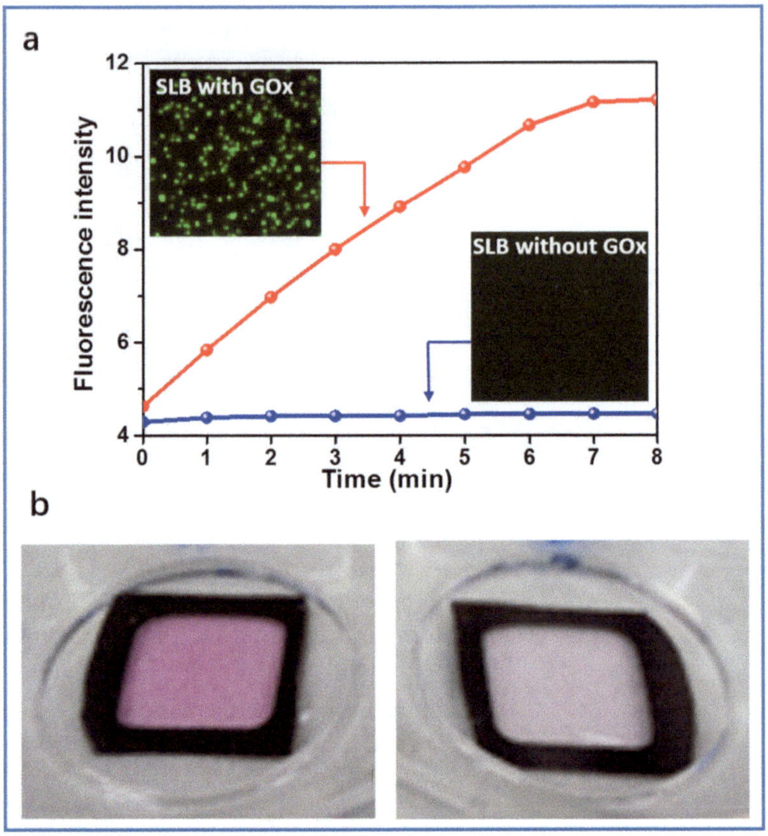

Fig. 3 Schematic for the activity study of GOx-triangle DNA origami. (**a**) The activity of GOx was indirectly observed with TIRF microscopy by monitoring the increased fluorescence of resorufin (fluorescent product of oxidized Amplex Red). Insets show the specific tethering of GOx on SLB. (**b**) After 30-min reaction, colorless Amplex Red solution was turned into pink by GOx (left) or remained colorless (right)

DNA, and incubate for 2 h at room temperature. Then, SA-HRP-alexa647-DNA is localized on SLB containing ssDNA as described above.

3.5 Imaging of Enzymatic Cascade Reactions

In this protocol, the moving trajectory and distribution of enzymes are observed by total internal reflection fluorescence (TIRF) microscopy (N-storm, Nikon). A 100× objective lens (NA 1.49) and electron multiplying charge-coupled devices (EMCCD) camera (Andor, iXon 3) are also used. Solid-state lasers (Coherent) at 488 nm (150 mW), 561 nm (200 mW), and 640 nm (200 mW) wavelengths are used for exciting atto488, Cy3, and alexa647, respectively.

1. Firstly, drop the immersion oil (refractive index $n = 1.518$) on a 100× numerical aperture (NA) 1.65 objective lens. Then, place

the cover glass on the stage of TIRF microscope, and optimize the focus, intensity, and incidence angle of the TIRF microscope (*see* **Note 13**).

2. The diffusion of catalase or HRP on SLB is tracked as follows: After tethering enzymes on SLB, a series of videos and pictures are acquired by using the capture software at a high frequency (10–50 Hz, 20–100 ms). Then, the videos are used for subsequent analysis (*see* **Note 14**).
3. Tracking of the diffusional motions of catalase in cascade reactions on SLB is obtained as follows: Firstly, 488 nm and 561 nm lasers are used for the GOx and DNA origami colocalization by observing the spot overlaps in an imaging region (40 μm × 40 μm). After that, the observed spot is moved to the center of imaging region, and 50 μL 10 mM glucose is added into the sample chamber. The dynamic video of catalase is immediately acquired with a 640 nm laser in a 30 ms time resolution. As shown in Fig. 4, the moving trajectories of catalase around immobilized GOx are extracted for subsequent analysis.
4. The distribution of catalase around GOx on SLB is obtained as follows: Firstly, an imaging region (40 μm × 40 μm) containing only one GOx spot is defined by using the method described above. After that, the image is immediately acquired with 640 nm laser by using the capture software at a high frequency (10–50 Hz, 20–100 ms). Besides, 3-min interval between images is used to avoid the influence of photobleaching. Finally, the images are used for subsequent localizing and analysis.

3.6 Fluorescence Recovery After Photobleaching (FRAP) Experiment

Fluorescence recovery after photobleaching (FRAP) experiments are used to measure the diffusion coefficient (D):

1. SLB containing NBD-PE is prepared as mentioned above (*see* Subheading 3.3). Before photobleaching, an image is captured for recording the initial fluorescence intensity. Then, a 30 μm circular bleach spot is generated by 5 min of irradiation with 488 nm laser.
2. Record the fluorescence intensity recovery after photobleaching in 20–30 min with a time interval of 20 s. Then, the FRAP curve obtained from the experiment is fitted by the equation reported in the literature:

$$f(t) = a + b \times \left(1 - e^{-\lambda t}\right)$$

3. After the parameters λ are obtained, the characteristic diffusion time ($\tau_{1/2}$) of fluorescence intensity recover to 50% is

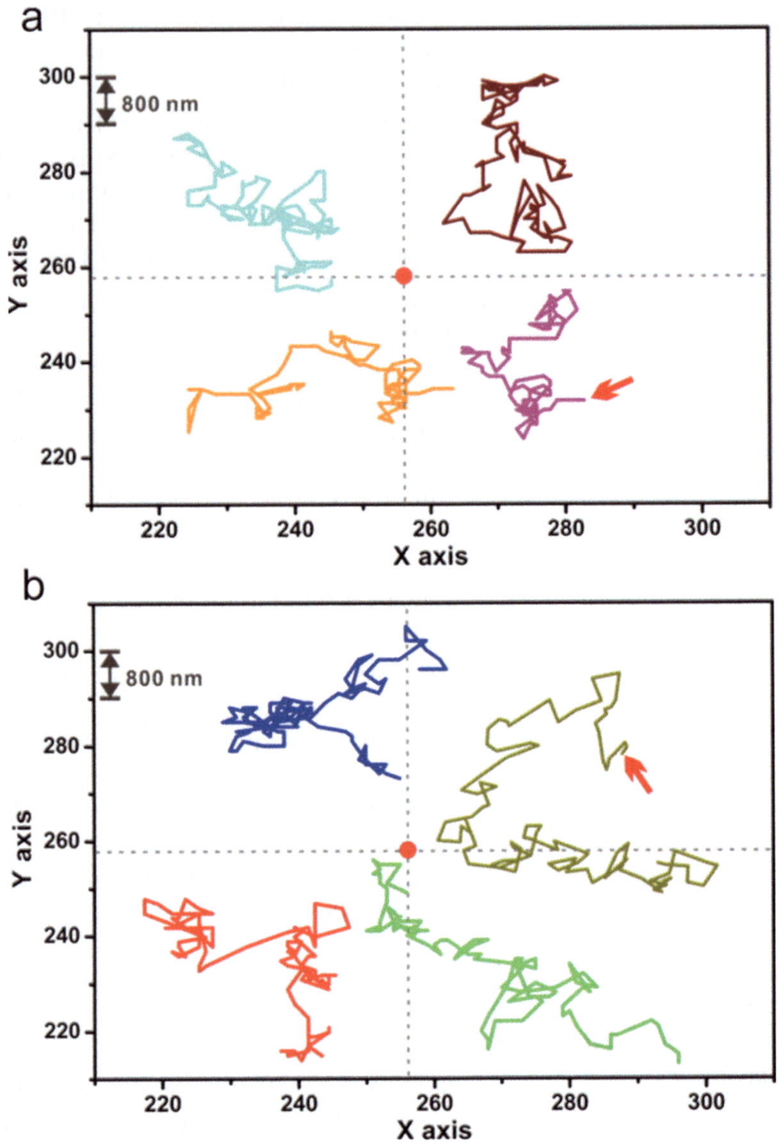

Fig. 4 Representative moving trajectories of catalase around immobilized GOx without (**a**) or with (**b**) glucose in 3 s. The red spot represents the position of GOx-triangle DNA origami

calculated from the equation $\tau_{1/2} = \ln 2/\lambda$. Then, the lateral diffusion of lipids (D) is calculated using the following equation:

$$D = 0.224\, W^2/\tau_{1/2}$$

(W is the radius of the photobleached spot. D is the lateral diffusion of lipids within the lipid bilayer.)

4. FRAP experiments are also used to confirm the successful insertion of cholesterol-DNA into SLB and the fluidity of cholesterol-DNA with Cy5-labeled DNA by 640 nm laser.

3.7 Fluorescence Anisotropy Experiment

Fluorescence anisotropy experiment is performed using a fluorescence spectrophotometer (F4500, HITACHI):

1. One mL of 4 nM catalase-DNA-atto488 or 30 nM atto488 is used for the fluorescence anisotropy measurement. The excitation wavelength is 480 nm, and the range of emission wavelength is 505–600 nm.
2. Add 1 μL of 1 M H_2O_2 into 1 mL of 4 nM catalase-atto488-DNA, and collect immediately the emission spectrum.

3.8 Analysis and Calculation of Experimental Data

The image analysis procedure was firstly done using ImageJ software. The "Manual Tracking" in ImageJ software is used for the trajectory tracking, and a program wrote with MATLAB is used for the calculation of mean square displacement (MSD) by the following formula:

$$\mathrm{MSD} = 4D\tau^{\alpha}$$

(D is the diffusion coefficient of target. τ is the observed lag time and α is related to target motion type.)

The diffusivity of individual catalases in the cascade reaction was also analyzed. The distribution of H_2O_2 around GO_x follows:

$$n(r,t) = \sum_{i=0}^{i=t/\tau-1} \frac{1}{\left[4\pi D_{H_2O_2}(t-i\tau)\right]^{3/2}} \exp\left[-\frac{r^2}{4D_{H_2O_2}(t-i\tau)}\right]$$

($n(r$; t) is the concentration of H_2O_2 at the distance (r, μm) from the initial position in the diffusion time (t, s). $D_{H_2O_2}$ is the diffusion coefficient of H_2O_2. ($D_{H_2O_2}$) = 1000 $\mu m^2/s$; k_{cat} (GOx) = 300 s^{-1}; $t = 3$ s.)

The translational move of catalase is calculated by the following formula:

$$s = v\tau_R$$

(s is the translational displacement of catalase. v is the translational speed of catalase. τ_R is the time scale of rotational diffusion of catalase.)

$$\mathrm{v} = \frac{\gamma}{\tau_f}, \gamma = \left(\frac{k_B T}{6\pi\varepsilon R} + \frac{2m}{3\phi^2\tau_f}\alpha Q\right) \cdot \frac{3\omega^2}{2RD}, D_O = \frac{k_B T}{6\pi\varepsilon R}, \tau_R = \frac{8\pi\omega R^3}{k_B T}$$

(a and R represent the radius of O_2 and catalase, respectively; ε is the dynamic viscosity of lipid membrane; ω is the diffusiophoretic; D_O is the diffusion coefficient of O_2 produced by catalytic reactions;

m is the mass of catalase; ϕ is the apparent friction coefficient; Q is the reaction heat released during individual reaction event; and α is the fraction of Q that is used for enhancing diffusion coefficient.)

$$\frac{1}{\tau_f} = \frac{V_M C_S}{K_M + C_S}$$

($\frac{1}{\tau_f}$ is the reciprocal of the time interval between two adjacent reactions. V_M is the maximum reaction rate, K_M is the Michaelis constant, and C_S is the substrate concentration.)

The plugin "thunder storm" 4 is used for localizing the spots of catalase, and the corresponding coordinate positions are extracted with a home-written software.

As shown in Fig. 5, the position of GOx is defined as the original point (x_0, y_0), and the relative position of catalase (xi, yi) toward GOx is analyzed by standard deviation (σ). With home-written software, σ_x and σ_y are defined as:

$$\sigma_x = \sqrt{\frac{1}{N}\sum_{i=1}^{N}(X_i - X_0)^2}$$

$$\sigma_y = \sqrt{\frac{1}{N}\sum_{i=1}^{N}(Y_i - Y_0)^2}$$

where σ_x is the x direction and σ_y is the y direction; σ is the discrete degree of the distribution of catalase around immobilized GOx.

Besides, a chemotaxis factor "κ" is further introduced for the relative position analysis of enzymes in the cascade reaction:

$$\kappa = \frac{\sum_{i=1}^{n}\sqrt{X\sigma i^2 + Y\sigma i^2}}{n}$$

$X\sigma i$ is the standard deviation of the X direction, $Y\sigma i$ is the standard deviation of the Y direction, and n is the number of measurement times.

A smaller κ would indicate a relatively concentrated distribution of catalase around GOx.

4 Notes

1. All solutions are filtered through a 0.22 μm filter and stored at 4 °C without light.
2. In order to achieve maximum DNA-enzyme conjugation efficiency, primary amines should be eliminated from enzyme storage solution. Reaction buffers like PBS or HEPES are used.
3. The temperature and pH value are crucial for SPDP modification on enzyme. The pH below 8 will cause the primary amines

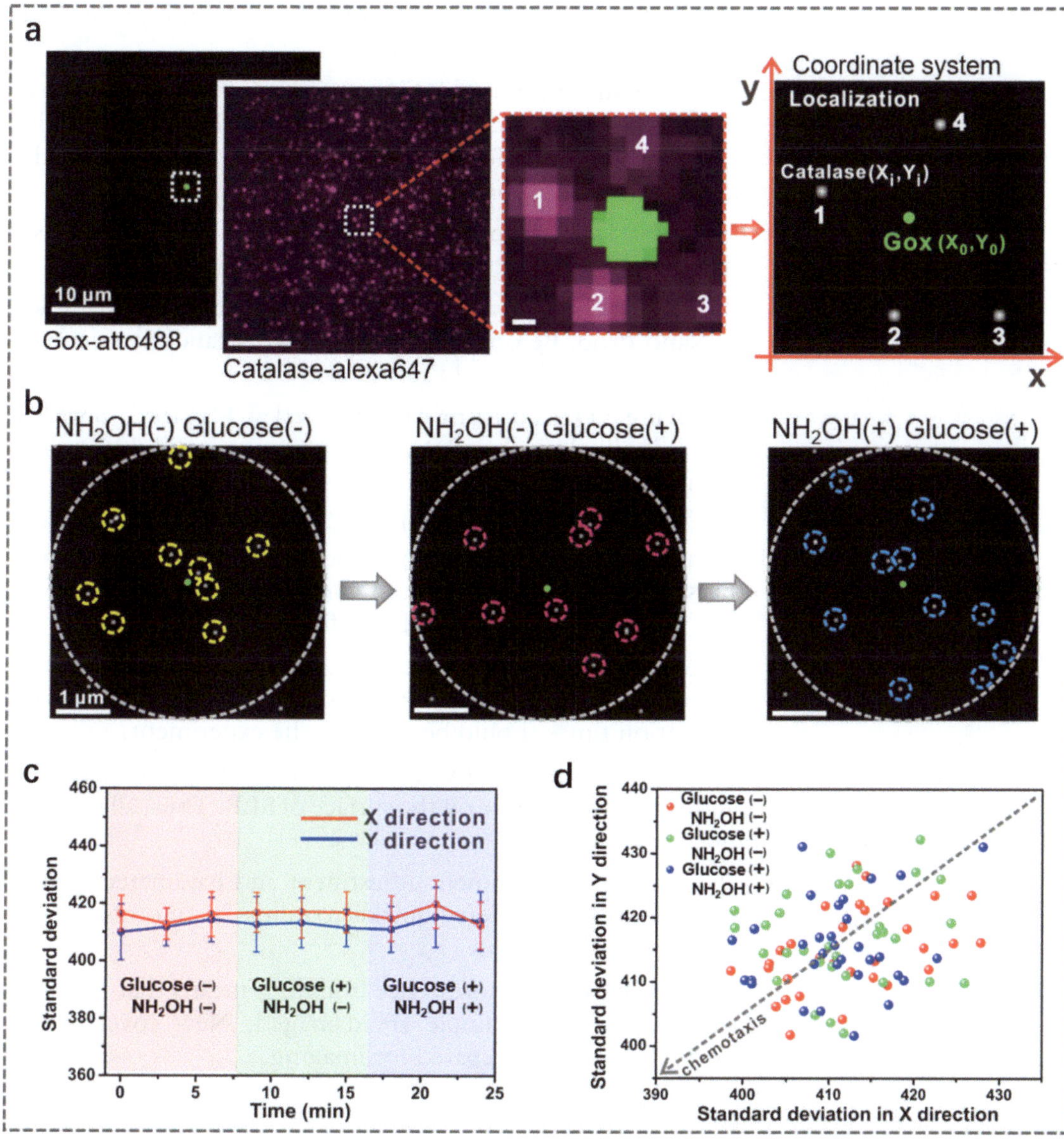

Fig. 5 Distribution analysis of catalase around immobilized GOx. (**a**) In situ imaging of the distribution of catalase around immobilized GOx. Scale bar is 200 nm. (**b**) TIRFM images showing the distribution of catalase around GOx upon different treatment. (*C*) Mean standard deviation (σ) analysis. (**d**) Scatter distribution of the standard deviation (σ) in the X-axis direction and Y-axis direction

protonation, and the pH value higher than 9 will cause NHS ester hydrolysis. Changes in temperature also cause a significant drop in conjugation efficiency.

4. The ratio of SPDP molecules per enzyme should be 1 or 2. More than three SPDP molecules per enzyme will significantly damage the activity of enzymes.

5. The sulfhydryl group of thiol-modified DNA is easily oxidized to disulfide bonds. Before DNA-enzyme conjugation, 20-fold

excess TCEP should be used to reduce the thiol groups and thus activate the thiolated DNA.

6. If the measured ratio of DNA to enzyme is still higher than three DNA molecules per enzyme, an additional wash with high salt concentration solution is needed to eliminate the nonspecific adsorption of DNA molecules on enzymes.
7. Excess of DNA staples ensures complete folding of DNA structure.
8. All DNA strands ordered from Takara should be re-quantified by Nano-drop. Before assembly, all DNA strands were cooled at 4 °C.
9. The triangle DNA origami can be stored at 4 °C for 1–3 weeks. The assembled enzyme-DNA nanostructures should be used up in one week.
10. The cover glass in contact with air will damage the supported membrane. In the preparation process of SLB, the solution in the sample chamber cannot be completely removed. In general, 20–30 μL buffer should be left to keep the SLB moist.
11. The enzymes at the interface of SLB must be monomolecular dispersed. Thus, lower enzyme concentration and shorter incubation times should be used for the experiment.
12. The Mg^{2+} ions in buffer will induce nonspecific absorption and aggregation of catalase on the surface of SLB. Thus, all solution must be free of Mg^{2+}.
13. The process of equipment adjustment and parameter optimization must be completed quickly to minimize photobleaching of fluorescent proteins.
14. If the background fluorescence is high, it means that coverslip isn't clean or the sample are damaged. New coverslips or samples should be prepared for imaging.

References

1. Biteen J, Willets KA (2017) Introduction: super-resolution and single-molecule imaging. Chem Rev 117:7241–7243
2. Weiss S (2000) Measuring conformational dynamics of biomolecules by single molecule fluorescence spectroscopy. Nat Struct Mol Biol 7:724–729
3. Joo C, Balci H, Ishitsuka Y, Buranachai C, Ha T (2008) Advances in single-molecule fluorescence methods for molecular biology. Annu Rev Biochem 77:51–76
4. Chen P, Zhou XC, Shen H, Andoy NM, Choudhary E, Han KS, Liu GK, Meng WL (2010) Single-molecule fluorescence imaging of nanocatalytic processes. Chem Soc Rev 39: 4560–4570
5. Axelrod D, Burghardt TP, Thompson NL (1984) Total internal reflection fluorescence. Ann Rev Biophys Bioeng 13:247–268
6. Schneckenburger H (2005) Total internal reflection fluorescence microscopy: technical innovations and novel application. Curr Opin Biotechnol 16:13–18
7. Zhao Z, Fu J, Dhakal S, Johnson-Buck A, Liu MH, Zhang T, Woodbury NW, Liu Y, Walter NG, Yan H (2016) Nanocaged enzymes with

enhanced catalytic activity and increased stability against protease digestion. Nat Commun 7: 10619
8. Su X, Ditlev JA, Hui EF, Xing WM, Banjade S, Okrut J, King DS, Taunton J, Rosen MK, Vale RD (2016) Phase separation of signaling molecules promotes T cell receptor signal transduction. Science 352:595–599
9. Varela JA, Rodrigues M, De S, Flagmeier P, Gandhi S, Dobson CM, Dobson CM, Klenerman D, Lee SF (2018) Optical structural analysis of individual α-Synuclein oligomers. Angew Chem Int Ed 57:4886–4890
10. Hong F, Zhang F, Liu Y, Yan H (2017) DNA origami: scaffolds for creating higher order structures. Chem Rev 117:12584–12640
11. Wang P, Meyer TA, Pan V, Dutta PK, Ke Y (2017) The beauty and utility of DNA origami. Chem 2:359–382
12. Endo M, Yang Y, Sugiyama H (2013) DNA origami technology for biomaterials applications. Biomater Sci 1:347–360
13. Madsen M, Gothelf KV (2019) Chemistries for DNA nanotechnology. Chem Rev 119:6384–6458
14. Zhou K, Dong J, Zhou Y, Dong J, Wang M, Wang Q (2019) Toward precise manipulation of DNA-protein hybrid nanoarchitectures. Small 2019:1804044
15. Chen Y, Ke G, Ma Y, Zhu Z, Liu M, Liu Y, Yan H, Yang CJ (2018) A synthetic light-driven substrate channeling system for precise regulation of enzyme cascade activity based on DNA origami. J Am Chem Soc 140:8990–8996
16. Li JM, Johnson-Buck A, Yang YR, Shih WM, Yan H, Walter NG (2018) Exploring the speed limit of toehold exchange with a cartwheeling DNA acrobat. Nat Nanotechnol 13:723–729
17. Endo M, Sugiyama H (2014) Single-molecule imaging of dynamic motions of biomolecules in DNA origami nanostructures using high-speed atomic force microscopy. Acc Chem Res 47:1645–1653
18. Rothemund PW (2006) Folding DNA to create nanoscale shapes and patterns. Nature 440: 297–302

Chapter 9

Single-Molecule Nanomechanical Genotyping with DNA Origami-Based Shape IDs

Qian Li, Jie Chao, Honglu Zhang, and Chunhai Fan

Abstract

Atomic force microscopy (AFM)-based nanomechanical imaging provides a sub-10-nm-resolution approach for imaging biomolecules under ambient conditions. Here we describe how to generate a set of DNA origami-based shape IDs (triangular and cross shape, with and without streptavidin) to site-specifically label target genomic DNA sequences containing two single-nucleotide polymorphisms (SNPs). Adjacent labeling sites separated by only 30 nucleobases (~10 nm) can be differentiated under AFM imaging. We can directly genotype single molecules of human genomic DNA.

Key words DNA origami, Atomic force microscopy, Genotyping, Nanomechanical imaging labels, Single-nucleotide polymorphisms, Single-molecule analysis

1 Introduction

Variations in DNA sequences are closely related to human diseases [1]. Imaging-based genotyping approaches provide powerful tools to perform rapid and high-throughput genotyping of single-nucleotide polymorphisms (SNPs) or haplotyping [2, 3]. The optical resolution limit (~200–300 nm) of fluorescent imaging restricts the genotyping resolution to ~1000 nucleobases [4]. The more advanced super-resolution fluorescence imaging improves the genotyping resolution to 100 nucleobases [5]. Atomic force microscopy (AFM) provides a nanomechanical imaging approach for genetic analysis with nanometer resolution [6–8]. Direct reading of genetic information using AFM has long been a dream since its invention [8]; however, its application in genetic analysis remains to be limited due to the lack of shape-specific labels for site-specific labeling of genomic DNA [9–11]. Here we demonstrate differentially shaped, highly hybridizable self-assembled DNA origami nanostructures that can serve as shape IDs for nanomechanical

Julián Valero (ed.), *DNA and RNA Origami: Methods and Protocols*, Methods in Molecular Biology, vol. 2639,
https://doi.org/10.1007/978-1-0716-3028-0_9,

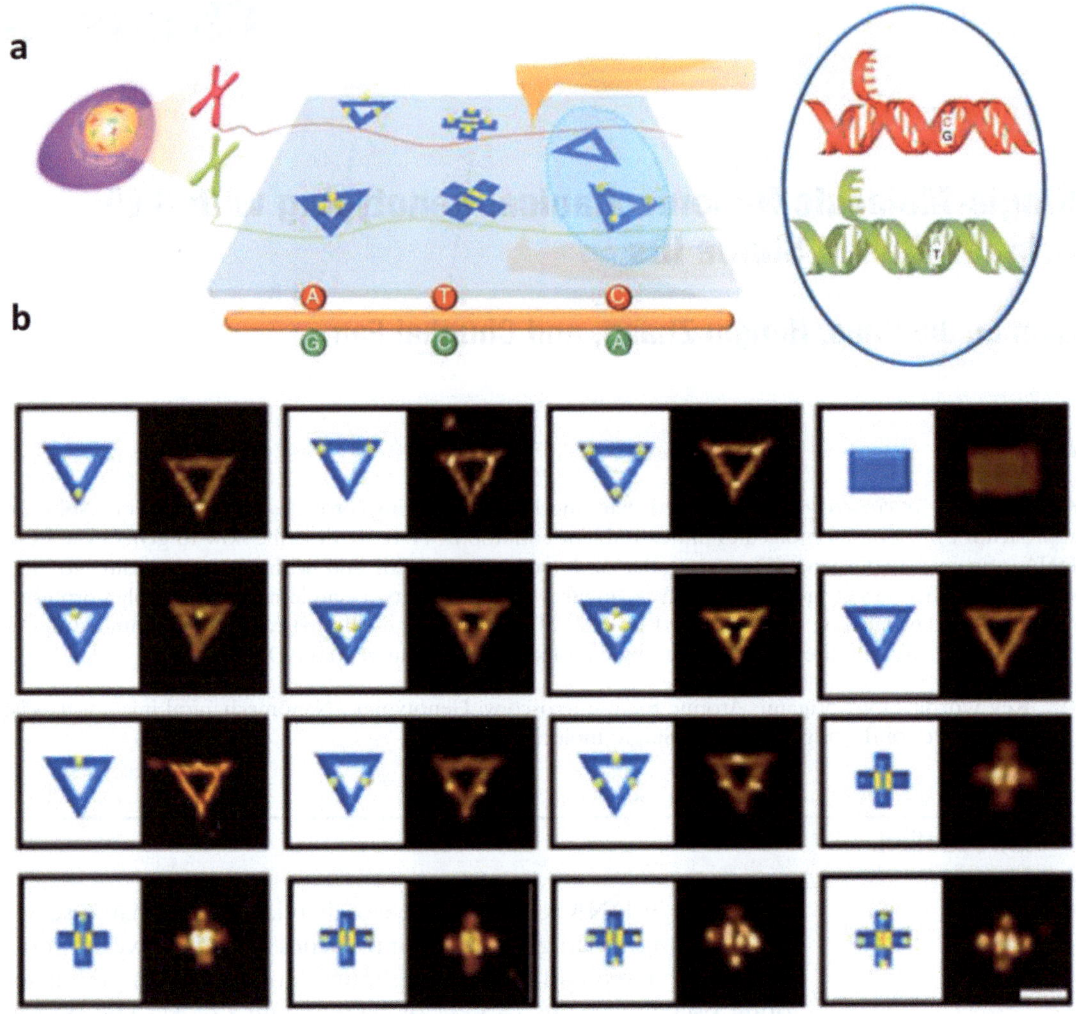

Fig. 1 (**a**) Schematic illustration of AFM-based single-molecule nanomechanical haplotyping with DNA origami shape IDs. Diploid genomic DNA was extracted from genetic samples, and the two alleles of each SNP are site-specifically labeled with different shape IDs. Consequently, the haplotype of this genomic DNA can be directly imaged under AFM. (**b**) Design and AFM images of 16 shape IDs using DNA origami decorated with or without STV. Scale bar, 100 nm. (Adapted from Ref. [12], with permission from Springer Nature)

imaging of SNPs (*see* Fig. 1a) [12–14]. With these origami shape IDs, we can directly genotype single molecules of human genomic DNA with an ultrahigh resolution of ~10 nm and determine types of disease-associated, long-range haplotypes.

2 Materials

All reagents should be stored and prepared according to the manufacturer's recommendations. Prepare all solutions using ultrapure water (18 MΩ·cm at 25 °C) and analytical grade reagents at room temperature, and store them at 4 °C (unless indicated otherwise). Diligently follow local or institutional regulations when recycling or disposal of waste materials. Use protective clothing (lab coats and gloves) throughout the procedure.

2.1 Preparation of DNA, Origami-Based Shape IDs

1. Milli-Q water (18 MΩ·cm).
2. Tris-EDTA (TE) buffer (1×): Mix 6.0 g of Tris base and 9.3 g of ethylenediaminetetraacetic acid (EDTA)-2Na in 90 mL of Milli-Q water. Use HCl and NaOH to adjust the solution to pH 8.0, and then add water to a final volume of 100 mL. The buffer can be stored at 4 °C for at least 6 months.
3. TBE buffer (10×): Add 108 g of Tris base, 55 g of boric acid, and 7.5 g of EDTA-2Na to 1 L of Milli-Q water. Use HCl and NaOH to adjust the solution to pH 8.0. The buffer can be stored at room temperature (25 °C) for up to 3 months.
4. TAE/Mg^{2+} buffer (10×): Add 48.5 g of Tris base, 11.4 mL of acetic acid, 25.4 g of $MgCl_2 \cdot 6H_2O$, and 7.5 g of EDTA-2Na to 1 L of Milli-Q water. Use HCl and NaOH to adjust the solution to pH 8.0. The buffer can be stored at room temperature for up to 3 months.
5. DNA staples and capturers (*see* **Note 1**).
6. M13mp18 ssDNA (250 μg/mL).
7. Biotin-modified ssDNA (optional).
8. Streptavidin (STV) (optional).

2.2 Labeling of ssDNA Template with Origami-Based Shape IDs

1. Mediator sequences (*see* **Note 2** and Fig. 2).
2. Genomic DNA samples.
3. Deoxynucleotide solution mix (10 mM each dNTP).
4. Vent (exo-) DNA polymerase (2000 U/mL).
5. ThermoPol reaction buffer.
6. Gold nanoparticles (spherical gold nanoparticles, 5 nm) (*see* **Note 3**) [15, 16].

2.3 Agarose Gel Electrophoresis and DNA Extraction

1. 0.5% (wt/vol) agarose gel: Add 0.25 g of agarose to 50 μL of 0.5 × TBE buffer. Heat and pour the mixture into the gel cassette. After filling the cassette, insert the comb until the teeth are completely immersed in the gel. Allow the gel to polymerize for 25–30 min (*see* **Note 4**).
2. Loading buffer (6×).
3. DNA ladder.
4. Gel running buffer: TBE (0.5×).
5. Gel staining solution: GelRed (1×).
6. Quantum Prep Freeze'N Squeeze DNA gel extraction spin column.

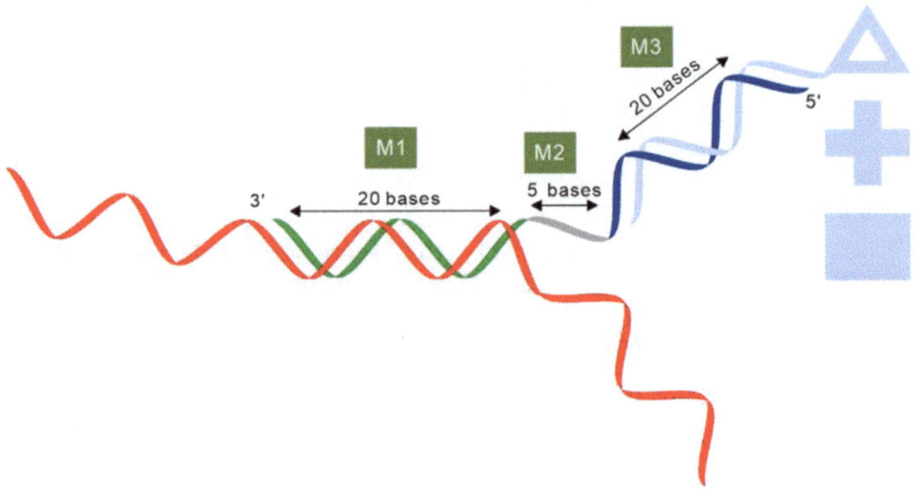

Fig. 2 Design of "Mediator" primer strands (M-strands). They consist of three blocks, M1, M2, and M3. M1 block is 20-base long, which is used to hybridize with the target ssDNA. M1 block is set at 3′-end of the M-strand, and the first base at 3′-end is the complementary base to the target SNP. M2 block is a spacer with five "T" bases. M3 block is also 20-base long, which can be recognized by M3′ on the DNA origami shape IDs [12–14]. (Reproduced from Ref. [13], with permission from Springer Nature)

2.4 AFM Imaging

1. Mica (1 × 1 cm^2).
2. Transparent tape.
3. Nitrogen gas.
4. Dilution buffer: TAE/Mg^{2+} (1×).

3 Methods

3.1 Preparation of DNA Stock Solutions

1. Centrifuge the EP tubes containing DNA sequences at "13,684 *g*" at room temperature for 10 min (*see* **Note 1**).
2. Dissolve all DNA sequences in 1× TE buffer, and dilute each DNA sample to 100 μM in 1× TE buffer (*see* **Note 5**).
3. Divide each sample into 10 μL aliquots. Label each tube and store the tubes at −20 °C until further use (*see* **Note 6**).

3.2 Preparation of DNA Origami Shape IDs

1. Mix the scaffold ssDNA and staple strands at a molar ratio of 1: 10 in a total volume of 100 μL in 1× TAE/Mg^{2+} buffer (*see* **Note** 7).
2. Heat the mixture solutions at 95 °C for 5 min, and then cool down to 4 °C in a PCR thermocycler at a cooling rate of 0.6 °C per min (*see* **Notes** 7 and **8**). Store the samples at 4 °C and use them within 1 week.

3.3 Purification of DNA Origami Shape IDs

1. Prepare samples for agarose gel electrophoresis by adding 1 μL of 6 × loading buffer to 5 μL of DNA origami shape IDs.
2. Load the DNA ladder and DNA origami shape IDs samples into the agarose gel (0.5% wt/vol), and conduct electrophoresis at 80 V (5 V/cm) for 100 min in 0.5 × TBE gel running buffer (*see* **Note 4**).
3. After 100 min, stop the electrophoresis. Place the gel into a container filled with 50 mL of 1× GelRed staining buffer for 15 min in the dark (*see* **Note 9**).
4. Image the gel and analyze the DNA bands with a gel image-analysis system (*see* **Note 10**). Excise the band of interest in the gel.
5. Chop the trimmed gel slice into smaller pieces, and place them into the filter cup of a Quantum Prep Freeze'N Squeeze DNA gel extraction spin column (*see* **Note 11**).
6. Place the Quantum Prep Freeze'N Squeeze DNA gel extraction spin columns (filter cup nested within the dolphin tube) at −20 °C for 5 min. Then centrifuge these spin columns at 13,000 *g* for 3 min at room temperature.
7. Collect the purified DNA origami shape IDs samples from the collection tube, and determine their concentrations by UV-visible spectroscopy measurement. Store the samples at 4 °C and use them within 1 week.

3.4 Preparation of DNA Origami Shape IDs Decorated with STV (Optional)

1. Add a tenfold molar excess of STV to purified DNA origami shape IDs containing biotin-modified ssDNA staples, and incubate the mixture at 37 °C for 4 h (*see* **Note 12**).

3.5 AFM Imaging of DNA Origami Shape IDs (With and Without STV)

1. Stick transparent tape to one side of the 1 × 1 cm^2 mica surface. Then tear it softly to eliminate mica layers to obtain a freshly cleaved mica surface for sample mounting (*see* **Note 13**).
2. Deposit 3 μL of 1.25 nM DNA origami ID solution onto the mica surface, and leave it to adsorb to the surface for 3–5 min.
3. Rinse the mica surface gently with Milli-Q water, and then dry it with nitrogen gas softly (*see* **Note 14**).
4. Image the samples by AFM in air under tapping mode with TESPA-V2 tips (*see* **Note 15**). See Fig. 1b for example images of DNA origami shape IDs with and without STV.

3.6 Direct Haplotyping of Genomic DNA by DNA Origami Shape IDs

1. Generate ssDNA template from genomic dsDNA samples (*see* **Note 16**).
2. Anneal the ssDNA template with the mediator primers (*see* Figs. 2 and 3). Initiate the extension at 95 °C (2 min) in a PCR thermal machine, allowing the strands to anneal by

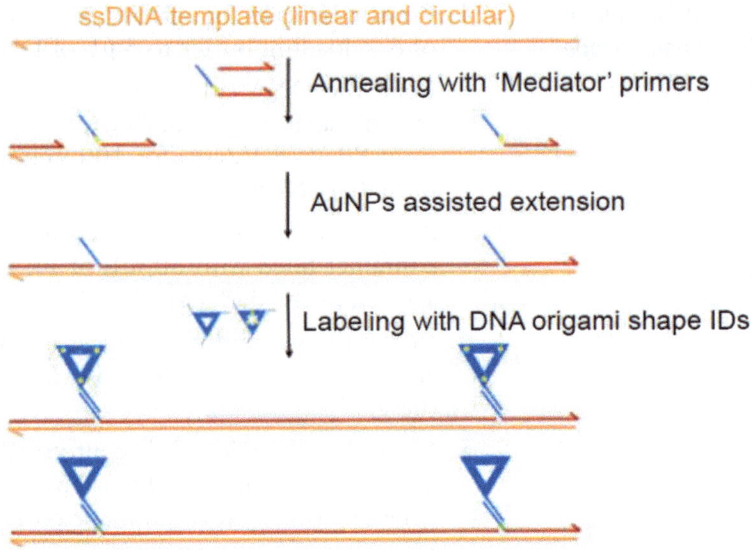

Fig. 3 Schematic illustration of labeling of ssDNA with DNA origami shape IDs. The desired ssDNA template generated from genomic dsDNA sample is generated, extended using the M-strand, and hybridized with the DNA origami shape IDs. (Reproduced from Ref. [13], with permission from Springer Nature)

gradually cooling to 25 °C at a rate of 0.6 °C/min. Then add 2 μL of Vent (exo-) DNA polymerase and 4 μL of dNTP, and incubate the sample at 60 °C for 45 min (*see* **Notes 3** and **17**).

3. Purify the extended products by agarose gel electrophoresis (*see* Subheading 3.3).
4. Determine the concentration of the purified samples (genomic dsDNA labeled with mediator primers) using UV-visible spectroscopy measurement (*see* **Note 18**).
5. Mix 2.5 μL of 1.0 nM genomic dsDNA labeled with mediator primers with 5 μL of 5.0 nM triangular DNA origami shape ID and 5 μL of 5.0 nM cruciform DNA origami shape ID (molar ratio of 1:10). Add 5 μL of 10× TAE buffer and 32.5 μL of ultrapure water (50 μL in total volume), and incubate the mixture at 37 °C for 30 min.
6. Perform AFM imaging (*see* Subheading 3.5). See Figs. 4 and 5 for example images of labeling the genomic DNA with DNA origami shape IDs.

4 Notes

1. To avoid loss of DNA powder, do not open the EP tubes before completing centrifugation.

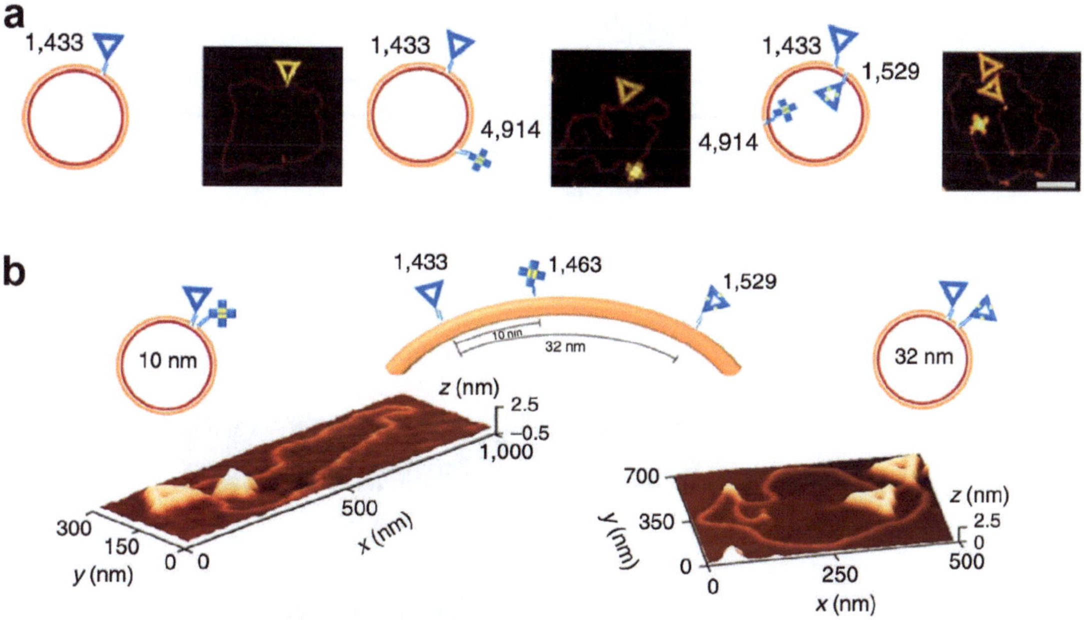

Fig. 4 Nanomechanical imaging and genotyping resolution of circular ssDNA with origami shape IDs. (**a**) Imaging of phiX174 with three different shape IDs (triangular corresponding to site 1433, triangular with STVs corresponding to site 1529, and cross corresponding to site 4914). Scale bar, 200 nm. (**b**) Schematic showing three-dimensional AFM topographic images of triangular- (corresponding to site 1433), cross- (corresponding to site 1463), and STV-decorated triangular (corresponding to site 1529)-shaped IDs for labeling phiX174. Left, the contour length between two labeled sites (1433and 1463) is ~10 nm (30 bp). Right, the contour length between two labeled sites (1433 and 1529) is ~32 nm (96 bp)

2. "Mediator" DNA strands consist of three blocks, M1, M2, and M3 (*see* Fig. 2). M1 block is 20-base long, which is used to hybridize with the target ssDNA. M1 block is set at 3′-end of the M-strand, and the first base at 3′-end is the complementary base to the target SNP. M2 block is a spacer with five "T" bases. M3 block is also 20-base long, which can be recognized by M3′ on the DNA origami shape IDs [12–14]. Design the sequences of the mediator DNA primers using software, such as Primer Premier 5.0 and NUPACK.
3. It is critical to employ AuNPs to facilitate extension of the ssDNA target annealed with "Mediator" DNA strands [15, 16].
4. Pour the gel slowly and smoothly. Before inserting the comb, make sure that there are no bubbles in the cassette.
5. To protect DNA from degradation, use 1× TE buffer to dissolve and dilute DNA. To ensure the formation of DNA origami structures, it is critical to determine the concentration of DNA samples with UV measurements.

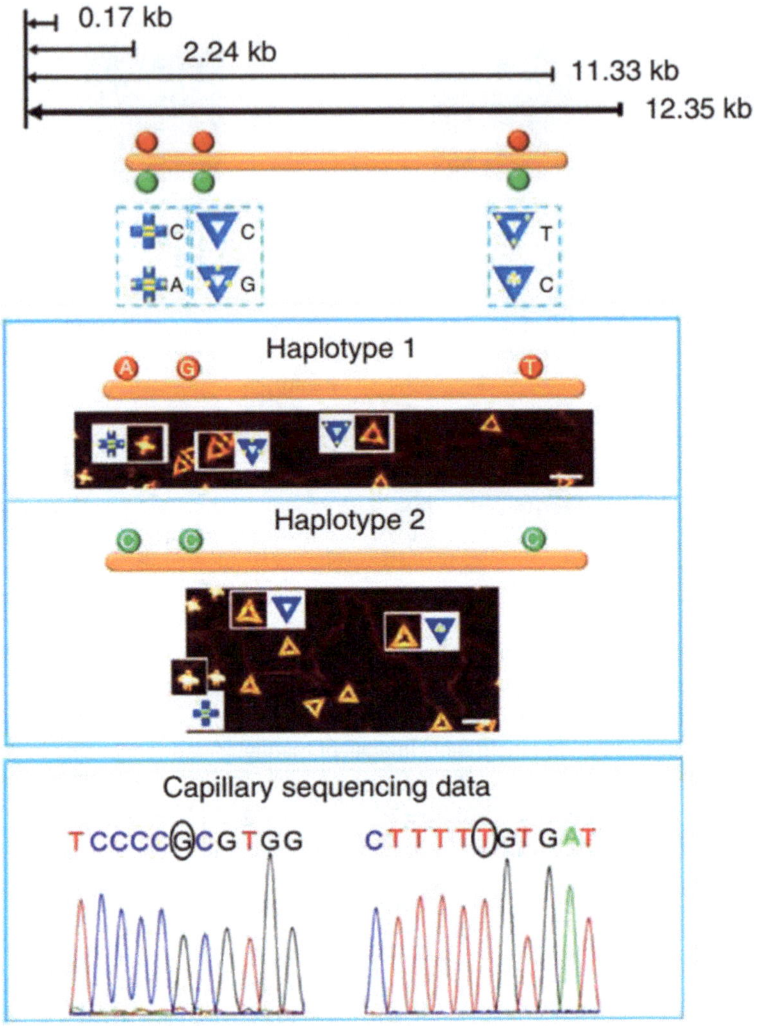

Fig. 5 Single-molecule haplotyping of P53 gene with origami shape IDs. Top, schematic illustration of a 12 kb region of the human p53 gene located on the chromosome 17. Each origami shape ID corresponds to a specific allele. Middle, AFM images for the haplotypes of these three SNPs. Haplotype 1 contains A-G-T and haplotype 2 contains C-C-C. Bottom, when the first SNP was A on one sequence, the other SNPs obtained from independent capillary sequencing were G and T, which was consistent with that from the nanomechanical imaging in haplotype 1. Scale bar, 200 nm

6. Dividing samples into aliquots is important to protect the DNA from repeated freeze-thaw cycles. Stock solutions can be stored at −20 °C for at least 1 year without freeze-thaw cycles.
7. Make sure correct staples and capturers are used. For the staples and capturers, and mixture components, used to fabricate different DNA origami shape IDs with and without streptavidin (STV), see references [12–14] for details, especially reference [13].

8. DNA origami shape IDs of different shapes have different optimal annealing procedures. See references [12–14] for detailed procedures, especially reference [13].
9. Remove the gel gently and make sure the gel is not damaged.
10. Typically, there are two bands in the lane: the upper one represents the target DNA origami and the lower one represents the redundant staples.
11. If the volume of the trimmed gel slice is too large to fit into a single filter cup, use two or more, and pool the recovered pure samples together.
12. For DNA origami shape IDs decorated with STV, use biotinylated mediator primers for genotyping experiments. Because of the addressability of DNA origami, different numbers of STV molecules can be site-specifically anchored on prescribed staples carrying biotin tags.
13. It is necessary to clean the mica surface with transparent tape before loading DNA origami samples.
14. Be careful not to damage the samples when rinsing and drying the surface.
15. The diameter of the tip is critical for clear imaging. The recommended pixel size is between 100 nm^2 and 3 μm^2, and the recommended imaging area is 512 × 512 pixels.
16. For detailed procedure of how to generate long ssDNA template from genomic dsDNA samples, refer to handbooks and instructions provided by reagent suppliers. The ssDNA samples can be stored at 4 °C for up to 1 week.
17. Make sure the solution is well mixed by vortexing the tube before annealing.
18. Measure the concentration of target DNA and DNA origami shape IDs by UV-visible spectroscopy.

Acknowledgments

This work was supported by the Ministry of Science and Technology of China (2017YFA0205302, 2016YFA0201200, and 2016YFA0400900), the NSFC (61771253 and 21390414), the Program for Changjiang Scholars and Innovative Research Team in University (IRT 15R37), the Natural Science Foundation of Jiangsu Province (BK20151504), and the Priority Academic Program Development of Jiangsu Higher Education Institutions (PAPD, YX03001).

References

1. Collins FS, Green ED, Guttmacher AE, Guyer MS (2003) A vision for the future of genomics research. Nature 422:835–847
2. Xiao M, Wan E, Chu C, Hsueh WC, Cao Y, Kwok PY (2009) Direct determination of haplotypes from single DNA molecules. Nat Methods 6:199–201
3. Lam ET, Hastie A, Lin C, Ehrlich D, Das SK, Austin MD et al (2012) Genome mapping on nanochannel arrays for structural variation analysis and sequence assembly. Nat Biotechnol 30:771–776
4. Hell SW (2007) Far-field optical nanoscopy. Science 316:1153–1158
5. Baday M, Cravens A, Hastie A, Kim H, Kudeki DE, Kwok PY et al (2012) Multicolor super-resolution DNA imaging for genetic analysis. Nano Lett 12:3861–3866
6. Müller DJ, Dufrêne YF (2008) Atomic force microscopy as a multifunctional molecular toolbox in nanobiotechnology. Nat Nanotechnol 3:261–269
7. Hansma HG, Sinsheimer RL, Li MQ, Hansma PK (1992) Atomic force microscopy of single- and double-stranded DNA. Nucleic Acids Res 20:3585–3590
8. Gross L, Mohn F, Moll N, Liljeroth P, Meyer G (2009) The chemical structure of a molecule resolved by atomic force microscopy. Science 325:1110–1114
9. Lu JH, Li HK, An HJ, Wang GH, Wang Y, Li MQ et al (2004) Positioning isolation and biochemical analysis of single DNA molecules based on nanomanipulation and single-molecule PCR. J Am Chem Soc 126:11136–11137
10. Kufer SK, Puchner EM, Gumpp H, Liedl T, Gaub HE (2008) Single-molecule cut-and-paste surface assembly. Science 319:594–596
11. Woolley AT, Guillemette C, Cheung CL, Housman DE, Lieber CM (2000) Direct haplotyping of kilobase-size DNA using carbon nanotube probes. Nat Biotechnol 18:760–763
12. Zhang HL, Chao J, Pan D, Liu HJ, Qiang Y, Liu K et al (2017) DNA origami-based shape IDs for single-molecule nanomechanical genotyping. Nat Commun 8:20170406
13. Chao J, Zhang HL, Xing YK, Lie Q, Liu HJ, Wang LH et al (2018) Programming DNA origami assembly for shape-resolved nanomechanical imaging labels. Nat Protoc 13:1569–1585
14. Liu K, Pan D, Wen YQ, Zhang HL, Chao J, Wang LH et al (2018) Identifying the genotypes of Hepatitis B Virus (HBV) with DNA origami label. Small 14:1701718
15. Chen P, Pan D, Fan CH, Chen JH, Huang K, Wang DF et al (2011) Gold nanoparticles for high-throughput genotyping of long-range haplotypes. Nat Nanotechnol 6:639–644
16. Li HK, Huang JH, Lv JH, An HJ, Zhang XD, Zhang ZZ et al (2005) Nanoparticle PCR: nanogold-assisted PCR with enhanced specificity. Angew Chem Int Ed 44:5100–5103

Chapter 10

Using Single-Molecule FRET to Evaluate DNA Nanodevices at Work

Nibedita Pal and Nils G. Walter

Abstract

The observation of DNA nanodevices at a single molecule (i.e., device) level and in real time provides rich information that is typically masked in ensemble measurements. Single-molecule fluorescence resonance energy transfer (smFRET) offers a means to directly follow dynamic conformational or compositional changes that DNA nanodevices undergo while operating, thereby retrieving insights critical for refining them toward optimal function. To be successful, smFRET measurements require careful execution and meticulous data analysis for robust statistics. Here we outline the elemental steps for smFRET experiments on DNA nanodevices, starting from microscope slide preparation for single-molecule observation to data acquisition and analysis.

Key words DNA nanotechnology, Fluorescence microscopy, FRET-based distance measurements, Single-molecule fluorescence resonance energy transfer, Surface immobilization

1 Introduction

The DNA duplex is inarguably one of the most fascinating examples of molecular self-assembly. Its programmability and structural predictability based on simple sequence-based rules are enabling the design of remarkable two- and three-dimensional structures with unprecedented molecular precision and diversity [1–7]. Structural DNA nanotechnology underwent a revolution when Paul Rothemund introduced "scaffolded DNA origami" wherein a long single-stranded DNA scaffold is folded intricately with the help of a large number of small staple strands segmentally complementary to the scaffold [2]. This pioneering work opened up entire new avenues for designing self-assembled DNA nanostructures. While from the very conception of DNA nanotechnology the design of intricate static DNA architectures has remained a focus, in recent years the engineering of nanoscale machines and actuated nanodevices has become increasingly popular. The strategies developed to power these nanodevices include strand displacement

Julián Valero (ed.), *DNA and RNA Origami: Methods and Protocols*, Methods in Molecular Biology, vol. 2639,
https://doi.org/10.1007/978-1-0716-3028-0_10,

through toehold exchange [8, 9]; the application of external stimuli such as ions [10], light [11], pH [12], etc.; or the exploitation of the catalytic activity of DNA [13, 14]. These DNA-based nanodevices have found interesting and versatile applications ranging from payload delivery [6] to biosensing of pH [15], ions [16], and enzymatic activity [17] across a living cell. Among the numerous DNA nanodevices designed so far, DNA walkers offer the prospect of cargo transport that simulates the function of biological motor proteins such as kinesin and myosin. One challenge in designing such walkers, or any other actuated DNA nanodevice, remains in their often-sluggish performance, taking significant time to perform just one operation. Recent advances have started to improve the design for faster translocation by cartwheeling [9] and transduction of external energy sources [18] to create DNA walkers of higher speed.

For the facile design of DNA nanodevices, several simulation software packages have been developed [19, 20]. Next, biochemical characterization and manipulation techniques are readily available to produce the DNA design. For example, restriction enzymes and ligases cut double-stranded DNA at specific sites and join the ends together, respectively [21–23]. Polymerase chain reaction (PCR) and cloning make it possible to generate DNA of a specific sequence in large quantity [24]. Long single-stranded DNA is commercially available as a scaffold strand in the form of M13 bacteriophage DNA [2, 4, 7] or can be generated by asymmetric PCR [25, 26]. Combining all these techniques makes it quite feasible to translate a computer-generated model of a DNA nanodevice into reality [27].

Once designed and synthesized, the characterization of DNA nanodevices remains a crucial step for advances in the field of DNA nanotechnology. Atomic force microscopy (AFM) is one of the standard tools with nanoscale spatial resolution and an ability to image under native conditions, albeit after firm adsorption to a mica mineral surface, that has been extensively applied to DNA nanodevices such as walkers [14, 28, 29]. Another important toolset entails either negative stain or cryogenic electron microscopy, which requires generating a thin layer of sample to be scanned by an electron beam in a vacuum [28, 29]. However, many DNA nanodevices require additional characterization to observe, ideally in real-time, dynamic events such as actuated shape changes or molecular locomotion. Here, bulk fluorescence spectroscopy offers a solution since it operates in real time and in solution under a broad set of native condition, is easy to detect and quantify, and is compatible with in situ manipulation such as the introduction of triggers and cofactors as well as with the use of microplates for parallelization and high-throughput screening [30].

DNA nanodevices often are actuated through conformational changes triggered by DNA strand displacement or changes in buffer additives. Fluorescence (or Förster) resonance energy transfer (FRET) offers a solution to observe in real time conformational changes between two judiciously placed fluorophores, which often proves critical toward understanding and optimizing functional performance [30, 31]. During FRET, the excitation energy of one fluorophore, the donor (e.g., Cyanine 3/Cy3), is non-radiatively transferred to the second fluorophore, or acceptor (e.g., Cyanine 5/Cy5), in a distance-dependent manner. As such, FRET enables one to monitor conformational changes in the 1–10-nm distance range that a DNA nanodevice undergoes during its operation. Importantly, modern single-molecule fluorescence microscopy has revolutionized how we perceive any molecular system as it reveals diversity of system behavior and malfunction [32, 33]. Single-molecule FRET (smFRET) of surface-immobilized DNA nanodevices allows us to monitor their actuation or locomotion over long periods of time when photobleaching and blinking of the fluorophores are minimized by removal of oxygen and speedy recovery from dark triplet states, respectively [34]. For example, smFRET has been employed recently for the real-time observation of the unidirectional rotation of a DNA nanoengine as an integral part of one of the fastest DNA walkers reported so far [18]. smFRET enabled the direct measurement of the rotation speed as an essential step in determining the nanoengine walking speed. Similarly, a recent application of smFRET to optimizing a cartwheeling DNA walker helped optimize its design parameters and thus its locomotion [9]. These examples show the synergistic interplay between design and characterization for optimizing the function of DNA nanodevices.

A total internal reflection fluorescence microscope (TIRFM) (*see* Fig. 1) is a powerful tool for smFRET detection with high signal-to-noise ratio [35]. In this chapter, we outline the steps necessary for efficient wide-field, camera-based, prism-type TIRFM for smFRET data acquisition, which allows one to image many single nanodevices simultaneously. We also describe general methodology for the analysis of smFRET data. As a representative example, we describe a DNA catenane-based nanoengine consisting of a catalytic stator that unidirectionally rotates against an interlocked rotor; a zinc finger protein fused to T7 RNA polymerase attached to the rotor harnesses the energy of NTP hydrolysis to fuel the continuous rotatory motion [18]. Although illustrated with this example, any fluorescently labeled nanodevice can be investigated using the smFRET protocols outlined here with only limited adjustments for specific needs. Similar protocols can be used for assays involving single-particle tracking [9] or single-molecule kinetic analysis of RNA transient structure (SiM-KARTS) strategies for monitoring nanodevice function [36]. Related super-resolution

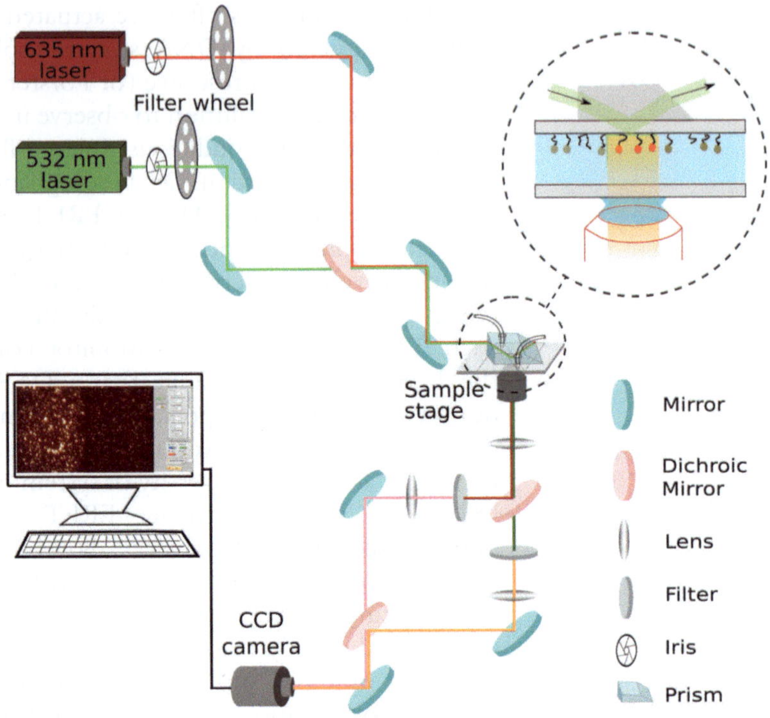

Fig. 1 Prism-type total internal reflection fluorescence microscope (TIRFM)

fluorescence techniques are another powerful toolset that can be used to monitor DNA nanodevice movement or compositional change over time, for the details of which the reader is referred to prior works [37, 38].

2 Materials

All solutions are prepared using autoclaved doubly deionized water (18 MΩ · cm at 25 °C). Chemicals are of highest commercially available purity and used without further purification.

2.1 Cleaning of Quartz Microscope Slides

1. Alconox.
2. Potassium hydroxide pellets (KOH).
3. Ammonium hydroxide.
4. Hydrogen peroxide.
5. Propane torch (14.1 OZ., Worthington).
6. Aqueous "base piranha": 20% v/v hydrogen peroxide, 20% v/v ammonium hydroxide, 60% v/v water.

2.2 Functionalization of Quartz Slide and Making Microfluidic Channel

1. Quartz microscope slide, 1 inch × 3 inches × 1 mm (G. Finkenbeiner).
2. Micro coverslip, rectangular, No. 1½, 24 × 30 mm.
3. 0.2″ ID × 0.6″ OD 100-80 microbore Tygon tubing (Cole-Parmer).
4. Double-sided sticky tape, ½″ wide.
5. Two-component epoxy resin (Double Bubble, Hardman Adhesives).
6. (3-Aminopropyl)triethoxysilane (APTES).
7. Acetone, HPLC grade.
8. Biotin-PEG-succinimidyl valerate (biotin-PEG-SVA, molecular weight 5000 Da) and methoxy-PEG-succinimidyl valerate (mPEG-SVA, molecular weight 5000 Da) (Laysan Bio Inc).
9. Biotinylated bovine serum albumin.
10. Disulfosuccinimidyl tartrate (Soltec Ventures).
11. Streptavidin.
12. Sodium bicarbonate.

2.3 Buffer Preparation (See Notes 1 and 2)

1. Trizma base, crystalline (≥99%).
2. Magnesium chloride hexahydrate ($MgCl_2$).
3. Sodium hydroxide pellets, anhydrous (NaOH).
4. Hydrochloric acid (HCl), 1 N.
5. Sodium chloride (NaCl, ≥99%).
6. Ethylenediaminetetraacetic acid (EDTA).
7. Acetic acid.
8. T50 buffer: 10 mM Tris-HCl, pH 8.0, 50 mM NaCl.
9. 1× TAE buffer: 40 mM Tris base, 20 mM acetic acid, pH 7.5, 1 mM EDTA.
10. 1× TAE-Mg^{2+} buffer: 40 mM Tris base, 20 mM acetic acid, pH 7.5, 12.5 mM $MgCl_2$, 1 mM EDTA.
11. 5× transcription buffer: 200 mM Tris–HCl, pH 7.9 at 25 °C, 50 mM DTT, 50 mM NaCl, and 10 mM spermidine.
12. Transcription mixture: rNTP set (2 mM each GTP, ATP, CTP, and UTP), 1× transcription buffer, 40 U RNasin ribonuclease inhibitor, 25 mM $MgCl_2$, 1× OSS, and 5% (v/v) DMSO.

2.4 Oxygen Scavenger System for smFRET Assay

1. Trolox (6-hydroxy-2,5,7,8-tetramethylchroman-2-carboxylic acid), 97% (Acros Organics).
2. Protocatechuate 3,4-dioxygenase from *Pseudomonas* sp. (PCD).
3. Protocatechuic acid (PCA).

4. Glycerol (≥99%).
5. 5 M KOH prepared in double distilled water.
6. PCD stock buffer: 100 mM Tris–HCl, pH 8.0, 50 mM KCl, 1 mM EDTA, and 50% (v/v) glycerol.
7. 1× OSS: 20 nM PCD, 5 mM PCA, and 2 mM Trolox.

2.5 Prism-Type Total Internal Reflection Fluorescence Microscopy

More detailed specifications of the opto-mechanical components required for assembling a prism-type TIRFM can be found somewhere else [35]. Briefly, the components required are as follows:

1. Continuous-wave 532 nm green laser (CrystalLaser, CL532-050-L, 50 mW).
2. Continuous-wave 635 nm red laser (Coherent CUBE, 635-25C, 25 mW).
3. Inverted fluorescence microscope (Olympus IX71).
4. Intensified charge-coupled device (I-CCD) camera (I-Pentamax, Princeton Instruments).
5. 60 × 1.2 NA water immersion objective (Olympus UPlanApo).
6. Pellin-Broca prism (Thorlabs).
7. Dichroic mirror (Chroma, cutoff 610 nm).
8. Neutral density variable filter (Edmund Industrial Optics).
9. Band-pass filter (HQ580/60 m, Chroma).
10. Long-pass filter (HQ655LP, Chroma).
11. Immersion oil, low fluorescence (Olympus).
12. Mirrors (protected silver mirror, 1″ diameter) and lenses (Plano-convex, BK7, AR coating, Thorlabs).
13. Vibration isolation optical table (ST-UT2, Newport).
14. Computer.

3 Methods

3.1 Cleaning of Quartz Microscope Slides

The quartz slides are surface functionalized following a standard protocol [37, 39]. Before surface functionalization, they are subjected to a thorough cleaning procedure, as follows:

1. Place the quartz microscope slides into the Coplin jar, and sonicate in water-alconox (100:1) mixture for 1 h.
2. Wash thoroughly with water to ensure that no residual detergent is left.
3. If the slides are being reused from prior experiments, they are manually rubbed with ethanol to get rid of residual glue and then thoroughly rinsed with water.

4. Sonicate the quartz slides in 1 M KOH for 20 min followed by thorough washing with water.
5. If the slides are being reused, flame the slides with a propane torch to burn off any residual contamination or glue.
7. Put the slides in aqueous "base piranha" and heat at 60–70 °C for 30 min.
8. Thoroughly rinse the slides with water and dry under nitrogen gas flow.

3.2 Surface Functionalization of Quartz Microscope Slides

After cleaning, the slides undergo surface functionalization. Before proceeding, ensure that the quartz slides are completely dry:

1. Incubate the quartz slides in 2% (v/v) 3-amino-propyltriethoxysilane (APTES) in acetone. After incubating for 20 min, sonicate for 1 min and incubate for an additional 10 min.
2. Wash the quartz slides thoroughly with water and dry under nitrogen flow.
3. Place the quartz slides in a clean pipette box, keeping the surface to be functionalized face up.
4. Prepare a 1:3 mixture of biotin-PEG-SVA and mPEG-SVA in freshly prepared 0.1 M sodium bicarbonate (*see* **Note 3**).
5. Centrifuge the mixture for 1 min at 10,000 rpm to remove any air bubbles.
6. Place 70 μL of the solution on the slide surface to be passivated, and then sandwich gently by placing a dried glass coverslip on top. Care is to be taken to ensure no air bubbles are trapped.
7. Incubate the slides at room temperature in a dark and moist environment for 3–4 h (or overnight for best results). Fill the bottom of the pipette box partially with water for overnight incubation.
8. Carefully disassemble the glass coverslip by sliding it off and disposing it in the proper waste.
9. Rinse the quartz slides thoroughly with water and dry under nitrogen flow.
10. Place the quartz slides back in the pipette box, keeping the functionalized surface face up.
11. Dissolve 12 mg sulfo-DST in 420 μL of a freshly prepared 1 M aqueous sodium bicarbonate solution. Centrifuge the solution at 10,000 rpm for 1 min to remove any air bubbles.
12. Place 70 μL of the solution on the PEG-functionalized surface of the slide and sandwich as before with a glass coverslip.
13. After 30 min of incubation in a moist environment, remove the glass coverslip by sliding it off and properly disposing of it.
14. Thoroughly rinse the slides with water and dry under nitrogen flow.

3.3 Assembling Microfluidic Sample Cells

For TIRFM experiments, a microfluidic sample cell is assembled on a PEG-functionalized quartz slide by following these steps (*see* Fig. 2):

1. For buffer exchange, drill holes into the quartz slide using a 1 mm diamond drill bit.
2. Place double-sided sticky tape diagonally on the PEGylated surface to make a channel of 5–6-mm width.
3. Gently place a dry and clean glass coverslip on top of the tape to complete the microfluidic channel. The surface of the coverslip can be PEGylated as described above for additional passivation.
4. Tightly seal the channel by gently pressing the coverslip with the help of a pipette tip over the entire area covering the double-sided tape.
5. Seal particularly the ends of the channel with epoxy resin.
6. Cut a 200 μL pipette tip at 8–9 mm from the tip. Affix that cut pipette tip to the holes with epoxy resin.
7. Attach Tygon tubing with epoxy glue to the pipette tips for constructing an inlet and an outlet for the microfluidic channel (*see* Fig. 2).

The complete sample chamber can be stored for 2–3 weeks in a dark environment at room temperature. For reuse, the quartz slides are boiled in water for 30 min or longer. Tape, cover glass, and adhesive are peeled off using a razor blade and the slides subjected to the above cleaning protocol.

3.4 Preparation of an Oxygen Scavenging System (OSS)

1. Prepare 1 μM PCD in PCD stock buffer.
2. Sterile filter (0.2 μm) and divide the PCD solution into 0.5 mL aliquots and store at −80 °C until needed.
3. Prepare 100 mM PCA in water aided by the dropwise addition of 5 M KOH. Adjust pH to ~8.3.
4. Sterile filter and divide the PCA solution into 1 mL aliquots. Store at −20 °C.
5. Dissolve Trolox in water to a final concentration of 100 mM. Slowly add 5 M KOH to aid dissolving Trolox. Vortex vigorously and check pH after each addition. Adjust pH to 10–11. Store at −20 °C until needed.
6. Prepare 1× OSS right before using it in a smFRET experiment.

3.5 Surface Immobilization of Single-DNA Nanodevices and smFRET Data Acquisition

In TIRFM, an evanescent wave is generated using either a quartz prism or the microscope objective itself. This evanescent wave illuminates surface-immobilized fluorescent molecules. In this chapter, we describe smFRET data acquisition using a prism-type TIRFM (*see* Fig. 1 and **Note 4**):

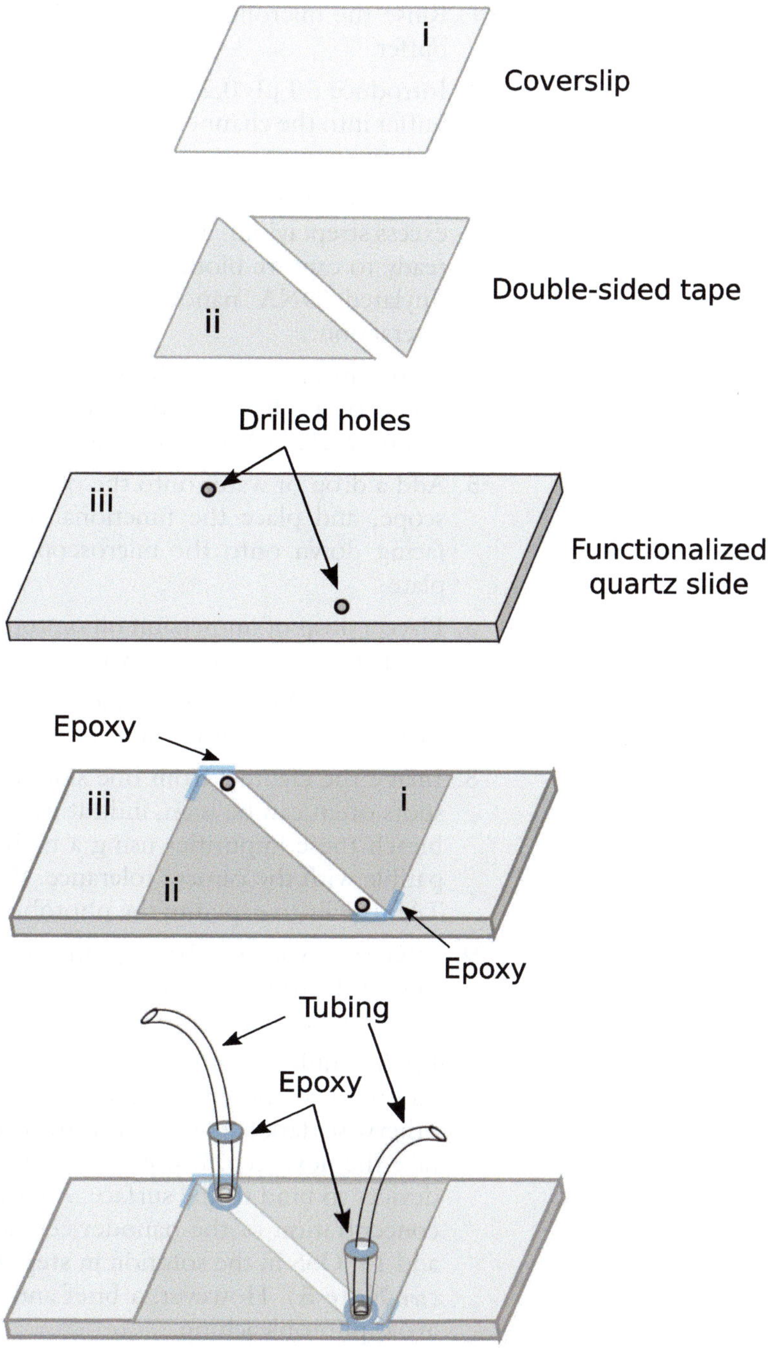

Fig. 2 Microfluidic channel assembly for single-molecule fluorescence microscopy

1. Rinse the microfluidic channel 2–3 times with ~200 μL T50 buffer.
2. Introduce 50 μL 0.2 mg/mL of streptavidin prepared in T50 buffer into the channel through the inlet tube, and incubate for 3–4 min.
3. Wash the channel with T50 buffer several times to flush out excess streptavidin. The slide surface within the channel is now ready to capture biotinylated DNA nanoengines or other biotinylated DNA nanodevices through a biotin-streptavidin interaction.
4. In the meantime, incubate the nanoengine with a threefold excess of T7 RNA polymerase-ZIF complex (T7RNAP-ZIF) (~0.6 nM:1.7 nM) at 37 °C for 30 min.
5. Add a drop of water onto the objective of the inverted microscope, and place the functionalized slide with the coverslip facing down onto the microscope stage held by an adapter plate.
6. Place a drop of immersion oil on top of the slide, and position the quartz prism carefully with the help of its holder.
7. Turn the 532 nm laser on at low power and adjust the objective focus to the glass-water interface.
8. Image the channel from one side to the other. Several bright spots often can be seen, indicating residual impurities. Photobleach these impurities using a high laser power that is compatible with the camera tolerance. Wash several times with 1× TAE buffer to expedite the photobleaching.
9. After removing the background, block the laser light with a shutter. Introduce ~100–300 μL of DNA nanodevices into the microfluidic chamber in the dark.
10. Briefly unblock the laser excitation and image using low excitation power. Several bright spots should be visible within the otherwise dark field of view. If the density of spots is sufficient, take the next step. If not, wait 3–4 min for more DNA nanodevices to bind to the surface. Alternatively or in addition, the concentration of the nanodevices can be increased. One may add 1× OSS in the solution in **step 9** to avoid photobleaching (*see* **Note 5**). However, a brief and low excitation power can avoid photobleaching.
11. Block the laser and wash away excess unbound catenane and T7RNAP-ZIF using 1× TAE-Mg^{2+} buffer.
12. Prepare enzymatic OSS at 1× concentration.
13. Prepare transcription mixture to relax the DNA double helix. Incubate on the slide for 3 min.
14. Seek a suitable field of view with a sufficient number of bright spots (*see* Fig. 3a).

15. Adjust the laser power to achieve a suitable signal-to-noise ratio while maintaining reasonably slow photobleaching.
16. Record fluorescence time traces in the form of movies for both donor and acceptor fluorophores using an I-CCD (or other single-molecule-sensitive camera) at the desired camera integration time. For example, to monitor the rotation of single nanoengines, we used 100 ms integration time. For our studies, the movie is saved in a data acquisition program written in MATLAB (scripts available upon request).
17. Either at the beginning or toward the end of the data acquisition, excite the acceptor fluorophore (Cy5 in our case) using the 638 nm red laser for a few frames to ensure the presence of acceptor (to distinguish from low-FRET states).
18. Repeat **steps 14–17** on different fields of view of the same slide to record a sufficient number of trajectories to achieve reliable statistics.

3.6 smFRET Data Analysis

The smFRET data collected as movies are typically analyzed using scripts written in MATALB or Interactive Data Language (IDL) (*see* **Note 6**):

1. Identify suitable molecules within the image that exhibit higher intensity than the surrounding pixels, and spatially match their signals in the donor and acceptor channels (Cy3 and Cy5 in our case (*see* **Note** 7)) (*see* Fig. 3a).
2. Extract time traces for the molecules identified using a suitable data analysis package. We use code written in the IDL package along with MATLAB scripts to extract and further analyze the trajectories (available upon request).
3. Set proper selection criteria. A significant signal-to-noise ratio, single-step photobleaching to reflect individual nanoengines, an acceptor intensity above a certain threshold, and other selection criteria can be applied, depending on the complexity of the trajectories. For our studies, we chose trajectories exhibiting an acceptor (Cy5) fluorescence intensity upon direct excitation well above background, anticorrelated donor (Cy3), and acceptor fluorescence intensities, as well as single-step photobleaching of both donor and acceptor. The FRET efficiency (E) at each time point is then calculated as $I_D/(I_A + I_D)$, where I_A and I_D are the apparent acceptor and donor fluorescence intensities, respectively (*see* Fig. 3b).
4. Generate a FRET efficiency histogram from the first 50 frames (or for the entire trajectory) of a sufficient number (typically >100) of trajectories, and fit it with a multi-peak Gaussian function as suitable to achieve low fit residuals (*see* Fig. 3c) [40]. The mean FRET value(s) represented by the center of

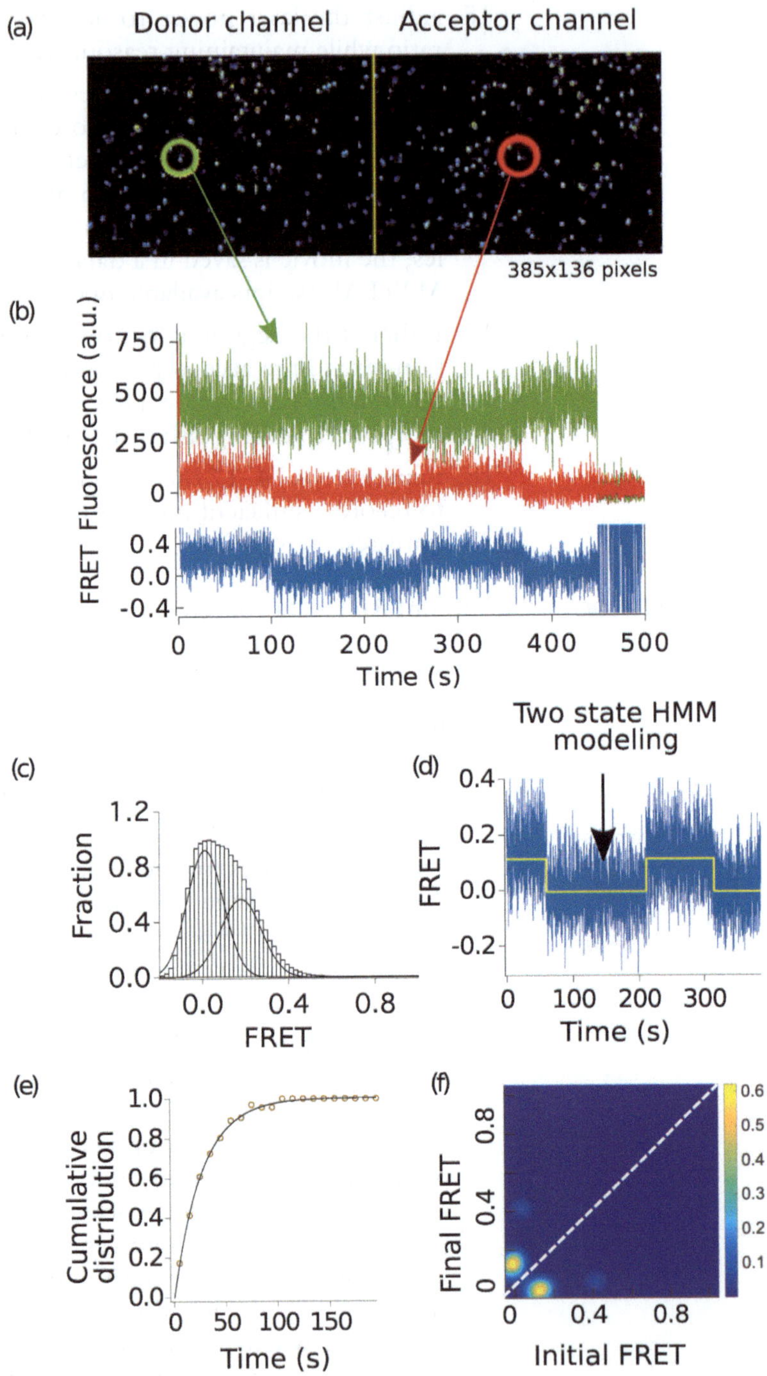

Fig. 3 smFRET data analysis steps of a DNA nanoengine [18] (**a**) Fluorescence signals from the fluorescently labeled nanoengines. (**b**) Extracted representative fluorescence intensity trajectories of donor (Cy3, green) and acceptor (Cy5, red) and corresponding FRET intensity trajectory (blue). (**c**) Single-molecule FRET efficiency histogram with multi-peak Gaussian fit. (**d**) Two-state hidden Markov (HMM) modeling of the FRET efficiency trajectory. (**e**) Cumulative dwell time distribution with suitable fit function. (**f**) Constructed transition occupancy density plot (TODP) from the idealized FRET trajectories

the peak(s) can be used to estimate the distance(s) between the fluorophores (R) using the following equation:

$E = \frac{1}{1+(R/R_0)^6}$, where R_0 is the Förster distance that is unique for a pair of donor and acceptor molecules; for Cy3 and Cy5, it is typically ~54 Å.

5. Fit the FRET intensity trajectories with proper idealizing models to extract the dwell times in the various FRET states. In the case of the nanoengine, we used two-state hidden Markov modeling, as implemented in the QuB package developed at the State University of New York at Buffalo, to extract the dwell times in the high- and low-FRET states (*see* Fig. 3d).
6. After idealization, extract the dwell times in each state for all trajectories.
7. Plot cumulative dwell time distributions, and fit them with suitable, often exponential, functions to extract the most probable dwell times using any suitable analysis software. For our purposes, we use OriginPro (*see* Fig. 3e).
8. From the idealized trajectories, construct the transition occupancy density plot (TODP) (*see* Fig. 3f). TODPs depict the fraction of molecules that populate a particular FRET state at least once, ensuring that fast transitions do not dominate this heat map [40]. Any population found on the diagonal represents static molecules (displaying no change in FRET state with time), whereas the off-diagonal population(s) represents dynamic molecules transitioning between FRET states during the observation time window (*see* Fig. 3d).

4 Conclusions and Future Outlook

In this chapter, we have illustrated a general procedure for how smFRET can be used as a technique to observe dynamic change in DNA nanodevices exemplified by a DNA catenane-based nanoengine. Application of smFRET of course is not limited to such comparably smaller structures of DNA nanodevices but additionally has been successfully employed investigating large, complex DNA origami-based nanodevices [28, 41–43]. These studies have demonstrated how the high temporal and spatial resolution offered by smFRET can be utilized to monitor origami-based DNA nanodevices whose response is often driven by conformational or compositional rearrangements triggered by environmental change. For example, smFRET measurements have been utilized to quantify depletion forces as low as ~100 fN in a DNA origami-based dynamic nanodevice [44]. Similarly, the mechanical properties of a DNA origami structure during voltage sensing can be monitored through smFRET signal intensity changes [41].

DNA origami-based nanostructures also have been used extensively as a canvas for functionalization with enzymes [7, 45], fluorescent dyes [46], DNA walkers [9], etc. with molecular scale precision. Single-molecule fluorescence microscopy has been used routinely as a technique to interrogate these systems, covering a broad spectrum of applications ranging from monitoring in real-time molecular kinetics [47, 48] and DNA walker locomotion [9] to enzyme catalytic activity inside a DNA cage [7]. However, the application of smFRET to monitoring conformational changes in the 3D structure of a DNA nanodevice during actuation is a relatively underexplored area and deserves more attention. As DNA nanodevices are becoming increasingly actuated, often going through intermediate configurations within sub-second time scales [49–51], the mapping of intra- and intermolecular distances in real time during dynamic actuation is becoming ever more important. Since smFRET provides a route to 3D distance triangulation by inclusion of multiple FRET pairs [51, 52], generally at the sub-second time scale [53], it can not only be used to complement existing characterization methods in DNA nanotechnology but also can open new possibilities for designing and optimizing rapidly actuated DNA origami of increasing size and complexity. Additionally, correlated measurements by combining several single-molecule techniques such as monitoring smFRET intensity while the DNA nanodevice is manipulated by atomic force microscopy (AFM) or optical tweezers, a detailed understanding of DNA nanodevice functionality in a complex environment can be achieved. Due to the necessary domain knowledge, implementing smFRET experiments for DNA nanodevices may appear a daunting barrier for newcomers to the field. Equipped with this protocol, however, we hope that scientists with little prior experience in single-molecule fluorescence microscopy will be able to adapt smFRET for their specific purposes.

5 Notes

1. Some of the reagents are potentially hazardous and should be handled with the manufacturer-prescribed precaution.
2. Freshly prepare all buffer solutions.
3. Biotin-PEG and mPEG should be stored at −20 °C and used within 6 months. Before opening the bottles, first warm them up to room temperature to avoid moisture condensation.
4. The steps described in this chapter follow the requirements for our DNA nanoengine [18] and DNA walker [9].
5. Sufficient care should be taken to protect the fluorescently labeled molecules from light during handling to prevent photobleaching. For example, dark Eppendorf tubes, aluminum foil wrapping, and low ambient light can be used for the purpose.

6. Numerous data analysis protocols have been developed for smFRET. In addition, good judgment should be used in interpreting all data to ensure that the conclusions are supported by the raw data.
7. Although Cy3 and Cy5 are used as FRET pair in our studies, different FRET pairs can be utilized instead. In that case, the excitation laser, dichroic filter, and emission filters need to be adjusted accordingly.

References

1. Winfree E, Liu F, Wenzler LA, Seeman NC (1998) Design and self-assembly of two-dimensional DNA crystals. Nature 394: 539–544
2. Rothemund PWK (2006) Folding DNA to create nanoscale shapes and patterns. Nature 440:297–302
3. Douglas SM, Dietz H, Liedl T et al (2009) Self-assembly of DNA into nanoscale three-dimensional shapes. Nature 459:414–418
4. Andersen ES, Dong M, Nielsen MM et al (2009) Self-assembly of a nanoscale DNA box with a controllable lid. Nature 459:73–76
5. Dietz H, Douglas SM, Shih WM (2009) Folding DNA into twisted and curved nanoscale shapes. Science 325:725–730
6. Douglas SM, Bachelet I, Church GM (2012) A logic-gated nanorobot for targeted transport of molecular payloads. Science 335:831–834
7. Zhao Z, Fu J, Dhakal S et al (2016) Nanocaged enzymes with enhanced catalytic activity and increased stability against protease digestion. Nat Commun 7:10619
8. Thubagere AJ, Li W, Johnson RF et al (2017) A cargo-sorting DNA robot. Science 357: eaan6558
9. Li J, Johnson-Buck A, Yang YR et al (2018) Exploring the speed limit of toehold exchange with a cartwheeling DNA acrobat. Nat Nanotechnol 13:723–729
10. Gerling T, Wagenbauer KF, Neuner AM, Dietz H (2015) Dynamic DNA devices and assemblies formed by shape-complementary, non--base pairing 3D components. Science 347: 1446–1452
11. You M, Chen Y, Zhang X et al (2012) An autonomous and controllable light-driven DNA walking device. Angew Chem Int Ed 51:2457–2460
12. Idili A, Vallée-Bélisle A, Ricci F (2014) Programmable pH-triggered DNA nanoswitches. J Am Chem Soc 136:5836–5839
13. Pei R, Taylor SK, Stefanovic D et al (2006) Behavior of polycatalytic assemblies in a substrate-displaying matrix. J Am Chem Soc 128:12693–12699
14. Lund K, Manzo AJ, Dabby N et al (2010) Molecular robots guided by prescriptive landscapes. Nature 465:206–210
15. Modi S, Swetha MG, Goswami D et al (2009) A DNA nanomachine that maps spatial and temporal pH changes inside living cells. Nat Nanotechnol 4:325–330
16. Saha S, Prakash V, Chakraborty K, Krishnan Y (2015) A pH-independent DNA nanodevice for quantifying chloride transport in organelles of living cells. Nat Nanotechnol 10:645–651
17. Dan K, Veetil AT, Chakraborty K, Krishnan Y (2019) DNA nanodevices map enzymatic activity in organelles. Nat Nanotechnol 14: 252–259
18. Valero J, Pal N, Dhakal S et al (2018) A bio-hybrid DNA rotor-stator nanoengine that moves along predefined tracks. Nat Nanotechnol 13:496–503
19. Castro CE, Kilchherr F, Kim D-N et al (2011) A primer to scaffolded DNA origami. Nat Methods 8:221–229
20. Douglas SM, Marblestone AH, Teerapittayanon S et al (2009) Rapid prototyping of 3D DNA-origami shapes with caDNAno. Nucleic Acids Res 37:5001–5006
21. Loenen WAM, Dryden DTF, Raleigh EA et al (2014) Highlights of the DNA cutters: a short history of the restriction enzymes. Nucleic Acids Res 42:3–19
22. Buckhout-White S, Person C, Medintz IL, Goldman ER (2018) Restriction enzymes as a target for DNA-based sensing and structural rearrangement. ACS Omega 3:495–502
23. Song IH, Shin SW, Park KS et al (2015) Enzyme-guided DNA sewing architecture. Sci Rep 5:1–9
24. Garibyan L, Avashia N (2013) Research techniques made simple: Polymerase Chain Reaction (PCR). J Invest Dermatol 133:20382

25. Veneziano R, Shepherd TR, Ratanalert S et al (2018) In vitro synthesis of gene-length single-stranded DNA. Sci Rep 8:1–7
26. Liu Q, Wang L, Frutos AG et al (2000) DNA computing on surfaces. Nature 403:175–179
27. Ke Y, Meyer T, Shih WM, Bellot G (2016) Regulation at a distance of biomolecular interactions using a DNA origami nanoactuator. Nat Commun 7:1–8
28. Birkedal V, Dong M, Golas MM et al (2011) Single molecule microscopy methods for the study of DNA origami structures. Microsc Res Tech 74:688–698
29. Jungmann R, Scheible M, Simmel FC (2012) Nanoscale imaging in DNA nanotechnology. Wiley Interdiscip Rev Nanomed Nanobiotechnol 4:66–81
30. Lakowicz JR (2006) Principles of fluorescence spectroscopy, 3rd edn. Springer, New York
31. Mao C, Sun W, Shen Z, Seeman NC (1999) A nanomechanical device based on the B-Z transition of DNA. Nature 397:144–146
32. Walter NG, Huang C, Manzo AJ, Sobhy MA (2008) Do-it-yourself guide: how to use the modern single-molecule toolkit. Nat Methods 5:475–489
33. Widom JR, Dhakal S, Heinicke LA, Walter NG (2014) Single-molecule tools for enzymology, structural biology, systems biology and nanotechnology: an update. Arch Toxicol 88:1965–1985
34. Aitken CE, Marshall RA, Puglisi JD (2008) An oxygen scavenging system for improvement of dye stability in single-molecule fluorescence experiments. Biophys J 94:1826–1835
35. Gibbs D, Kaur A, Megalathan A et al (2018) Build your own microscope: step-by-step guide for building a prism-based TIRF microscope. Methods Protoc 1:40
36. Rinaldi AJ, Lund PE, Blanco MR, Walter NG (2016) The Shine-Dalgarno sequence of riboswitch-regulated single mRNAs shows ligand-dependent accessibility bursts. Nat Commun 7:1–10
37. Michelotti N, de Silva C, Johnson-Buck AE et al (2010) A bird's eye view. Tracking slow nanometer-scale movements of single molecular nano-assemblies. Methods Enzymol 475: 121–148
38. Johnson-buck A, Nangreave J, Kim D et al (2013) Super-resolution fingerprinting detects chemical reactions and idiosyncrasies of single DNA pegboards. Nano Lett 13:728–733
39. Chandradoss SD, Haagsma AC, Lee YK et al (2014) Surface passivation for single-molecule protein studies. J Vis Exp 86:e50549
40. Blanco M, Walter NG (2010) Analysis of complex single-molecule FRET time trajectories. Methods Enzymol 472:153–178
41. Hemmig EA, Fitzgerald C, Maffeo C et al (2018) Optical voltage sensing using DNA origami. Nano Lett 18:1962–1971
42. Saccà B, Ishitsuka Y, Meyer R et al (2015) Reversible reconfiguration of DNA origami nanochambers monitored by single-molecule FRET. Angew Chem Int Ed 54:3592–3597
43. Li CY, Hemmig EA, Kong J et al (2015) Ionic conductivity, structural deformation, and programmable anisotropy of DNA origami in electric field. ACS Nano 9:1420–1433
44. Hudoba MW, Luo Y, Zacharias A et al (2017) Dynamic DNA origami device for measuring compressive depletion forces. ACS Nano 11: 6566–6573
45. Fu J, Liu M, Liu Y et al (2012) Interenzyme substrate diffusion for an enzyme cascade organized on spatially addressable DNA nanostructures. J Am Chem Soc 134:5516–5519
46. Stein IH, Steinhauer C, Tinnefeld P (2011) Single-molecule four-color FRET visualizes energy-transfer paths on DNA origami. J Am Chem Soc 133:4193–4195
47. Jungmann R, Steinhauer C, Scheible M et al (2010) Single-molecule kinetics and super-resolution microscopy by fluorescence imaging of transient binding on DNA origami. Nano Lett 10:4756–4761
48. Johnson-Buck A, Nangreave J, Jiang S et al (2013) Multifactorial modulation of binding and dissociation kinetics on two-dimensional DNA nanostructures. Nano Lett 13:2754–2759. https://doi.org/10.1021/nl400976s
49. Lauback S, Mattioli KR, Marras AE et al (2018) Real-time magnetic actuation of DNA nanodevices via modular integration with stiff micro-levers. Nat Commun 9:1446
50. Song J, Li Z, Wang P et al (2017) Reconfiguration of DNA molecular arrays driven by information relay. Science 357:371
51. Kopperger E, List J, Madhira S et al (2018) A self-assembled nanoscale robotic arm controlled by electric fields. Science 359:296–301
52. Hu J, Liu MH, Zhang CY (2019) Construction of tetrahedral DNA-quantum dot nanostructure with the integration of multistep Förster resonance energy transfer for multiplex enzymes assay. ACS Nano 13:7191–7201
53. Jeong C, Cho WK, Song KM et al (2011) MutS switches between two fundamentally distinct clamps during mismatch repair. Nat Struct Mol Biol 18:379–385

Part IV

Applications of DNA and RNA Origami

Chapter 11

Parallel Functionalization of DNA Origami

Rasmus P. Thomsen, Rasmus S. Sørensen, and Jørgen Kjems

Abstract

DNA origami enables the creation of large supramolecular structures, with precisely defined features at the nanoscale. The concept thus naturally lends itself to the concept of molecular patterning, i.e., the positioning of molecular moieties and functional features. Creation of nanoscale patterns was already disseminated by Rothemund in 2006, in which DNA hairpins were used to produce nanoscale patterns on the flat origami canvases (Rothemund PWK, Nature 440(7082):297–302, 2006). For this type of application, it is often desired to produce multiple different patterns using the same origami canvas by reusing existing origami staple strands, rather than ordering new, custom oligonucleotides for each unique pattern. This chapter presents a method where the enzyme terminal deoxynucleotidyl transferase (TdT) is used in a parallelized reaction to add functional moieties to the end of a selected pool of unmodified staple strand oligonucleotides, which are then incorporated at precisely defined positions in the DNA origami canvas. Introducing arrays of functional features using this enzymatic functionalization of origami staple strands offers a very high degree of flexibility, versatility, and ease of use and can often be obtained faster than custom synthesis. For small synthesis scales, typically employed during initial functional screening of many different molecular patterns, the method also offers a significant advantage in terms of cost. During the past years, we have utilized this to incorporate a large variety of molecules including bulky proteins (Sørensen RS, Okholm AH, Schaffert D, Kodal ALB, Gothelf KV, Kjems J, ACS Nano 7:8098–8104, 2013) in designed patterns from modified nucleotide triphosphate (NTP) building blocks (Jahn K, Tørring T, Voigt NV, Sørensen RS, Kodal ALB, Andersen ES, Bioconjug Chem 22:819–823, 2011). The near-quantitative yields obtained by enzymatic functionalization allow synthesis of a large set of oligonucleotides in a one-pot reaction from commercial starting materials without the need for individual post-purification. Based on the chosen subset of staple strand, it is possible to create any designed functionality, array, or pattern. Here we describe the process going from an idea/design of a DNA origami-specific molecular pattern to nucleotide synthesis and subsequent parallel functionalization of the DNA origami, assembly, and the final characterization.

Key words DNA, Functionalization, Enzymatic, DNA origami, Nanotechnology

1 Introduction

With the advancement of structural DNA nanotechnology, development has naturally shifted towards a more functional-centric focus [4, 5]. With methods like DNA origami [1], tailor-made nanostructures can be constructed, and functional features can be

Julián Valero (ed.), *DNA and RNA Origami: Methods and Protocols*, Methods in Molecular Biology, vol. 2639, https://doi.org/10.1007/978-1-0716-3028-0_11,

introduced at chosen spatial positions with sub-10 nm precision. Often, DNA nanostructure modification is done by introducing external functional entities. Examples include small molecules like ligands, lipids [6, 7], haptens, dyes [8], peptides and proteins [9], interesting polymers [10], and inorganic nanoparticles [11]. This large array of possibilities allows researchers to engineer and elaborate functional DNA nanostructures from nanorobots interacting with cells [12, 13] to plasmonic devices [11] and scaffolds for patterning purposes.

With DNA, being a rather chemically inert molecule, such modifications require specific handles. As a popular solution, useful chemical moieties like nucleophilic amines or thiols are often introduced during the synthesis of synthetic oligonucleotides. Another highly popular method is the use of biotin-streptavidin bridges [9] (for a more in-depth review *see* [14]). Such methods are however not very flexible as each new design requires de novo synthesis of the synthetic oligonucleotides. This can be especially expensive and laborious for DNA origami structures as specific subsets of staple strands are used to vary design patterns in order to explore and optimize the spatial positioning for the given purpose.

A more flexible and parallel alternative is post-synthetic labeling of plain DNA oligonucleotides using enzymes [3]. Selected oligonucleotides can be individually attached non-templated with the (a) specific nucleotide triphosphate(s) to the 5′ or the 3′ end using the Vaccinia Capping Enzyme (Vce) or (b) the terminal deoxynucleotidyl transferase respectively (TdT) [15]. While the Vce only accepts GTP as a substrate with limited alterations and caps at the 5′ end of an oligonucleotide, the TdT enzyme is much more versatile and accepts any available natural nucleotide triphosphate and attaches it to the unmodified 3′ hydroxyl group. Highly interestingly, TdT is a rather promiscuous enzyme which can accept an impressive variety of modified nucleotide triphosphates as substrate for 3′ elongation [2]. This even includes larger biomolecules like dendrimers. As a result, a subset of DNA strands in a given nanostructure design, being a DNA origami or other method, can be modified in batches of specific patterns. This is in particular useful regarding DNA origamis, being molecular pegboards, as subsets can be selected in chosen modules, functionalized in parallel, and their effect studied [3]. As an alternative approach, 3′-end tailing of protruding DNA staple strands is possible on assembled DNA origamis in solution [16]. Importantly, with enzymatic TdT, labeling of the 3′-end with a single nucleotide extension is possible and preferred in many cases using ddNTPs. A more in-depth discussion about the strategic choice of oligonucleotide staple strand modification strategy is found in **Note 1**.

Important to notice, TdT is a processive enzyme, and if a natural nucleotide triphosphate is used, a homopolymeric tail of the given nucleotide of various length is made depending on the substrate concentration and incubation time. We routinely use TdT to attach 3′ hexynyl-ddUTP, Aminoallyl-dNTP, 5-ethynyl-dUTP, NTPs pre-coupled with large biomolecules, biotinylated NTPs, sialic acids, lipids, carbohydrates, biotin, fluorophores, PEG of various sizes, and array of click and NHS chemistry-ready modifications in poly-N-tails.

2 Materials

All stock concentrations and additions of water are done with Milli-Q grade water with a resistivity of 18.2 MΩ·cm at 25 °C. Store all stocks and reagents at −25 °C unless otherwise stated. For DNA nanostructures, oligonucleotides can be ordered from IDT (Integrated DNA Technologies). Various non-natural triphosphate nucleotides can be ordered from several commercial sites like Jena Bioscience, Trilink Biotechnologies, Metkinen Chemistry, Base-click GmbH, and others.

2.1 Designing Modular DNA Origami

1. Cadnano (https://cadnano.org) or other DNA origami design software.

2.2 Pipetting of Modular Staple Strand Pools

1. DNA single strands are synthesized by Integrated DNA Technologies. Typically for DNA origami, >200 single DNA strands are needed and ordered in 96-well plates and diluted to 100 μM in desired buffer (RNase free water or IDTE buffer: 10 mM Tris, 0.1 mM EDTA at pH 8.0 or 7.5).
2. A clean space for pipetting or preferably a pipetting robot.

2.3 Copper(I)-Catalyzed Alkyne-Azide Cycloaddition (CuAAC)

1. (Di)deoxynucleotide triphosphate precursor:
   R1. Aminoallyl-dUTP.
   R2. 5-propargylamino-ddUTP.
   R3. 5-Ethynyl-dUTP.
   R4. Other useful: Biotin-modified (d)dNTP, CuAAC ready (d)dNTP, and fluorophore-modified (d)dNTP.
2. Materials to be attached to nucleotide:
   R1. 5kPEG-NHS.
   R2. Azido-dPEG8-NHS, NHS-DBCO bifunctional linker, streptavidin.
   R3. Benzyl azide.
   R4. Other useful: Hetero and homobifunctional linkers, CuAAC-modified chemical groups, and NHS-activated modification.

3. Buffer components:
   R1. HEPES-buffer (1 M, pH 8.0).
   R2. HEPES-buffer (50 mM, pH 7.8), DMSO, HEPES (37.5 mM, pH 8.0), 250 mM NaCl solution.
   R3. CuAAC buffer: 50 mM $CuSO_4$, 100 mM BTTAA, 200 mM Ascorbic acid. HEPES-buffer (250 mM, pH 7.5), DMSO.
4. Other materials:
   R1. Thermomixer shaker/temperature control.
   R2. 10 kDa cut-off Amicon spin-dialysis filters, reverse phase HPLC column.
   R3. HPLC with reverse phase HPLC column.
5. HPLC purification:
   R1. MQ-water (H_2O).
   R2. Acetonitrile (MeCN).
   R3. 1 M triethylammonium acetate (TEAA).
   R4. 1 L Buffer A: 900 ml H_2O with 50 ml 50 mM TEAA and 50 ml MeCN, pH 7.
   R5. 1 L Buffer B: 950 ml MeCN with 50 ml 1 M TEAA, pH 7.
   R6. HPLC RP-column (Phenomenex Kinetex C18-Evo or the like).
   R7. HPLC setup (Agilent 1200 series or the like).

2.4 TdT 3′-End Labeling of Selected Staple Strands

1. Terminal deoxynucleotidyl transferase kit is acquired from Sigma Aldrich (previously Roche) which contains the TdT enzyme (400 U/μl), 5× $CoCl_2$ (25 mM), and 5× TdT buffer.
2. 0.4 M EDTA solution pH 8.2, to chelate Co^{2+}.
3. Heated block/shaker for incubation at 37 °C.
4. Oligonucleotides.
5. 3 M Sodium acetate (NaOAC) at pH 5.8.
6. Ice-cold 96% ethanol.
7. Optionally: 70% ethanol.
8. SequaGel or other acrylamide gel system (National Diagnostics) for denaturing polyacrylamide gel electrophoresis (denaturing PAGE): UreaGel Concentrate and UreaGel Diluent.
9. 10% Ammonium Persulfate (W/V in water) + TEMED.
10. SYBR Gold Nucleic Acid Gel Stain.
11. GeneRuler Ultra Low Range ladder.

2.5 Assembly of DNA Origami with TdT-Functionalized Staple Strands and Gel Electrophoresis

1. 10× TAE pH 8.3 (1× TAE: 40 mM Tris-acetate with 1 mM EDTA).
2. Magnesium chloride, $MgCl_2$, 0.2 M stock solution.
3. Assembly buffer (1× TAE/Mg^{2+} buffer): 40 mM Tris-acetate and 1 mM EDTA at pH 8.3 with 12.5 mM $MgCl_2$.

2.6 Characterization of Labeled DNA Origami

Agarose gel characterization:

1. 10× TBE buffer. 1× TBE is 100 mM Tris-base, 90 mM boric acid and 1 mM EDTA at pH 8.4.
2. 6× native loading dye: 40% glycerol with Bromophenol Blue or other loading dye.
3. SYBR Safe Nucleic Acid Gel Stain (or ethidium bromide).

AFM:

1. Asylum Research cantilevers model TR400PSA or the like with a spring constant of in the 0.05 Nm^{-1} range.
2. Washing buffer: 1× TAE with appropriate Mg^{2+} concentration.
3. Optional: NiCl solution.
4. Tweezers of choice.
5. Mica.
6. Gwyddion or similar software for AFM image analysis.

TEM:

1. Carbon film coated 400 mesh Cu grids (CF-400-Cu_UL, EMS).
2. 2% Uranyl formate (UF) ($UO_2(CHO_2)_2{\cdot}H_2O$) solution, 2% uranyl formate (UF) solution:
   - UF powder.
   - 10 M KOH.
   - Vortexer and table spin centrifuge.
   - Liquid nitrogen (LN2).
   - 0.2 μm cellulose syringe filters.
   - 1.5 ml Eppendorf tubes and PCR tubes (optional).
3. Whatman filter paper (GE HealthCare, #1001-085).
4. Timer.
5. Tweezers of choice to handle grids.
6. While most images can be viewed with ImageJ, recommended processing software are Relion [17, 18], cisTEM [19, 20], Scipion [21, 22], and CryoSPARC [23, 24].

3 Methods

3.1 Designing Modular DNA Origami

DNA origami design can be facilitated by using any design software of choice. We recommend caDNAno (version 2 [25] or 2.5 [26]). To ease the selection of multiple staples for "one-pot" labeling, we suggest color-coding the staple strands in caDNAno before exporting to CSV. This makes it easy to arrange staple strands in the desired modular pools. We also suggest mapping each hex color code (e.g., #FF3311) to a module name (e.g., "Red-pattern") – this is easily done using, e.g., Python or Excel. Each modular pool can be individually functionalized to create patterns and/or patches of specific functionality. Importantly, to create a plain DNA origami, combine stoichiometric amounts of all module pools to create a "complete-design" master pool; then mix some of the master pool with the M13 scaffold strand.

3.2 Pipetting of Modular Staple Strand Pools

1. Prepare and label clean Eppendorf tubes for each of the modular pools of DNA staple strands (*see* **Note 2**).
2. For each pool, transfer equal amount/volume of all staple strands into one tube. When all staple strands have been pipetted, mix the solution by pipetting several times (or vortexing), and then spin down any liquid from the sides of the tube using a benchtop centrifuge.
3. Optional: As DNA origami typically contains more than 200 smaller staple strands, it can be beneficial to use an automated pipetting robot. Using a pipetting robot is not necessarily faster than manually pipetting, but the automation increases reproducibility and frees the scientist's time.

3.3 Synthesis of Modified Nucleotide Triphosphates

Selection of nucleotide for labeling of oligonucleotides with TdT is highly dependent on intended use. If the specific dideoxynucleotide triphosphate is commercially available, this step can be skipped. We recommend using either deoxynucleotide triphosphates (dNTPs), dideoxynucleotide triphosphates (ddNTPs), or ribonucleotide triphosphates (rNTPs) as substrate to obtain oligonucleotides with specific properties (*see* **Note 3**). A few commonly encountered troubleshooting steps can be found in **Note 4**.

The synthesis strategy is usually customized depending on the kind of functionalization you desire and what is commercially available. The modification is usually positioned on the nucleobase but can also be placed on the 3′-position. The following description will outline the synthesis using amine-NHS or alkyne-azide reactions. Here we will describe three examples to create a custom nucleotide from a commercially available substrate using simple and efficient biochemistry methods. Figure 1 visualizes the schemes used to create the modified nucleotides.

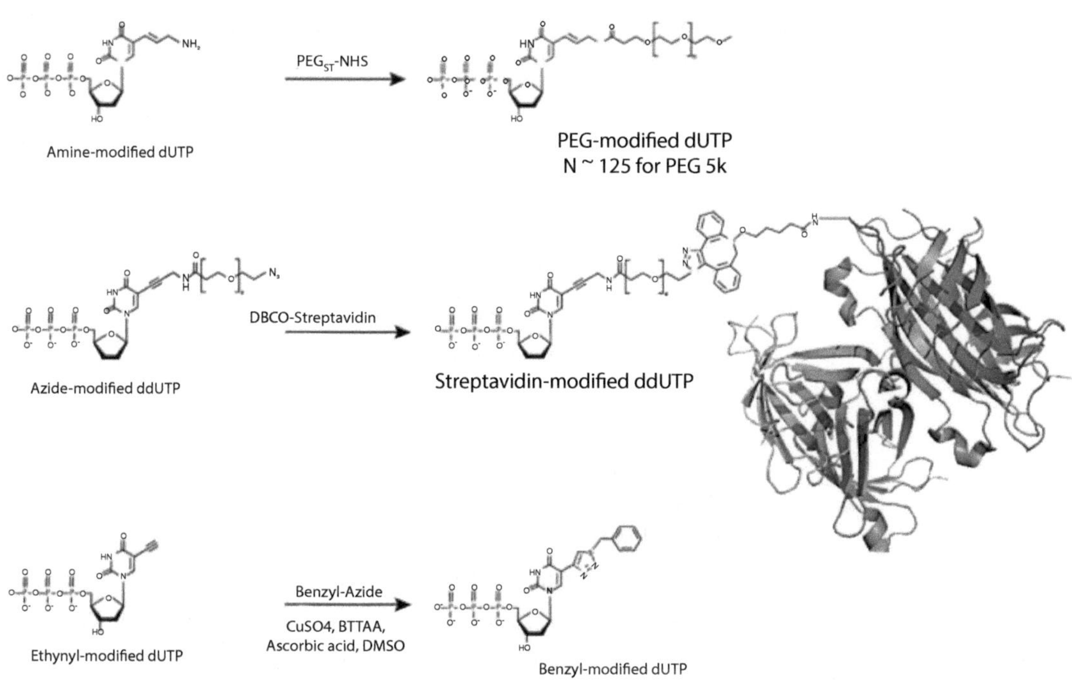

Fig. 1 Reaction examples. (Figure adapted from Ref. [2])

3.3.1 Synthesizing a dUTP with a 5 kDa PEG Polymer on the Base [2]

Reaction 1 NHS-activated conjugation of 5 kDa PEG to aminoallyl-dUTP: This method has been used to prepare 70 nmol of modified nucleotide using NHS-amine reaction on a base modified primary amine (aminoallyl).

1. Mix 60 μl water with 14 μl 5 mM aminoallyl-dUTP (70 nmol) and 10 μl 1 M HEPES (pH 8.0, 10 μmol).
2. Add 500 nmol (2.5 mg) dry 5 kDa PEG-NHS to get a final concentration of about 6 mM.
3. Incubate at 25 °C and 1300 rpm shake for 2 h.
4. The crude product can be purified on a reversed-phase HPLC column, as outlined below. The unmodified aminoallyl-dUTP will elute with the void volume, while the conjugated PEG-dUTP will elute at around 25–50% MeCN, depending on column.

3.3.2 Synthesizing a ddUTP with a Streptavidin (STV) Protein on the Base [2]

Reaction 2 Cu-free click conjugation of STV to azide-ddUTP to produce Azido-dPEG8-propargylamino-ddUTP (ap-ddUTP) in three steps: 1. Synthesizing ap-ddUTP by conjugating 5-propargylamino-ddUTP to azido-dPEG8-NHS. 2. Global protein labeling of primary amines on STV with DBCO using DBCO-NHS (DBCO-NHS (dibenzylcyclooctyne-N-hydroxysuccinimide ester). 3. Cu-free click reaction between purified products from **step 1** and

2. Utilizing CuAAC reaction was not possible due to solubility/precipitation issues (*see* **Note 5**).

1. First, react nucleotide triphosphate derivative with a bifunctional NHS ester azide (ap-ddUTP). To that end, mix 150 nmol 5-propargylamino-ddUTP in 80 μL of HEPES buffer (50 mM, pH 7.8) with 600 nmol of azido-dPEG8-NHS in 6 μL of DMSO.
2. Stir the reaction for 16 h at RT.
3. Purify using RP-HPLC (See commonly used procedure below).
4. For unspecific labeling of STV protein with NHS-DBCO, mix 24 nmol of STV with 600 nmol of 100 mM DBCO-NHS in DMSO in 480 μL 37.5 mM HEPES buffer (pH 8.0) and 25% v/v DMSO.
5. Incubate the reaction for 2 h at RT.
6. Dilute the product tenfold and purify using 10 kDa MWCO Amicon spin-dialysis filters (10,000 g, 20 min, 4 C) while washing three times with water. Dilute the final 25 μL of recovered product 1:1 with water.
7. Finally, conjugate azido-ddUTP (**steps 1–3**) with DBCO-labeled STV (**steps 4–6**). For that, mix 30 μL diluted STV-DBCO with 7.5 nmol of ap-ddUTP in 55 μL 50 mM HEPES buffer (pH 7.8) for 16 h at RT.
8. Purify STV-ddUTP using 10 kDa MWCO Amicon spin-dialysis filters as above but washed with 250 mM NaCl twice and finally in water.

3.3.3 Synthesizing a dUTP with a Benzyl on the C5 Position on the Base

Reaction 3 CuAAC of 5-Ethynyl-dUTP with Benzyl azide to prepare 100 nmol alkyne (Ethynyl)-modified nucleotides.

1. Prepare the CuAAC "click" buffer by mixing 2 μl 50 mM $CuSO_4$ with 2 μl 100 mM BTTAA, and finally add 16 μl of 200 mM ascorbic acid to the total volume of 20 μl. Incubate on the table for 10–15 min before use (*see* **Note 6**).
2. To produce a final reaction volume of 50 μl, mix 5 μl of a 250 mM HEPES-buffer (pH 7.5). Add 120 nmol of 5-Ethynyl-dUTP (1.2 μl of 100 mM). To get the highest yield of modified nucleotides, add at least around 5× excess of Benzyl azide (6 μl of 100 mM in DMSO). Adjust the final DMSO content to 20% by adding 4 μl DMSO. Add 16 μl water to get a final volume of 50 μl, and add the 20 μl CuAAC buffer. Reaction can be optimized by (i) varying the

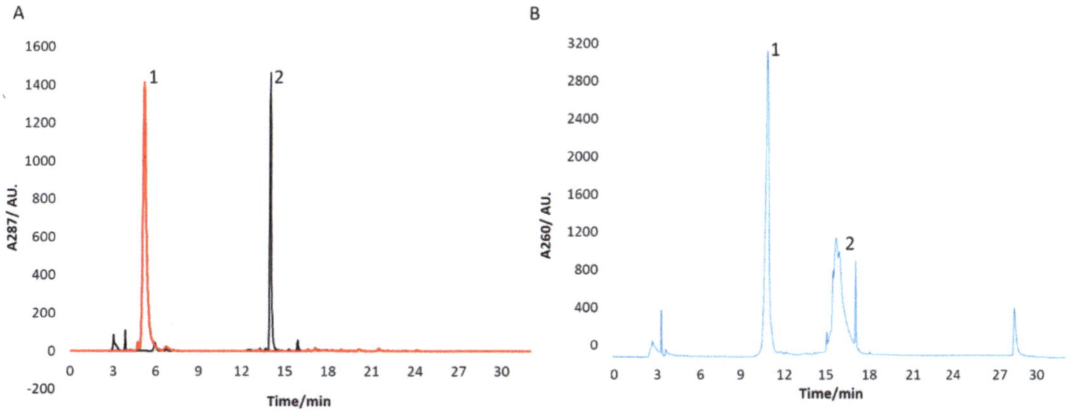

Fig. 2 HPLC chromatogram of functionalized nucleotide and oligonucleotides. (**a**) Benzyl-dUTP before (1) and after (2) coupling to benzyl azide (R3). (**b**) Chromatogram of 28 oligonucleotides enzymatically labeled with an alkyne nucleotide using TdT and post-chemical modification with palmitoyl azide. Unreacted oligonucleotides are found in peak (1) and palmitoylated in peak (2)

final DMSO concentration (*see* **Note 7**) or (ii) optimizing the click buffer: Cu-to-alkyne ratio (Here 1:1), alkyne-to-azide ratio (Here 1:6), or amount of stabilizing agent (Here 200 nmol BTTAA) (*see* **Note 8**).

3. Mix by vortexing and incubate at RT for 2 h or 4 °C overnight (*see* **Note 9**). Shaking is recommended, especially for high amounts.
4. Purify product using RP-HPLC. Product will typically elute at higher acetonitrile concentrations than input, depending on modification.

3.3.4 Common RP-HPLC Purification Approach

1. To analyze and purify the (oligo)nucleotide product(s), running a RP-HPLC is recommended in cases where it is possible. Due to the hydrophilic nature of the negatively charged (oligo) nucleotide phosphates, the conjugated counterparts often elute in later fractions when purified by RP-HPLC (*see* Fig. 2). After fraction collection, the nucleotide is lyophilized and resuspended into the desired concentration (typically in the 20 mM range for nucleotides and 100 μM for oligonucleotides) using a preferred buffer.
2. Run typically a linear gradient from 100% buffer A to 95% buffer B/5% buffer A over a period of 15–20 min (*see* **Notes 10** and **11**).
3. Set up either a peak-based or time-based collection of the desired fractions to collect product (and reactants).
4. After running the gradient, restore the HPLC column by running about 2 volumes of buffer A through (typically 5–8 min at 0.5 ml/min flow on Agilent 1200-series).

5. Identify peaks by running a denaturing page of fraction or do mass spectroscopy.
6. For HPLC analysis of reactions, load 10 nmol (if nucleotide) or 200 fmol (if oligonucleotide) of product to get unsaturated chromatogram, and analyze reaction yield (*see* **Note 12**).

3.4 TdT 3′-End Labeling of Selected Staple Strands (See Notes 13 and 14)

The following protocol has been used to produce 1 nmol of TdT-functionalized oligonucleotides with biotin-dUTPs or biotin-ddUTPs. This scheme can easily be scaled up by keeping the stoichiometry, or optimized by adjusting the excess and type (*see* **Note 15**) of nucleotides, amount of TdT, and/or the incubation time (*see* **Notes 16** and **17**). While the TdT enzyme is quite promiscuous, some nucleotide modifications are not tolerated (this includes hydrophobic moieties like palmitoyl). In these cases, it may be desirable to reverse the scheme and attach the precursor prior chemical modification (*see* **Note 18**).

To prepare 20 μl reaction, in a 1.5 ml microcentrifuge tube:

1. Add 3 μl water (adjust to a final volume of 20 μl final). Add 4 μl 100 μM DNA oligonucleotides. Add 4 μl 1 mM modified nucleotide triphosphate, 4 μl 25 mM $CoCl_2$ (to 5 mM final), and 4 μl 5× TdT buffer (to 1× final).
2. Mix by vortexing the solution before adding the enzyme.
3. Add 1 μl of TdT enzyme (400 U/μl, *see* **Note 19**).
4. Gently mix by flicking the tube and incubate the reaction for 30 min at 37 °C while shaking at 400–600 rpm.
5. Add 1 μl of 0.4 M EDTA to terminate the reaction (*see* **Notes 20** and **21**).

To remove excess nucleotides and enzyme from the oligo product, purify the crude reaction mix by ethanol precipitation (*see* **Note 22**):

6. Add 2 μl 3 M NaOAc (0.1× volume) and 60 μl ice-cold 96% ethanol (3× volumes).
7. Incubate on ice or in the freezer for at least 30 min.
8. Centrifuge the tube at 16,000 × *g* at 4 °C for 30 min.
9. Remove the supernatant, gently add about 1× volume 70% ice-cold ethanol, and spin 16,000 × *g* at 4 °C for 10 min.
10. Remove the supernatant and dry the pellet by leaving the tube open in the fume hood for about 5–30 min (depending on remaining volume).
11. Resuspend the pellet in water or your buffer of choice.
12. Quantify the oligonucleotide concentration by UV absorbance using the extinction coefficient at 260 nm.

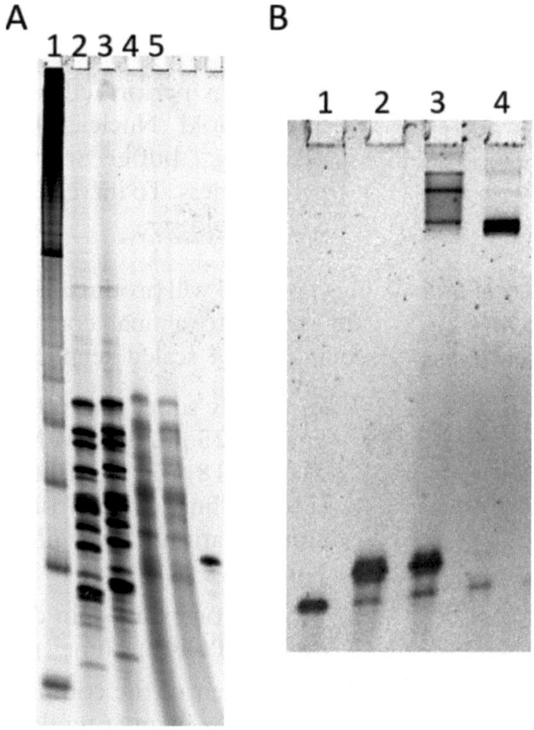

Fig. 3 Analysis of TdT labeling using denaturing PAGE. (**a**) Functionalization of 28 oligonucleotides of various sizes using TdT. lane 1: Generuler ULR ladder, lane 2: Unlabeled oligonucleotides, lane 3: TdT alkyne-functionalized oligonucleotides, and palmitoyl-conjugated oligonucleotides before (lane 4) and after (lane 5) RP-HPLC purification (Peak 2 in Fig. 2b). (**b**) Functionalization of two oligonucleotides of similar size using TdT with biotinylated-ddTTP. Lane 1: unlabeled oligonucleotide, lane 2 TdT-functionalized oligonucleotide with biotin-ddUTP, lane 3 and 4: biotinylated oligonucleotides incubated with increasing amount of streptavidin

3.4.1 PAGE Characterization of TdT-Labeled Oligonucleotides

1. Characterization of the strands is optimally performed by running the reactions on 12–16% denaturing PAGE depending on the length of DNA staple strands. Examples of denaturing PAGE characterization of TdT-labeled staple strands are shown in Fig. 3.
2. Cast a gel by mixing concentrated and diluent acrylamide SequaGel with at 1× TBE buffer.
3. Solidify gel by adding 80 μl APS and 4 μl TEMED per 10 ml. Add comb and let sit for about 30 min.
4. Prepare a gel chamber with buffer reservoirs and pre-run the gel at 20 W for about 20–30 min.
5. Rinse the wells using a syringe and load the prepared wells with a 1–2 pmol of each staple strand per well. If a mixture, e.g., contains 10 staple strands, the final load of each strand should be 1 pmol each, not 0.1 pmol.

6. Run the gel for about 1 h, dependent of the oligonucleotide length.
7. Staining is performed by incubating the gel in 5 μl 10,000× SYBR Gold Nucleic Acid Gel Stain diluted into 50 ml TAE/Mg^{2+} buffer. Stain for about 15 min depending on the gel thickness. To increase stain intensity, increase the amount of SYBR Gold.

3.5 Assembly of DNA Origami with TdT-Labeled Staple Strands

This protocol will produce 100 μl of 10 nM of a DNA origami with 200 staples to a final concentration of 50 nM each (5×). This scheme can be scaled keeping the stoichiometry constant.

1. In a PCR tube, mix 68 μl water, 10 μl 10× TAE buffer (1× final), 6.25 μl 200 mM $MgCl_2$ (12.5 mM final), 10 μl 100 nM M13Mp18/p7249 (10 nM final), and 10 μl staple strand pool (50 nM final of each oligo.; a pool of 200 staples at a total concentration of 100 μM contains 0.5 μM of each oligo) (*see* **Note 23**).
2. Mix by vortexing, spin down, and place the tube in a thermocycler ("PCR machine").
3. Dependent of the complexity of the origami, anneal the DNA origami by incubating the mixture at 80 °C for 5 min followed by a temperature ramp from 65 °C to 45 °C at a slow rate of 40 min/°C and a temperature ramp of 45 °C to 20 °C at 1 min/°C before storage at 10–4 °C (*see* **Note 24**).

3.6 Characterization of DNA Origami

To characterize nanoscale patterns and origami assemblies, nanocharacterization methods like atomic force microscopy (AFM) for planar structures and transmission electron microscopy (TEM) for 3D structures are highly useful techniques as demonstrated in Fig. 4.

3.6.1 Agarose Characterization of Labeled DNA Origami

1. Prepare a 1% agarose gel with 1× TBE containing 5 mM $MgCl_2$: Mix 1 gram of agarose in 97.5 ml 1× TBE buffer, and add 2.5 ml 200 mM $MgCl_2$ (to 5 mM final) in a volumetric flask. Heat the solution in the microwave at 800 W for 2–3 min until the agarose is fully dissolved.
2. Prepare 1× TBE with 5 mM $MgCl_2$ to use as running buffer: Mix 975 mL 1× TBE with 25 ml 200 mM $MgCl_2$.
3. Mix 5–10 μl of the folded DNA origami product with loading buffer, and load in a well of the agarose gel. Run the gel for about 3 h at 60 V (typically 4–6 W).
4. Image the gel to verify successful annealing and incorporation of modified staple strands.

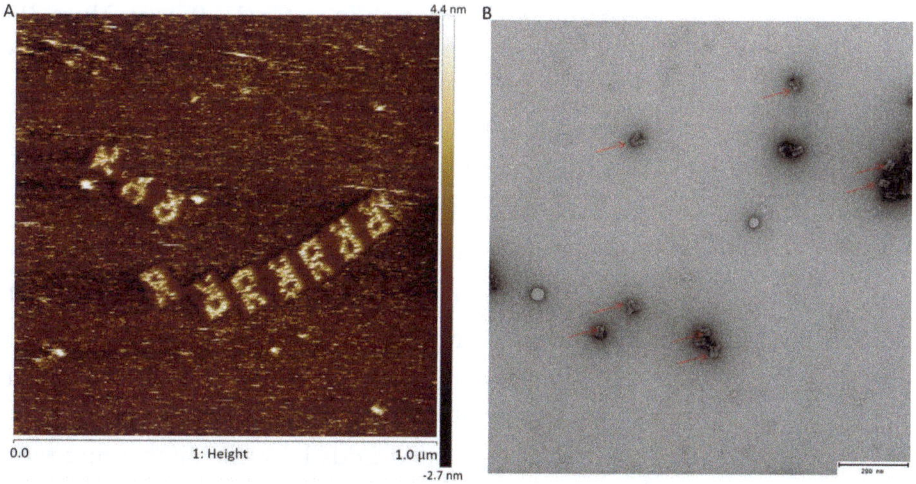

Fig. 4 Nano-characterization of biotin-streptavidin functionalized DNA origamis using liquid AFM (**a**) and TEM (**b**). (**a**) Using AFM, patterns on planar origami with bulky molecular structures can be observed. Here the pattern corresponding to the letter R has been modified in parallel with biotin-dUTP using TdT and incubated with streptavidin on the mica. (**b**) Using TEM, positions of large and bulky macromolecules can be observed in 3D DNA origami structures. Here two staple strands have been functionalized in parallel and incubated with streptavidin prior TEM characterization. Arrows indicated the presence of streptavidin inside the nanopore lumen

*3.6.2 AFM Characterization of Planar DNA Origami Structures (See **Notes 25** and **26**)*

Protocol presented here is based on published articles: [2, 3, 27].

1. Prepare a 10 nM concentration of DNA origami (*see* **Note 27**).
2. Using Scotch tape, cleave off a sheet of mica from a pre-cut piece of mica silicate of an appropriate size to get a clean and flat surface before depositing the origami sample.
3. Add 10 µl DNA origami to the freshly cleaved mica (*see* **Note 28**).
4. Incubate mica with sample at RT for a few minutes before washing the sample in an appropriate buffer to remove excess material. In a flow cell, flow buffer until several volumes has passed. In an open setup, wash 5–10 times in water and blow dry gently with N2. Then, add buffer before imaging.
5. Load mica on AFM/flow cell and attach the AFM probe to AFM, find the tip to be used, and calibrate photodiode.
6. Scan the surface using appropriate, instrument-specific AFM parameters. Example in Fig. 4a demonstrates characterization of a designed streptavidin pattern on a planar DNA origami.

*3.6.3 TEM Characterization of Three-Dimensional DNA Origami Structures (See **Note 29**)*

The protocol presented here is based on published articles: [28–30].

1. Prepare a 10 nM concentration of DNA origami. If needed, dilute sample in TAE buffer with $MgCl_2$ (specific concentration is structure dependent).

2. Prepare a 2% UF solution in water (W/V) (*see* **Note 30**). First, degas H_2O by applying vacuum or boil (if boiled, slowly cool down to room temp.).
3. Per 0.1 gram of UF, add 5 mL of diH_2O to an empty beaker.
4. Vortex solution for 10 min in the dark.
5. Add 2 μl 10 M KOH per 1 mL UF solution.
6. Vortex solution for 10 min in the dark.
7. Spin solution at max speed on a table centrifuge for 10 min.
8. Filter the solution with a 0.2 μm syringe filter in the dark.
9. Make 50 μl aliquots and snap freeze immediately by dumping into the liquid nitrogen.
10. Prepare carbon-coated grids by glow discharging samples, to clean the carbon, and make the surface hydrophilic (*see* **Note 31**).
11. Set up workstation by preparing tweezers, a timer, and Whatman filter paper.
12. Incubate 5 μl of sample onto the carbon site of the grid for about 60–180 s depending on desired grid coverage, sample concentration, etc. Quickly blot the grid from the side and immediately transfer 5 μl 2% UF stain in a stain/wash step (incubation of about 5 s). Blot the grid again from the side and add another 5 μl 2% UF solution to the grid, and incubate for about 15–20 s. Blot again from the side and let the grid dry on the table (*see* **Note 32**).
13. Store the dried grid in a grid box and load into TEM to image.
14. Image the grid using about 120 kV scopes or the like. Example in Fig. 4b demonstrates the characterization of the capture of streptavidin inside the pore of a 3D-DNA origami nanopore [28].

4 Notes

1. *Enzymatic* vs *chemical functionalization*. There are, obviously, many ways of incorporating modifications into DNA structures. The primary alternative to enzymatic labeling is to simply incorporate the required modification or chemical moieties directly onto the desired staple strand(s) during oligonucleotide synthesis or, alternatively, to add a single-stranded extension to either end, which can be used to hybridize to another chemically modified oligonucleotide (often described as a single-stranded *handle*). If only a few oligonucleotides need to be modified, chemical modification is often the most cost- and labor-efficient approach. However, since the price of chemically

modified oligonucleotides is still 5–10 times the cost of unmodified nucleotides, more extensive modifications are significantly cheaper to introduce by enzymatic labeling, or by hybridization to single-stranded protrusions from stable stranded oligo handles. Hybridization to a common handle is often the most scalable solution, since only a single chemically modified oligo needs to be synthesized/purchased. The downside by using a hybridization handle is that it does require either an additional purification step before adding the complementary modified oligonucleotide, or alternatively a very large excess of the modified oligonucleotide. The hybridization-handle approach also introduces the addition of a 15-nt double-helix protruding from the structure. Whether such a protrusion is acceptable depends on the specific use of the modified origami. Enzymatic labeling, on the other hand, adds only a single nucleotide protrusion to the structure, which may be desired or required for some projects. Enzymatic labeling is cost-efficient when only a small amount of each oligonucleotide is needed, which is often the case for projects involving DNA origami. DNA origami is typically assembled with a concentration of each staple strand in the range of around 100 nM, at a volume below 100 μl, totaling only 10 pmol of each oligo. Enzymatic functionalization at this scale, or any scale below 1 nmol, is very inexpensive in material cost. However, if scales above 10 nmol of each modified strand are required, it becomes more cost-effective to introduce the functional moiety during chemical oligonucleotide synthesis. In terms of production time, enzymatic functionalization is much faster than commercial oligonucleotide synthesis, if both the enzyme, nucleotide, and the unmodified staple strand are available. In that case, synthesis of a functionalized oligonucleotide can be completed in 2–3 h. Thus, having this basic 3′-end labeling toolkit consisting of TdT enzyme and frequently used nucleotides can really help bootstrap a DNA origami labeling project.

2. When pipetting staple strands, we suggest ordering all staple strands in well plates with the same initial concentration. In the examples below, we assume all staple strand oligos have been ordered at a fixed concentration of 100 μM. This way, the concentration of each strand in a pool of staple strands can be calculated simply as 100 μM divided by the number of staple strands, and the combined concentration of all staple strands remains fixed at 100 μM.
3. *Selection of nucleotide triphosphate, ddNTP* vs *dNTP* vs *rNTP:* Terminal transferase requires a DNA oligo of at least six nucleotides in length as substrate. However, terminal transferase can use a range of nucleotide triphosphates in the tailing reaction

including dNTPs, ddNTPs, and rNTPs (or just NTPs). If ddNTPs are used, the product will not contain any 3′-OH, ensuring that only a single nucleotide is incorporated. This is generally the most optimal option, as it ensures a single, homogenous product, with exactly one modification at the 3′ end. If dNTPs are used in the reaction, a string of modified nucleotides will be incorporated until all dNTPs in the solution have been incorporated (or until the reaction is stopped, e.g., quenched with EDTA). If rNTPs are used, the tailing reaction is less effective and will usually only introduce a few nucleotides. Since the terminal transferase enzyme has little to no affinity for RNA, the reaction will come to a halt as more rNTPs are incorporated. This typically results in a product with 2–6 rNTPs at the 3′-end of the oligo. However, because of the 2'-OH moiety, subsequent rNTPs may be susceptible to hydrolysis (the first rNTP is not). Both deoxyribonucleotides and ribonucleotide triphosphates are available with a range of different chemically reactive moieties, e.g., amino, alkyne, azide, as well as several fluorophores and haptens including biotin and digoxigenin. For dideoxynucleotides, the selection is currently limited to amino and biotin moieties.

4. *Troubleshooting, synthesis of Streptavidin-ddUTP.* You may experience issues with precipitation during protein conjugation. For instance, streptavidin may precipitate in the presence of DMSO. If you have problems with solubility, we recommend taking the following steps:
   - Try reversing the scheme so you prepare DBCO-ddUTP and azide-streptavidin, which is then clicked together. The azide is typically more soluble in aqueous solutions required by streptavidin, while the DBCO moiety may provide additional hydrophobicity that helps with purification by reversed-phase HPLC.
   - The DBCO-NHS is now also available as DBCO-sulfo-NHS, e.g., Sigma-Aldrich cat. no. 762040. The sulfate group makes the reagent more hydrophilic, enabling aqueous reactions with little or no DMSO.

5. Initial activation studies using TBTA catalysis led to severe precipitation problems.
6. Always prepare ascorbic acid solution fresh!
7. The optimal DMSO concentration is substrate dependent, and it is an advantage to optimize the conditions using a single "dummy" oligonucleotide as test bench material.
8. The efficiency of the CuAAC highly depends on the amount of available Cu(I) ions present; the more, the higher the efficiency. Thus, a stabilizing agent (here used THTPA, TBTA, or

BTTAA) is a vital component. The specific choice of stabilizing agent is often a lesser concern, but in certain cases, a specific agent might be beneficial or even crucial compared to others. Studies to compare the use of different agents have been performed under specific conditions [31]. In our hands, TBTA dissolved in DMSO works best at high DMSO concentrations in CuAAC reactions, while THPTA and BTTAA, which are soluble in water and DMSO, are preferred in low DMSO concentrations.

9. For certain substrates, the solution may appear hazy, which often disappears over time.
10. Optimization of TEAA content, gradient profile, and length can be highly advantageous in separating closely related hydrophobic molecules.
11. Increased TEAA generally makes the oligonucleotides elute later from the RP-HPLC column material due to increased ion pair interaction with the backbone.
12. It can be an advantage to make a UV-Vis of the unmodified nucleotide to know the best absorption peak beforehand.
13. TdT kits can be purchased from several vendors which each provide a unit measure. For this protocol, we have used the Roche recombinant terminal transferase (now sold by Sigma).
14. This procedure can be done for individual strands or for an ensample of staple strand in a one-pot reaction.
15. The TdT-enzyme enables three different products, depending on whether dideoxynucleotide (ddNTPs), dNTP, or rNTP is used. For ddNTP, only a single nucleotide will be attached to the 3′ end of the oligonucleotide substrate. Using dUTP, a stochastic number of nucleotides will be attached, and the length will depend on reaction time and stoichiometric ratio (dUTP:oligo). Finally, using ribonucleotides (UTPs) reduces the efficiency, and a shorter tail of 6–8 nucleotides will be produced.
16. *Avoid extended reaction times.* If the reaction is left at 37 °C for extended periods after the reaction has completed, the TdT enzyme may start to degrade the product, by removing nucleotides from the 3′ end of the oligo. This can be observed as a ladder of N-[1, 2, 3, …] products on the denaturing PAGE gel.
17. To avoid longer tails, short incubation, and high TdT, nucleotide concentration is desired. Alternatively, use pure ddNTPs or use a fraction (e.g., ¼ ddNTP) for stochastic 3′-end termination.

18. In the method described here, the modification is first attached to the nucleotide, which subsequently is conjugated to the oligodeoxynucleotides using the TdT enzyme. While this method works extremely well for most modifications, even large proteins and polymers, there may be instances where reversing the scheme is more optimal, especially for highly hydrophobic modifications, e.g., cholesterol and palmitoyl moieties. In this case, first end-labeling the oligodeoxynucleotides with an alkyne- or amine-functionalized ddNTP is preferred, producing a batch of reactive staple strands, and then conjugate the desired modification to the reactive staple strands.
19. The enzyme can often be reduced by up to a factor of 10, depending on the activity and quality of the enzyme. Also, note that the unit definition differs between vendors. Finally, keep enzyme on ice or in the freezer until use for optimal conservation and reuse.
20. *Quenching with EDTA is important.* Even if you intend to either heat-inactivate the enzyme, or proceed directly with ethanol precipitation, you should still add EDTA to the reaction in order to chelate the Co^{2+}. If the cobalt ions are not chelated, they will co-precipitate during ethanol precipitation (observed as a slightly red, purple, or pink hue in the pellet). This could make the pellet harder to resuspend and cause unexpected issues later.
21. The reaction can optionally be heated to 80 °C for 10 min to heat inactivate the TdT enzyme. However, it is important to still add EDTA to chelate the cobalt ions.
22. If further purification is needed, RP-HPLC is recommended as demonstrated in Fig. 2b.
23. Optional: If, e.g., 40 of the 200 strands is TdT-labeled and purified as a staple strand batch, add 40/200 = 1/5 volume (2.0 μl) of the labeled strands (2.5 μM each; 100 μM/40 staples) and 160/200 = 4/5 volume (8 μl) of the unlabeled background staple strands (0.625 μM each; 100 μM/160 staples).
24. Troubleshooting the folding of DNA origami: Make sure the PCR tubes are not overfilled if using a thermocycler with heated lid, as this may change the effective temperature during annealing.
25. The specified concentrations and incubation steps are only indicative and serve as a good starting point for initial studies. Often, more ideal, sample-specific conditions require further optimization.
26. Scan using liquid tapping mode.

27. Optional: Dilute origami sample to 1–2 nM in annealing buffer (TAE-Mg, TAE buffer supplemented with 5–40 mM $MgCl_2$; optimal $MgCl_2$ concentration depends on the origami structure).
28. Optional: Mixing/diluting the DNA origami sample with a small amount of nickel (e.g., to a final concentration 1–10 mM $NiCl_2$) will increase adsorption of DNA origami to the negatively charged mica surfaces.
29. The specified concentrations and incubation steps are only indicative and serve as a good starting point for initial studies. Often, more ideal, sample-specific conditions require further optimization.
30. Minimize exposure to light, work fast, and snap freeze right away. Store in the dark at −80 freezer. Ensure UF powder is yellow, not green. Final solution color should look yellowish.
31. Always freshly clean the grids by glow discharging only the ones you are going to need within a 20 min time to avoid contamination building up and drop in hydrophilicity.
32. Keep the stain in the dark as much as possible to avoid precipitation of the uranyl salt. A total stain time of about 25 s (stain wash + 2nd stain) is desired (rather do 20 s than 30 s).

References

1. Rothemund PWK (2006) Folding DNA to create nanoscale shapes and patterns. Nature 440(7082):297–302
2. Sørensen RS, Okholm AH, Schaffert D, Kodal ALB, Gothelf KV, Kjems J (2013) Enzymatic ligation of large biomolecules to DNA. ACS Nano 7(9):8098–8104
3. Jahn K, Tørring T, Voigt NV, Sørensen RS, Kodal ALB, Andersen ES et al (2011) Functional patterning of DNA origami by parallel enzymatic modification. Bioconjug Chem 22(4):819–823
4. Seeman NC (2010) Nanomaterials based on DNA. Annu Rev Biochem 79(1):65–87
5. Seeman NC, Sleiman HF (2018) DNA nanotechnology. Nat Rev Mater 3(1):17068
6. Langecker M, Arnaut V, Martin TG, List J, Renner S, Mayer M et al (2012) Synthetic lipid membrane channels formed by designed DNA nanostructures. Science 338(6109): 932–936
7. Krishnan S, Ziegler D, Arnaut V, Martin TG, Kapsner K, Henneberg K et al (2016) Molecular transport through large-diameter DNA nanopores. Nat Commun 7:12787
8. Tyagi S, Kramer FR (1996) Molecular beacons: probes that fluoresce upon hybridization. Nat Biotechnol 14(3):303–308
9. Niemeyer CM, Koehler J, Wuerdemann C (2002) DNA-directed assembly of Bienzymic complexes from in vivo biotinylated NAD (P)H:FMN oxidoreductase and luciferase. Chembiochem 3(2–3):242–245
10. Knudsen JB, Liu L, Kodal ALB, Madsen M, Li Q, Song J et al (2015) Routing of individual polymers in designed patterns. Nat Nanotechnol 10:892–898
11. Kuzyk A, Schreiber R, Fan Z, Pardatscher G, Roller EM, Högele A et al (2012) DNA-based self-assembly of chiral plasmonic nanostructures with tailored optical response. Nature 483(7389):311–314
12. Douglas SM, Bachelet I, Church GM (2012) A logic-gated nanorobot for targeted transport of molecular payloads. Science 335(6070): 831–834
13. Shaw A, Lundin V, Petrova E, Fördos F, Benson E, Al-Amin A et al (2014) Spatial control of membrane receptor function using ligand nanocalipers. Nat Methods 11(8): 841–846

14. Madsen M, Gothelf KV (2019) Chemistries for DNA nanotechnology. Chem Rev 119(10): 6384–6458
15. Dong Y, Liu D, Yang Z (2014) A brief review of methods for terminal functionalization of DNA. Methods 67(2):116–122
16. Okholm AH, Aslan H, Besenbacher F, Dong M, Kjems J (2015) Monitoring patterned enzymatic polymerization on DNA origami at single-molecule level. Nanoscale 7(25): 10970–10973
17. Scheres SHW (2012) RELION: implementation of a Bayesian approach to cryo-EM structure determination. J Struct Biol 180(3): 519–530
18. Relion 2.0. Cloud computing tools for cryo-EM – a resource for the cryo-EM community. [Online]. Available: http://cryoem-tools.cloud/. Accessed: 18-Feb-2020
19. Grant T, Rohou A, Grigorieff N (2018) cisTEM, user-friendly software for single-particle image processing. elife 7:e35383
20. cisTEM. cisTEM | Computational imaging system for transmission electron microscopy. [Online]. Available: https://cistem.org/software. Accessed: 18-Feb-2020
21. de la Rosa-Trevín JM, Quintana A, Del Cano L, Zaldívar A, Foche I Gutiérrez J, *et al.* (2016) Scipion: a software framework toward integration, reproducibility and validation in 3D electron microscopy. J Struct Biol 195(1): 93–99
22. Scipion. [Online]. Available: http://scipion.i2pc.es/
23. Punjani A, Rubinstein JL, Fleet DJ, Brubaker MA (2017) cryoSPARC: algorithms for rapid unsupervised cryo-EM structure determination. Nat Methods 14(3):290–296
24. cryoSPARC. cryoSPARC | Leading Cryo-EM Software Solutions. [Online]. Available: https://cryosparc.com/. Accessed: 18-Feb-2020
25. Cadnano. cadnano. [Online]. Available: http://cadnano.org/. Accessed: 17-Feb-2020
26. cadnano/cadnano2.5. cadnano, 2020
27. Klein WP, Thomsen RP, Turner KB, Walper SA, Vranish J, Kjems J et al (2019) Enhanced catalysis from multienzyme cascades assembled on a DNA origami triangle. ACS Nano 13(12): 13677–13689
28. Thomsen RP, Malle MG, Okholm AH, Krishnan S, Bohr SS-R, Sørensen RS et al (2019) A large size-selective DNA nanopore with sensing applications. Nat Commun 10(1):1–10
29. Manuguerra I, Grossi G, Thomsen RP, Lyngsø J, Pedersen JS, Kjems J et al (2017) Construction of a polyhedral DNA 12-arm junction for self-assembly of wireframe DNA lattices. ACS Nano 11(9):9041–9047
30. Nielsen TB, Thomsen RP, Mortensen MR, Kjems J, Nielsen PF, Nielsen TE et al (2019) Peptide-directed DNA-templated protein labelling for the assembly of a pseudo-IgM. Angew Chem Int Ed 58(27):9068–9072
31. Besanceney-Webler C, Jiang H, Zheng T, Feng L, Soriano del Amo D, Wang W et al (2011) Increasing the efficacy of bioorthogonal click reactions for bioconjugation: a comparative study. Angew Chem Int Ed 50(35): 8051–8056

Chapter 12

Protein Coating of DNA Origami

Heini Ijäs, Mauri A. Kostiainen, and Veikko Linko

Abstract

DNA origami has emerged as a common technique to create custom two- (2D) and three-dimensional (3D) structures at the nanoscale. These DNA nanostructures have already proven useful in development of many biotechnological tools; however, there are still challenges that cast a shadow over the otherwise bright future of biomedical uses of these DNA objects. The rather obvious obstacles in harnessing DNA origami as drug-delivery vehicles and/or smart biodevices are related to their debatable stability in biologically relevant media, especially in physiological low-cation and endonuclease-rich conditions, relatively poor transfection rates, and, although biocompatible by nature, their unpredictable compatibility with the immune system. Here we demonstrate a technique for coating DNA origami with albumin proteins for enhancing their pharmacokinetic properties. To facilitate protective coating, a synthesized positively charged dendron was linked to bovine serum albumin (BSA) through a covalent maleimide-cysteine bonding, and then the purified dendron-protein conjugates were let to assemble on the negatively charged surface of DNA origami *via* electrostatic interaction. The resulted BSA-dendron conjugate-coated DNA origami showed improved transfection, high resistance against endonuclease digestion, and significantly enhanced immunocompatibility compared to bare DNA origami. Furthermore, our proposed coating strategy can be considered highly versatile as a maleimide-modified dendron serving as a synthetic DNA-binding domain can be linked to any protein with an available cysteine site.

Key words Nucleic acids, DNA nanotechnology, DNA nanostructures, Self-assembly, Electrostatic interaction, Protein, Dendron, Drug delivery, DNA origami stability

1 Introduction

Owing to the fully predictable Watson-Crick base pairing, DNA molecules can be used as programmable building blocks for creating accurate nanostructures from the bottom up using various different techniques [1, 2]. DNA origami technique [3–5] is based on assembling a long DNA strand (scaffold) into a desired shape by folding it using dozens of short strands (staples). The utility of the method emerges from its accuracy and robustness; one can create arbitrary-shaped DNA nanostructures with custom properties, position molecular components with them at the (sub-)nanometer scale [6], and simultaneously achieve high

Julián Valero (ed.), *DNA and RNA Origami: Methods and Protocols*, Methods in Molecular Biology, vol. 2639, https://doi.org/10.1007/978-1-0716-3028-0_12,

fabrication yields [7]. With the help of the novel design techniques [8], user-friendly software [9–12], and automation [11–13], the structural DNA nanotechnology has reached the enabled state [14] at which new biotechnological applications come increasingly into view [1, 2, 5, 8, 9, 14]. Indeed, during the past decade, vast numbers of implementations based on high addressability and modularity of DNA nanostructures have been reported. Such applications include, for example, templates for nanoelectronics [15] and nanopatterning [16, 17], plasmonic and nanophotonic devices [18–21], chemical reactors [22–25], artificial ion channels [26], nanoscale measurement and imaging tools [27, 28], and rulers for super-resolution microscopy [29, 30].

Besides these intriguing examples, to use DNA nanostructures as smart drug-delivery vehicles and biomedical devices has recently drawn a lot of attention [31, 32]. The advantages of the DNA objects in such applications are rather obvious; they are made of natural polymers [33], and they can be loaded with a variety of drugs [34–37]; moreover, they are easily modified for targeting [36, 37], and they can perform (multiple) programmed tasks [36–39]. However, there are several drawbacks, including relatively poor transfection rates due to their high polarizability [40–42], varying stability of the structures in different media [33, 43–48], and their proneness to degradation by nucleases [33, 44, 47–52]. One of the obstacles is that despite the intrinsic biocompatibility of DNA, these structures tend to cause an undesired inflammatory response [51, 53]. Nevertheless, there are ways to overcome these issues by rational design [52]; by lipid [53], protein [51, 54–56], and polymer coatings [49, 50, 57, 58]; by protein functionalization [59, 60]; and by covalent crosslinking of neighboring DNA strands [61].

Here, we present a versatile technique to improve transfection, stability, and immunocompatibility of DNA nanostructures by coating them with inert proteins. To demonstrate the method, we use a 60-helix bundle (60HB) [43] and a capsule-like DNA origami [25] as example objects and shield them with bovine serum albumin (BSA) proteins (*see* Fig. 1). BSA is covalently linked through its single surface cysteine to maleimide-modified synthetic dendron that acts as a positively charged binding domain [51]. The major procedures in the protocol are origami assembly and purification, the conjugation of BSA and dendrons, chromatographic purification of the assembled conjugates, and electrostatic assembly of protein-coated origami structures. Here, we describe the coating protocol with a second-generation dendron (G2). The synthesis of maleimide-modified dendrons, including G2, has been previously described in detail by Kostiainen et al. [62, 63] and will not be included in this protocol.

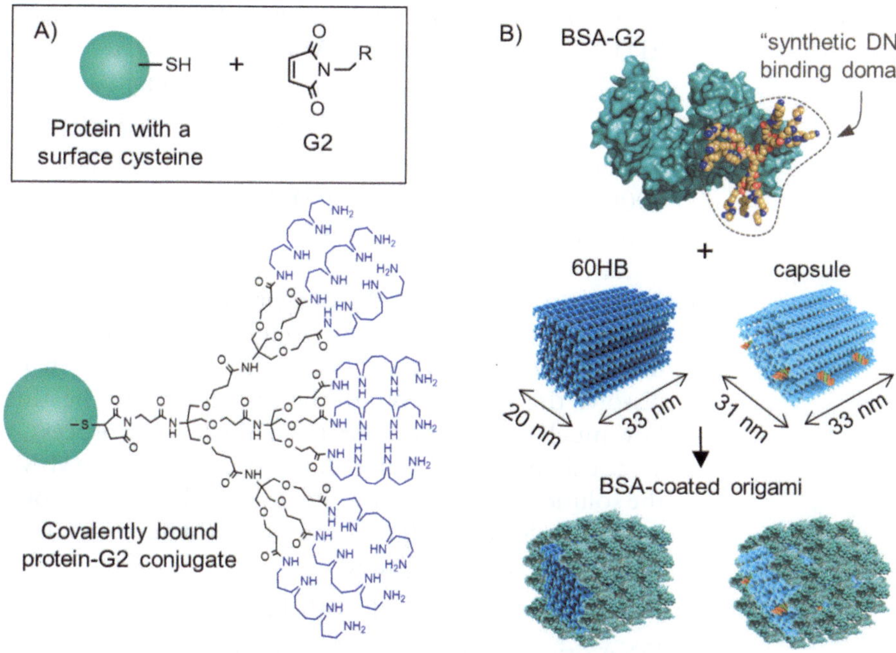

Fig. 1 (**a**) Principle of conjugation between a surface cysteine-containing protein and a maleimide-modified G2 dendron. *Inset:* Conjugation between a cysteine site and a maleimide. (**b**) G2 acts as a synthetic DNA-binding domain for bovine serum albumin (BSA) due to its protonated amine groups (BSA and G2 in scale). G2-modified BSA (BSA-G2) conjugates form a protecting layer onto the negatively charged DNA origami surface through the electrostatic interaction. BSA-G2 coating is demonstrated for the 60-helix bundle (60HB) and the capsule-like DNA origami. (Figure has been adapted from Ref. [51] and completely modified)

2 Materials

2.1 Assembly and Purification of DNA Origami

2.1.1 DNA Origami

1. 7249 nucleotides long viral single-stranded DNA (ssDNA), M13mp18, at 100 nM concentration for 60HB origami.
2. 8064 nucleotides long ssDNA, at 100 nM concentration for capsule origami.
3. Set of short staple strands (Integrated DNA Technologies), at 100 μM concentration, for both the 60HB and the capsule structure. 60HB includes 141 different staple strands, and the full list of sequences can be found in Ref. [43]. A closed capsule structure is annealed by using 264 staple strands listed in Ref. [35].
4. 2.5× 60HB folding buffer (FOB): 100 mM Tris-base, 47.5 mM acetic acid, 2.5 mM EDTA, 50 mM $MgCl_2$, and 12.5 mM NaCl, pH ~8.3.
5. 2.5× capsule FOB: 100 mM Tris-base, 47.5 mM acetic acid, 2.5 mM EDTA, 37.5 mM $MgCl_2$, and 12.5 mM NaCl, pH ~8.3.

2.1.2 Purification of DNA Origami

1. 1× 60HB FOB: 40 mM Tris-base, 19 mM acetic acid, 1 mM EDTA, 20 mM $MgCl_2$, 5 mM NaCl, pH ~8.3.
2. 1× capsule FOB: 40 mM Tris-base, 19 mM acetic acid, 1 mM EDTA, 15 mM $MgCl_2$, 5 mM NaCl, pH ~8.3.
3. Polyethylene glycol (PEG) precipitation buffer: 15% PEG 8000 (w/v), 1× TAE (40 mM Tris-base, 19 mM acetic acid, 1 mM EDTA), 505 mM NaCl.

2.1.3 Agarose Gel Electrophoresis

1. 2% agarose gel: dissolve 2 g of agarose into 100 mL of 1× TAE buffer with 11 mM $MgCl_2$. Weigh the flask, mix gently, and heat the solution shortly in a microwave oven in order to dissolve the agarose. Weigh the flask again after heating to see how much water has evaporated in the heating process, and add an equal amount of water. Add 10 mL of 110 mM $MgCl_2$ to the solution. Allow the solution to cool down to 50–60 °C, and stain with 80 μL of 0.625 mg/mL ethidium bromide (EtBr) solution. Cast immediately.
2. Running buffer: 1× TAE with 11 mM $MgCl_2$.
3. DNA origami samples (*see* Subheadings 3.1 and 3.2).
4. 6× DNA gel loading dye.
5. M13mp18 ssDNA.
6. 1× 60HB FOB and 1× capsule FOB.

2.2 Cationic Protein-Dendron Conjugates

2.2.1 Conjugation

1. ~250 mg/mL BSA solution in water.
2. 0.2 M sodium phosphate buffer, pH 7.0.
3. 0.5 M EDTA, pH 7.0.

2.2.2 Purification of Protein-Dendron Conjugates with Cation Exchange Chromatography

1. Unpurified BSA-G2 conjugates (*see* Subheading 3.4).
2. Buffers and solutions for cation exchange chromatography: filter and degas all solutions before use.
   (a) Start buffer: 50 mM sodium phosphate buffer, pH 7.0.
   (b) Elution buffer: 50 mM sodium phosphate buffer, 1 M NaCl, pH 7.0.
   (c) 20% ethanol and distilled water (for washing and storing the column).
3. Fast liquid chromatography (FPLC) system equipped with a cation exchange column.
4. Centrifugal filters with a 10 kDa molecular weight cut-off.

2.3 Protein Coating and Electrophoretic Mobility Shift Assay

1. Purified BSA-G2 conjugates in 50 mM phosphate buffer, pH 7.0.
2. PEG-purified 60HB DNA origami solution at ~20 nM concentration.

3. 10 kDa molecular weight cut-off centrifugal filters.
4. 150 mM NaCl solution.
5. 1 M NaCl solution.
6. 2% agarose gel (*see* Subheading 2.1.3).
7. Running buffer: 1× TAE with 11 mM $MgCl_2$.
8. 6× DNA gel loading dye.

3 Methods

3.1 DNA Origami Assembly

1. Prepare stock solutions of the staple strands needed both for the 60HB and capsule origami: mix together an equal volume of each DNA oligonucleotide strand. The initial concentration of each strand should be 100 μM. The concentration of each staple strand in the capsule stock solution will be *ca.* 379 nM after mixing together all 264 staple strands and does not need any further dilution before use. After mixing together the 141 staple strands for the 60HB stock solution, dilute the mixture with water to yield a final 500 nM concentration of each strand.
2. Prepare DNA origami folding reaction mixtures in 100 μL quantities by mixing the following components in a 0.2 mL PCR tube:
   (a) 40 μL of 2.5× FOB
   (b) 40 μL of staple strand stock solution (yields a 10× molar excess of staples compared to the scaffold in the 60HB reaction mixture and ~7× in the case of capsule origami).
   (c) 20 μL of ssDNA scaffold DNA at 100 nM concentration (select scaffold according to the used design).
3. Anneal the reaction mixtures in a thermal cycler with the following thermal folding ramp (*see* **Note 1**):
   (a) From 65 °C to 60 °C: −1 °C/15 min.
   (b) From 60 °C to 40 °C: −0.25 °C/45 min.
   (c) Store at 12 °C until the program is manually stopped.
4. Transfer 10 μL of annealed DNA origami solution into a separate container for later analysis with agarose gel electrophoresis (AGE). Purify the rest of the sample using PEG precipitation procedure (*see* Subheading 3.2). Store all samples at 4 °C if not continuing directly to the purification procedure.

3.2 DNA Origami Purification

Because an excess amount of staple strands is used in the folding reaction, free staple strands remain in the sample after annealing. These strands should be removed as they will otherwise interfere with the coating procedure by binding to the cationized proteins.

Here, this is carried out with PEG precipitation, which is a non-destructive method for DNA origami purification. The protocol has been adapted from Stahl et al. [64].

1. Dilute the DNA origami solution with 1× FOB with a 1:4 dilution factor. Mix the diluted solution with an equal volume of PEG precipitation buffer, and mix thoroughly by pipetting back and forth.
2. Centrifuge the mixture at 14,000× *g* for 30 min at room temperature.
3. Remove the supernatant with a pipette. Add the original sample volume of 1× FOB on the pellet. Mix gently with a pipette and incubate the sample overnight at room temperature to allow a complete resuspension of the DNA origami pellet.
4. Estimate the DNA origami concentration in the purified sample by measuring sample absorbance at 260 nm wavelength with a UV/Vis spectrophotometer. Calculate the DNA origami concentration from the A_{260} value, extinction coefficient ε_{260}, and path length according to the Beer-Lambert law. Approximated extinction coefficient for the 60HB is $\varepsilon_{260} = 0.9 \times 10^8\ \mathrm{M}^{-1}\mathrm{cm}^{-1}$ and for the capsule $\varepsilon_{260} = 1.2 \times 10^8\ \mathrm{M}^{-1}\mathrm{cm}^{-1}$, based on the amount of double-stranded and single-stranded DNA residues in the structures [25, 51, 65].

3.3 Characterization of DNA Origami Folding and Purification Quality with AGE

After purification of excess staple strands, it is necessary to ensure that the folding quality and assembly yield, as well as the level of removal of excess staple strands is sufficient for the protein coating procedure. Characterization with AGE is based on a comparison of the electrophoretic mobility of unfolded scaffold DNA, DNA origami samples before purification of staple strands, and DNA origami samples after PEG purification.

1. Prepare AGE samples from unpurified and PEG-purified DNA origami solutions: mix 10 μL of DNA origami solution with 2 μL of 6× loading dye.
2. Prepare a reference sample from scaffold DNA by first diluting 100 nM of scaffold DNA to 20 nM concentration with 1× FOB, and mix 10 μL of the solution with 2 μL of 6× loading dye. Use the 7249 nt long scaffold as a reference to the 60HB and the 8046 nt scaffold for the capsule origami.
3. Place a gel running chamber on an ice bath, place a 2% agarose gel stained with ethidium bromide into the running chamber (*see* Subheading 2.1.3), and fill the chamber with running buffer. Load 10 μL of each sample in the wells of the gel. Run the gel with a constant voltage of 90 V for 45–60 min and image under UV light.

A properly assembled DNA origami sample should migrate resulting in a clear, narrow band and usually have a different migration speed on the gel than the scaffold reference sample (*see* **Note 2**). Excess staple strands migrate as a broad, blurry band near the dye front in the unpurified origami sample. An absence of the band in the PEG purified sample indicates efficient purification.

3.4 Preparation of BSA-G2 Conjugates

1. To prepare the conjugation reaction mixture, mix together 400 μL of ~250 mg/mL BSA solution, 100 μL of 10 mg/mL G2 dendron, 400 μL of 0.2 M sodium phosphate buffer (pH 7.0), and 30 μL of 0.5 M EDTA (pH 7.0). This results in a mixture with roughly 1.6 mM BSA, 0.34 mM G2, 86 mM sodium phosphate buffer and 16 mM EDTA, and an approximately 4.6-fold molar excess of BSA in respect to G2. Incubate the mixture on shaking at room temperature for 16 h.

The conjugation reaction yields covalently bound conjugates when the maleimide group in the G2 dendron reacts with the sulfhydryl group of a protein surface cysteine (*see* **Notes 3** and **4**).

3.5 Conjugate Purification with Cation Exchange Chromatography

Prior to use, the properly formed BSA-dendron conjugates should be separated from the unconjugated fraction based on the altered physical properties of the conjugates (*see* **Note 5**). In cation exchange chromatography, the separation is based on the positive charge of the conjugates. Use a suitable FPLC system equipped with a cation exchange column.

1. Wash the system and column with water and balance the column into 50 mM phosphate buffer at pH 7.0.
2. Inject and load the BSA-G2 conjugate sample into the column. Wash the column with several column volumes of 50 mM phosphate buffer. Follow the A_{280} value in the chromatogram and ensure that all unconjugated BSA flows through the column during the column wash.
3. Elute the BSA-G2 conjugates with a NaCl gradient by gradually increasing the volume fraction of 50 mM phosphate buffer + 1 M NaCl (pH 7.0) in the column from 0% to 100%. Collect the fractions that contain protein according to the A_{280} value.
4. Combine the collected fractions. If the volume of the pooled fractions is considerably higher than in the sample before purification, concentrate the solution back to roughly the original volume by using 10 kDa cutoff centrifugal filters.
5. Remove excess amounts of NaCl from the sample. This can be done, *e.g.*, with 10 kDa cutoff centrifugal filters by repeatedly diluting and concentrating the sample with 50 mM phosphate buffer, until the estimated NaCl concentration in the sample is $\leq$150 mM.

6. After the run, wash the system with 50 mM phosphate buffer and water, and balance the column into 20% ethanol for storage.
7. Determine the protein concentration in the product by measuring sample absorbance at wavelength 280 nm with a UV/Vis spectrophotometer. Calculate the BSA concentration from the A_{280} value, extinction coefficient ε_{280}, and path length according to the Beer-Lambert law using the extinction coefficient of BSA at 280 nm, $\varepsilon_{280} = 43{,}824\ M^{-1}cm^{-1}$.
8. Store the sample in aliquots at −20 °C. Optionally, measurements for determining the purity and composition of the conjugate can be performed (*see* **Note 6**).

3.6 Protein Coating of DNA Origami

An optimal molar ratio of BSA-G2 and DNA origami in the coating procedure can be found by analyzing samples prepared with various ratios with an electrophoretic mobility shift assay (EMSA) (*see* **Note 7**).

1. Before starting, exchange the buffer of BSA-G2 to 150 mM NaCl with 10 kDa spin filters.
2. Prepare a set of samples with a constant DNA origami concentration and a range of BSA-G2 concentrations: mix together a constant volume of PEG-purified DNA origami, BSA-G2 in 150 mM NaCl to yield a 0–4000× molar excess to the origami, 1 M NaCl to adjust the NaCl concentration of the sample at 150 mM, and distilled water to fill the volume of all samples as the same (*e.g.*, 20 μL). See Table 1 for pipetting guidelines for preparing a set of samples using 20 nM capsule origami in 1× capsule FOB and 25 μM BSA-G2 in 150 mM NaCl.

Table 1
Pipetting scheme for preparing BSA-G2 – capsule origami samples with a constant (6 nM) origami concentration and a 0–4000× BSA-G2 molar excess

c(BSA-G2)/*c*(origami)	*V*(origami) [μL]	*V*(BSA-G2) [μL]	*V*(1 M NaCl) [μL]	*V*(H_2O) [μL]
0	6	0	2.97	11.03
250	6	0.67	2.87	10.46
500	6	1.33	2.77	9.90
1000	6	2.67	2.57	8.76
1500	6	4.00	2.37	7.63
2000	6	5.33	2.17	6.50
3000	6	8.00	1.77	4.23
4000	6	10.67	1.37	1.96

The volumes have been calculated to yield the indicated molar ratio and 150 mM NaCl concentration when prepared using 20 nM capsule origami in 1× capsule FOB and 25 μM BSA-G2 in 150 mM NaCl

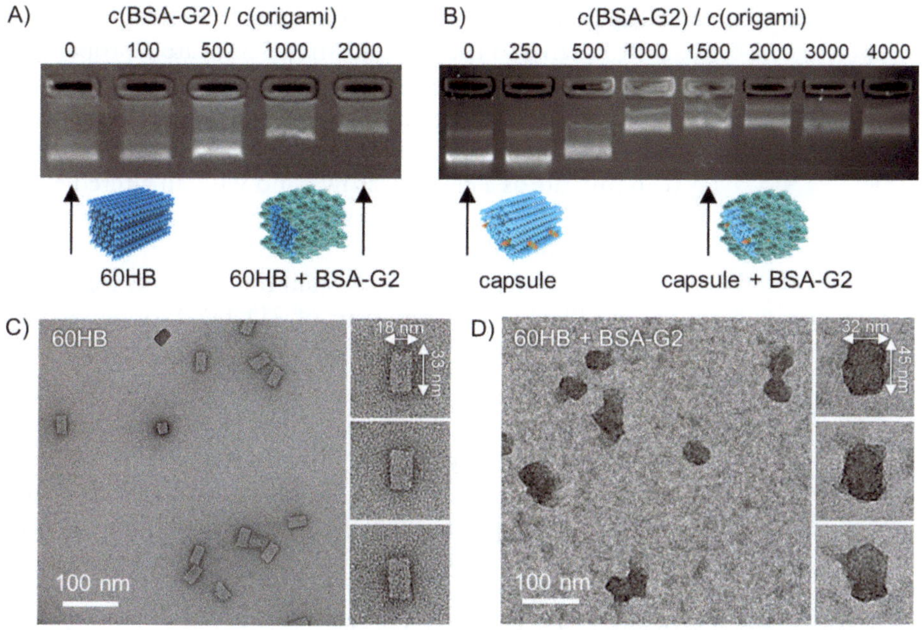

Fig. 2 Electrophoretic mobility shift assays for (**a**) 60HB and (**b**) capsule origami when complexed with different amounts of BSA-G2. (**c**) Transmission electron microscopy (TEM) images of bare 60HB. (**d**) TEM images of BSA-G2 coated 60HB. The scale bars in subfigures (**c**) and (**d**) are 100 nm and the size of the insets are 80 nm × 80 nm. (Subfigures (**a**), (**c**), and (**d**) have been adapted from Ref. [51] and modified)

3. Incubate the samples for 30 min at room temperature.
4. Mix 4 μL of 6× loading dye into 20 μL of each sample and load the samples on a 2% agarose gel. Run the gel with a constant voltage of 90 V for 45–60 min and image under UV light.
5. Select the optimal BSA-G2:DNA origami molar ratio by comparing the migration of the samples on a gel. Association of the cationic proteins with the origami surface will cause a decreased electrophoretic mobility (*see* Fig. 2). Select a molar ratio where the amount of BSA-G2 is enough to decrease the mobility to a minimum.

4 Notes

1. The described folding conditions and procedure are used for folding the 60HB and the capsule DNA origami, which are both three-dimensional structures designed in a honeycomb lattice [10]. Other designs may require adjustments to the FOB components, namely to the concentration of $MgCl_2$ and NaCl, as well as to the thermal ramp [66] for proper folding.

2. Usually, a bright, narrow band and an electrophoretic mobility shift relative to the scaffold strand indicate a proper folding of the DNA origami shape. In the reported conditions, the 60HB and capsule are expected to migrate faster than the scaffold strand (as in Fig. 2). In some cases, mis- of unfolded structures can migrate as a clear, narrow band with an altered electrophoretic mobility in comparison to the unfolded scaffold. For this reason, additional structural characterization with imaging techniques such as negative stain transmission electron microscopy (TEM), or in the case of 2D DNA origami shapes and atomic force microscopy (AFM), is recommended.
3. Optimal reaction conditions for the BSA-G2 conjugation. Thiol-maleimide reactions should be performed at pH 6.5–7.5 according to the instructions of the manufacturer (ThermoFisher Scientific). At this pH range, the maleimide group reacts specifically with sulfhydryl groups on the protein surface, and its rate of hydrolysis to non-reactive maleamic acid is low.
4. The method is modular allowing the use of any protein containing either a natural or engineered cysteine at the protein surface. For example, a hydrophobin II and an anti-HER2 fragment with an artificial C-terminal cysteine residue ($scFv^{C}$, Cys257) have been successfully conjugated with G2 by Auvinen et al. [51] and Seitz et al. [67], respectively.
5. There are other possible purification methods for BSA-G2. For instance, semi-preparative reversed-phase high-performance liquid chromatography (HPLC) can be used to separate the BSA-G2 conjugates from the unconjugated BSA based on the high hydrophobicity of the G2 [63].
6. The purity and composition of the BSA-G2 conjugate can be evaluated, for example, by measuring the mass-to-charge (*m/Z*) ratios of unconjugated BSA and BSA-G2 using matrix-assisted desorption/ionization/time-of flight mass spectrometry (MALDI-TOF). A *m/Z* ratio of 69,552.5 has been measured for BSA-G2, while the ratio for pure BSA is smaller, 66,444.3 [63].
7. The molar ratio between BSA-G2 and the DNA origami structure needs to be determined individually for each DNA origami structure. The size and shape of the DNA origami will affect the amount of BSA-G2 being able to bind to the DNA origami surface. Additionally, the ratio depends on the purity of the BSA-G2 conjugate; lower purity conjugates containing more unconjugated BSA require higher apparent concentration for an efficient coverage of the DNA origami surface. For the 60HB, optimal coating is achieved using 2000× BSA-G2 excess, and for the capsule with 1500× excess

Acknowledgments

We acknowledge the provision of facilities and technical support by Aalto University Bioeconomy Facilities, OtaNano – Nanomicroscopy Center (Aalto-NMC) and Micronova Nanofabrication Center. Financial support from the Academy of Finland (projects 286845, 308578, 303804, 267497), the Jane and Aatos Erkko Foundation, the Emil Aaltonen Foundation, the Sigrid Jusélius Foundation, and ERA Chair MATTER from the European Union's Horizon 2020 research and innovation programme under grant agreement No 856705 is gratefully acknowledged. This work was carried out under the Academy of Finland Centers of Excellence Programmes (2014–2019) and (2022–2029) in Life-Inspired Hybrid Materials (LIBER), project number 346110.

References

1. Jones MR, Seeman NC, Mirkin CA (2015) Programmable materials and the nature of the DNA bond. Science 347:1260901
2. Seeman NC, Sleiman HF (2018) DNA nanotechnology. Nat Rev Mater 3:17068
3. Rothemund PWK (2006) Folding DNA to create nanoscale shapes and patterns. Nature 440:297–302
4. Douglas SM, Dietz H, Liedl T, Högberg B, Graf F, Shih WM (2009) Self-assembly of DNA into nanoscale three-dimensional shapes. Nature 459:414–418
5. Hong F, Zhang F, Liu Y, Yan H (2017) DNA origami: scaffolds for creating higher order structures. Chem Rev 117:12584–12640
6. Funke JJ, Dietz H (2016) Placing molecules with Bohr radius resolution using DNA origami. Nat Nanotechnol 11:47–52
7. Praetorius F, Kick B, Behler KL, Honemann MN, Weuster-Botz D, Dietz H (2017) Biotechnological mass production of DNA origami. Nature 552:84–87
8. Bathe M, Rothemund PWK (2017) DNA nanotechnology: a foundation for programmable nanoscale materials. MRS Bull 42:882–888
9. Nummelin S, Kommeri J, Kostiainen MA, Linko V (2018) Evolution of structural DNA nanotechnology. Adv Mater 30:1703721
10. Castro CE, Kilchherr F, Kim D-N, Shiao EL, Wauer T, Wortmann P, Bathe M, Dietz H (2011) A primer to scaffolded DNA origami. Nat Methods 8:221–229
11. Benson E, Mohammed A, Gardell J, Masich S, Czeizler E, Orponen P, Högberg B (2015) DNA rendering of polyhedral meshes at the nanoscale. Nature 523:441–444
12. Veneziano R, Ratanalert S, Zhang K, Zhang F, Yan H, Chiu W, Bathe M (2016) Designer nanoscale DNA assemblies programmed from the top down. Science 352:1534
13. Linko V, Kostiainen MA (2016) Automated design of DNA origami. Nat Biotechnol 34: 826–827
14. Linko V, Dietz H (2013) The enabled state of DNA nanotechnology. Curr Opin Biotechnol 24:555–561
15. Maune HT, Han S, Barish RD, Bockrath M, Goddard WA III, Rothemund PWK, Winfree E (2010) Self-assembly of carbon nanotubes into two-dimensional geometries using DNA origami templates. Nat Nanotechnol 5:61–66
16. Ramakrishnan S, Subramaniam S, Stewart AF, Grundmeier G, Keller A (2016) Regular nanoscale protein patterns via directed adsorption through self-assembled DNA origami masks. ACS Appl Mater Interfaces 8:31239–31247
17. Shen B, Linko V, Tapio K, Pikker S, Lemma T, Gopinath A, Gothelf KV, Kostiainen MA, Toppari JJ (2018) Plasmonic nanostructures through DNA-assisted lithography. Sci Adv 4: eaap8978
18. Kuzyk A, Schreiber R, Fan Z, Pardatscher G, Roller E-M, Högele A, Simmel FC, Govorov AO, Liedl T (2012) DNA-based self-assembly of chiral plasmonic nanostructures with tailored optical response. Nature 483:311–314
19. Gopinath A, Miyazono E, Faraon A, Rothemund PWK (2016) Engineering and mapping

nanocavity emission via precision placement of DNA origami. Nature 535:401–405
20. Kuzyk A, Jungmann R, Acuna GP, Liu N (2018) DNA origami route for nanophotonics. ACS Photonics 5:1151–1163
21. Shen B, Kostiainen MA, Linko V (2018) DNA origami nanophotonics and plasmonics at interfaces. Langmuir 34:14911–14920
22. Grossi G, Jepsen MDE, Kjems J, Andersen ES (2017) Control of enzyme reactions by a reconfigurable DNA nanovault. Nat Commun 8:992
23. Linko V, Nummelin S, Aarnos L, Tapio K, Toppari JJ, Kostiainen MA (2016) DNA-based enzyme reactors and systems. Nanomaterials 6:139
24. Zhao Z, Fu J, Dhakal S, Johnson-Buck A, Liu M, Zhang T, Woodbury NW, Liu Y, Walter NG, Yan H (2016) Nanocaged enzymes with enhanced catalytic activity and increased stability against protease digestion. Nat Commun 7: 10619
25. Ijäs H, Hakaste I, Shen B, Kostiainen MA, Linko V (2019) Reconfigurable DNA origami nanocapsule for pH-controlled encapsulation and display of cargo. ACS Nano 13:5959–5967
26. Langecker M, Arnaut V, Martin TG, List J, Renner S, Mayer M, Dietz H, Simmel FC (2012) Synthetic lipid membrane channels formed by designed DNA nanostructures. Science 338:932–936
27. Castro CE, Dietz H, Högberg B (2017) DNA origami devices for molecular-scale precision measurements. MRS Bull 42:925–929
28. Rajendran A, Endo M, Hidaka K, Sugiyama H (2014) Direct and single-molecule visualization of the solution-state structures of G-hairpin and G-triplex intermediates. Angew Chem Int Ed 53:4107–4112
29. Steinhauer C, Jungmann R, Sobey TL, Simmel FC, Tinnefeld P (2009) DNA origami as a nanoscopic ruler for super-resolution microscopy. Angew Chem Int Ed 48:8870–8873
30. Graugnard E, Hughes WL, Jungmann R, Kostiainen MA, Linko V (2017) Nanometrology and super-resolution imaging with DNA. MRS Bull 42:951–959
31. Surana S, Shenoy AR, Krishnan Y (2015) Designing DNA nanodevices for compatibility with the immune system of higher organisms. Nat Nanotechnol 10:741–747
32. Linko V, Ora A, Kostiainen MA (2015) DNA nanostructures as smart drug-delivery vehicles and molecular devices. Trends Biotechnol 33: 586–594
33. Bila M, Kurisinkal EE, Bastings MMC (2019) Engineering a stable future for DNA origami as a biomaterial. Biomater Sci 7:532–541
34. Zhao X-Y, Shaw A, Zeng X, Benson E, Nyström AM, Högberg B (2012) DNA origami delivery system for cancer therapy with tunable release properties. ACS Nano 6:8684–8691
35. Kollmann F, Ramakrishnan S, Shen B, Grundmeier G, Kostiainen MA, Linko V, Keller A (2018) Superstructure-dependent loading of DNA origami nanostructures with a groove-binding drug. ACS Omega 3:9441–9448
36. Douglas SM, Bachelet I, Church GM (2012) A logic-gated nanorobot for targeted transport of molecular payloads. Science 335:831–834
37. Li S, Jiang Q, Liu S, Zhang Y, Tian Y, Song C, Wang J, Zou Y, Anderson GJ, Han JY, Chang Y, Liu Y, Zhang C, Chen L, Zhou G, Nie G, Yan H, Ding B, Zhao Y (2018) A DNA nanorobot functions as a cancer therapeutic in response to a molecular trigger in vivo. Nat Biotechnol 36:258–264
38. Marras AE, Zhou L, Su H-J, Castro CE (2015) Programmable motion of DNA origami mechanisms. Proc Natl Acad Sci USA 112: 713–718
39. Ijäs H, Nummelin S, Shen B, Kostiainen MA, Linko V (2018) Dynamic DNA origami devices: from strand-displacement reactions to external-stimuli responsive systems. Int J Mol Sci 19:2114
40. Okholm AH, Nielsen JS, Vinther M, Sørensen RS, Schaffert D, Kjems J (2014) Quantification of cellular uptake of DNA nanostructures by qPCR. Methods 67:193–197
41. Ora A, Järvihaavisto E, Zhang H, Auvinen H, Santos HA, Kostiainen MA, Linko V (2016) Cellular delivery of enzyme-loaded DNA origami. Chem Commun 52:14161–14164
42. Bastings MMC, Anastassacos FM, Ponnuswamy N, Leifer FG, Cuneo G, Lin C, Ingber DE, Ryu JH, Shih WM (2018) Modulation of the cellular uptake of DNA origami through control over mass and shape. Nano Lett 18:3557–3564
43. Linko V, Shen B, Tapio K, Toppari JJ, Kostiainen MA, Tuukkanen S (2015) One-step large-scale deposition of salt-free DNA origami nanostructures. Sci Rep 5:15634
44. Hahn J, Wickham SFJ, Shih WM, Perrault SD (2015) Addressing the instability of DNA nanostructures in tissue culture. ACS Nano 8: 8765–8775

45. Kielar C, Xin Y, Shen B, Kostiainen MA, Grundmeier G, Linko V, Keller A (2018) On the stability of DNA origami nanostructures in low-magnesium buffers. Angew Chem Int Ed 57:9470–9474
46. Xin Y, Piskunen P, Suma A, Li C, Ijäs H, Ojasalo S, Seitz I, Kostiainen MA, Grundmeier G, Linko V, Keller A (2022) Environment-dependent stability and mechanical properties of DNA origami six-helix bundles with different crossover spacings. Small 18:2107393
47. Ramakrishnan S, Ijäs H, Linko V, Keller A (2018) Structural stability of DNA origami nanostructures under application-specific conditions. Comput Struct Biotechnol J 16:342–349
48. Li Y, Schulman R (2019) DNA nanostructures that self-heal in serum. Nano Lett 19:3751–3760
49. Ponnuswamy N, Bastings MMC, Nathwani B, Ryu JH, Chou LYT, Vinther M, Li WA, Anastassacos FM, Mooney DJ, Shih WM (2017) Oligolysine-based coating protects DNA nanostructures from low-salt denaturation and nuclease degradation. Nat Commun 8: 15654
50. Agarwal NP, Matthies M, Gür FN, Osada K, Schmidt TL (2017) Block copolymer micellization as a protection strategy for DNA origami. Angew Chem Int Ed 56:5460–5464
51. Auvinen H, Zhang H, Nonappa, Kopilow A, Niemelä EH, Nummelin S, Correia A, Santos HA, Linko V, Kostiainen MA (2017) Protein coating of DNA nanostructures for enhanced stability and immunocompatibility. Adv Healthcare Mater 6:1700692
52. Ramakrishnan S, Shen B, Kostiainen MA, Grundmeier G, Keller A, Linko V (2019) Real-time observation of superstructure-dependent DNA origami digestion by DNase I using high-speed atomic force microscopy. ChemBioChem 20:1900369
53. Perrault SD, Shih WM (2014) Virus-inspired membrane encapsulation of DNA nanostructures to achieve in vivo stability. ACS Nano 8: 5132–5140
54. Mikkilä J, Eskelinen A-P, Niemelä EH, Linko V, Frilander MJ, Törmä P, Kostiainen MA (2014) Virus-encapsulated DNA origami nanostructures for cellular delivery. Nano Lett 14:2196–2200
55. Lacroix A, Edwardson TGW, Hancock MA, Dore MD, Sleiman HF (2017) Development of DNA nanostructures for high-affinity binding to human serum albumin. J Am Chem Soc 139:7355–7362
56. Hernandez-Garcia A, Estrich NA, Werten MWT, van der Maarel JRC, LaBean TH, de Wolf FA, Stuart MAC, de Vries R (2017) Precise coating of a wide range of DNA templates by a protein polymer with a DNA binding domain. ACS Nano 11:144–152
57. Kiviaho JK, Linko V, Ora A, Tiainen T, Järvihaavisto E, Mikkilä J, Tenhu H, Nonappa, Kostiainen MA (2016) Cationic polymers for DNA origami coating – examining their binding efficiency and tuning the enzymatic reaction rates. Nanoscale 8:11674–11680
58. Ahmadi Y, de Llano E, Barišić I (2018) (Poly)cation-induced protection of conventional and wireframe DNA origami nanostructures. Nanoscale 10:7494–7504
59. Saccà B, Niemeyer CM (2011) Functionalization of DNA nanostructures with proteins. Chem Soc Rev 40:5910–5921
60. Schaffert DH, Okholm AH, Sørensen RS, Nielsen JS, Tørring T, Rosen CB, Kodal AL, Mortensen MR, Gothelf KV, Kjems J (2016) Intracellular delivery of a planar DNA origami structure by the transferrin-receptor internalization pathway. Small 12:2634–2640
61. Gerling T, Kube M, Kick B, Dietz H (2018) Sequence-programmable covalent bonding of designed DNA assemblies. Sci Adv 4:eaau1157
62. Kostiainen MA, Szilvay GR, Smith DK, Linder MB, Ikkala O (2006) Multivalent dendrons for high-affinity adhesion of proteins to DNA. Angew Chem Int Ed 45:3538–3542
63. Kostiainen MA, Szilvay GR, Lehtinen J, Smith DK, Linder MB, Urtti A, Ikkala O (2007) Precisely defined protein–polymer conjugates: construction of synthetic DNA binding domains on proteins by using multivalent dendrons. ACS Nano 1:103–113
64. Stahl E, Martin TG, Praetorius F, Dietz H (2014) Facile and scalable preparation of pure and dense DNA origami solutions. Angew Chem Int Ed 53:12735–12740
65. Hung AM, Micheel CM, Bozano LD, Osterbur LW, Wallraff GM, Cha JN (2010) Large-area spatially ordered arrays of gold nanoparticles directed by lithographically confined DNA origami. Nat Nanotechnol 5:121–126
66. Ijäs H, Liedl T, Linko V, Posnjak G (2022) A label-free light-scattering method to resolve assembly and disassembly of DNA nanostructures. Biophys J 121:4800–4809
67. Seitz I, Ijäs H, Linko V, Kostiainen MA (2022) Optically responsive protein coating of DNA origami for triggered antigen targeting. ACS Appl Mater Interfaces 14:38515–38524

Chapter 13

Cellular Uptake of DNA Origami

Maartje M. C. Bastings

Abstract

This chapter discusses the methods involved in achieving and analyzing cellular uptake of DNA origami. While cells naturally internalize substances from their surroundings, more than a simple addition of DNA origami in the surrounding cell medium is necessary to ensure DNA origami particles successfully enter the intracellular environment. Starting with the folding of the DNA, careful handling of sterile buffers and tools is essential, as well as the use of an endotoxin free scaffold. We explain how DNA origami needs a certain form of stabilization or protection to survive the degrading low-salt and high-nuclease environment of common cell culture media. Depending on the preferred method of post-uptake analysis (confocal), microscopy, or flow cytometry, we elaborate on the full protocols and crucial steps to prepare cell uptake experiments. Finally, notes are added on the intracellular fate (*see* **Notes 14** and **15**), and cellular retention of DNA origami (*see* **Note 16**) is discussed.

Key words DNA origami, Cell uptake, Stability, Nanomaterials, Nanotechnology, Nanoparticles, Nanotherapeutics

1 Introduction

DNA origami nanostructures can be programmed to fold into prescribed spatial configurations and can be functionalized site specifically with a wide variety of guests, and induced to undergo conformational changes [1]. Their unique properties as controlled shape, addressability on the nanoscale, and responsiveness can be transferred to other nanomaterials – e.g., liposomes, polymeric or metallic particles, and proteins – and thereby augment their functionality [2]. Furthermore, DNA-based nanoparticles are biodegradable and biocompatible and thus form an interesting class of materials for the development of nano-diagnostics and therapeutics [3]. A pivotal requirement, however, is their successful cellular uptake, which requires careful sample preparation, handling, and analysis. Crucial to realize is the ongoing battle to find correct methods to truly analyze and quantify the integrity of DNA-based materials in cellular context that can be used independent of cell type and DNA origami design. Thus far, it has become clear that

Julián Valero (ed.), *DNA and RNA Origami: Methods and Protocols*, Methods in Molecular Biology, vol. 2639,
https://doi.org/10.1007/978-1-0716-3028-0_13,

uptake and cellular fate differ significantly with cell type as well as DNA nano-shape, size, and type of labeling and functionalization [4, 5]. While an ideal nano-tool in test-tube conditions, far from living material, the true potential of DNA origami structures as intracellular tools remains unclear due to missing fundamental insights into their uptake pathway, integrity, and intracellular fate. The protocol presented here offers guidelines to uptake experiments of DNA origami; however, it should be mentioned that each cell type and each DNA origami design behaves differently, and thus careful controls and modifications might be needed.

1.1 Uptake of Particles into Cells

Cells come in many sorts and flavors, and their behavior regarding to uptake differs greatly. The first decision to make when uptake experiments are concerned, independent on the DNA origami structure of choice, is the selection of the target cell type. Endothelial, epithelial, and even certain cancer cell lines show very little particle uptake, whereas macrophages and dendritic cells function as "professional eaters" to ensure our immune system works adequately. The uptake differences between for instance dendritic cells and HEK293 (human epithelial kidney cells, an epithelial cell line) or HUVEC (human umbilical vein endothelial cells, an endothelial cell line) can differ up to 15× after 12 h of incubation (*see* Fig. 1) [4].

Besides differences between cell types in absolute uptake quantities, uptake kinetics can also largely vary. It is therefore advisable to perform a kinetic experiment in order to map out the uptake kinetics of the cell type of choice. For example, endothelial and epithelial cells saturated after 2 h, whereas dendritic cells continued to endocytose up to 12 h (*see* Fig. 2). Additionally, also the structural design of the DNA origami influences uptake kinetics, with compact structures being taken up faster than hollow or elongated ones (*see* Fig. 2) [4]. Insights into the kinetic profile of the target cells will also impact decisions on stability requirements for the DNA origami.

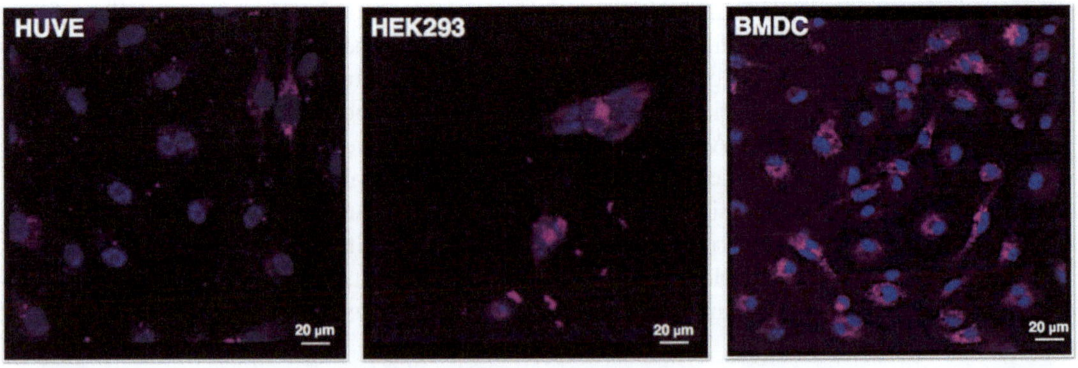

Fig. 1 Confocal imaging of the qualitative differneces of the uptake of DNA origami (Labeled in pink) in 3 different cell lines, HUVE cells, HEK293 cells and dendritic cells (BMDC) (Nuclei visible in blue). (Figure adapted with permission from Ref. [4])

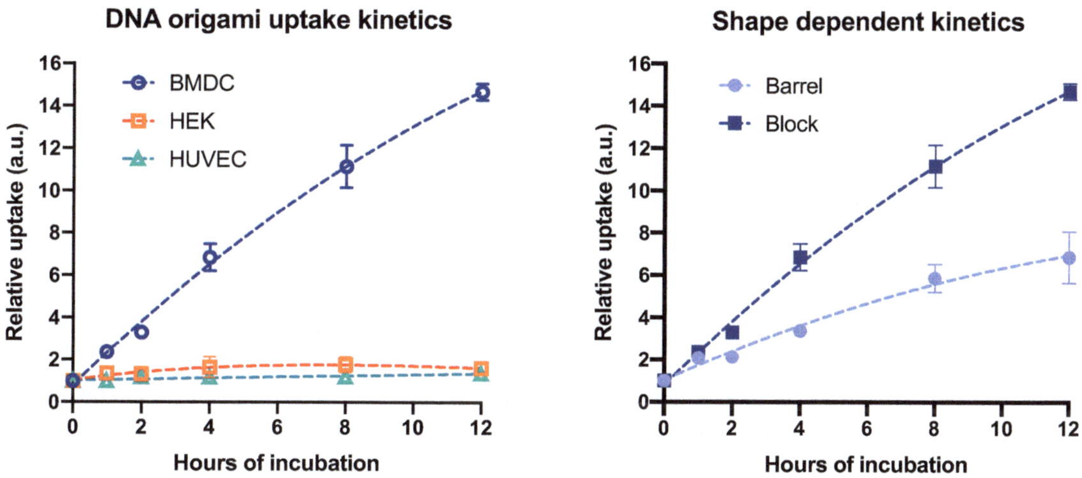

Fig. 2 (left) DNA origami uptake in 3 different cell types shows a significant difference in kinetics. (right) Kinetics of DNA-origami uptake is influenced by structural parameters of aspect ratio and compactness. Graph shows the comparison of an open barrel shape versus a compact block shape. (Figure adapted with permission from Ref. [4])

Since dendritic cells show the most pronounced DNA origami uptake, all exemplary figures demonstrated in this chapter will be based on this cell type.

1.2 Design Parameters for DNA Origami Compatible for Cell Uptake

1.2.1 Endotoxin Free Scaffold and Buffers

DNA origami that is meant to be internalized by cells needs to be designed and synthesized with some important changes from regular DNA origami. The most important difference is that the structure needs to be folded with an endotoxin free scaffold, as well as all staples dissolved in endotoxin free ultrapure water. Endotoxins are lipopolysaccharides (LPS) that are located within a cell wall of the bacteria that cannot be heat destroyed. These molecules can easily contaminate labware and as such can significantly impact both in vitro and in vivo experiments [6]. In vitro, there is increasing evidence that endotoxin causes a variety of problems for cell culture research, and in vivo, endotoxins provoke inflammatory responses. Bacteria present endotoxin in large amounts upon cell death and when they are actively growing and dividing. As the M13 phage scaffold is produced with the help of *E. coli*, the DNA origami scaffold is at risk of endotoxin contamination. Hence, any M13 scaffold used for cell experiments needs to be rendered endotoxin free before use in a DNA origami folding reaction. Current FDA limits state that endotoxin levels are <0.5 EU/mL unless there is contact with cerebrospinal fluid; then the limit is <0.06 EU/mL [7]. The presence of endotoxins can be measured via a selective LAL assay. Subheading 3.1.1 discusses the purification of the M13 phage scaffold from endotoxins as preparation for cellular uptake experiments.

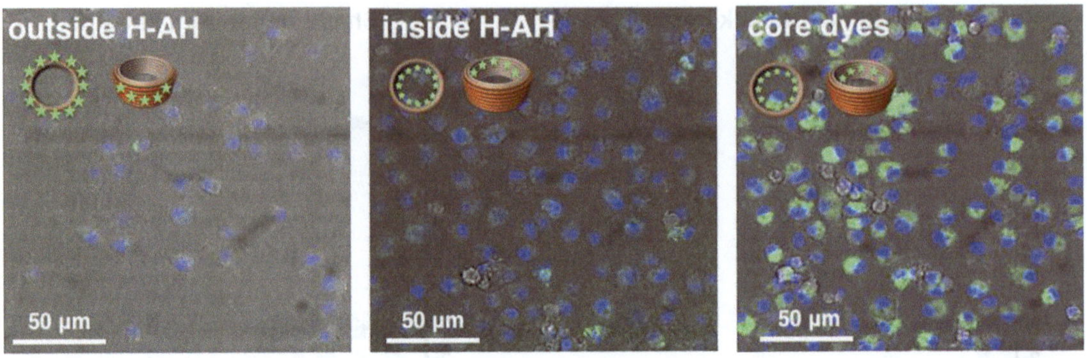

Fig. 3 Qualitative differences of cellular internalization of a barrel-shaped DNA-origami structure, decorated with fluorescent labels via a handle-antihandle approach (left and middle panels) and covalent core modification of staples (right panel). (Figure adapted with permission from Ref. [8])

1.2.2 Fluorescent Labeling

Dependent on the analysis technique used to follow cellular uptake, most likely the DNA origami needs a certain dye functionalization. In our experience, direct functionalization of staples that are part of the origami structure yields the most robust results. A handle-antihandle approach has a higher risk of degradation or destabilization which can result in a reduction of dyes per structure over time, negatively affecting the quantification of post-uptake analysis (*see* Fig. 3). Additionally, the presence of handles on the outside of a DNA origami structure negatively affects the uptake efficiency (*see* **Note 1**). Per DNA origami, three covalently attached dyes have demonstrated to be sufficiently visible in cell assays using 1 nM origami structures [8]. When multiple designs or batches of DNA origami are compared, it is crucial to calibrate the fluorescence per structure before performing uptake experiments. Without this step, an adequate comparison and quantification of uptake yield is impossible. Lastly, the type of dye can influence cellular uptake and localization and/or vary in intensity depending on the cellular environment, which should be taken into careful consideration in the experimental design [5, 9] (*see* **Note 2**).

1.2.3 Stability

Since the stability of DNA origami in cellular medium is not evident due to low salt concentrations and the presence of nucleases, a suitable protection strategy is required to guarantee the structures remain intact upon internalization [10]. Unprotected DNA origami will already (partially) fall apart when placed in cell medium conditions, and this significantly affects their uptake (*see* Fig. 4). Multiple strategies have been developed over the past years, reviewed here [11]. For each application and design, a careful choice of protection method needs to be made. One of the more general methods is the coating of DNA origami with a cationic, pegylated polymer [8]. Since this strategy has been demonstrated

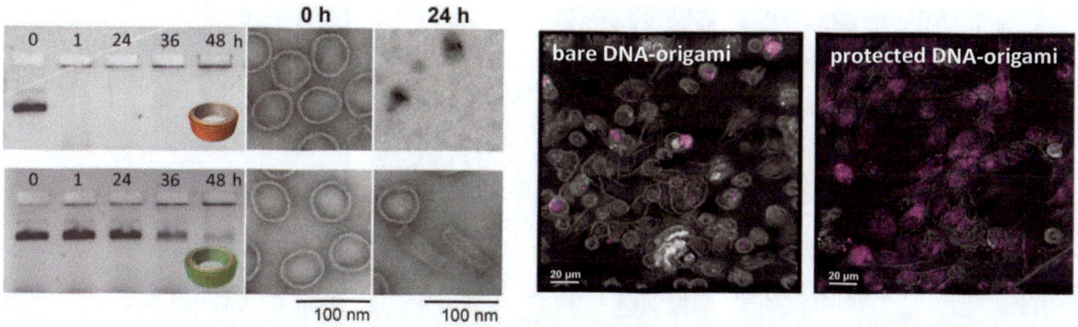

Fig. 4 (Left) Agarose gel and TEM analysis of the structural integrity of bare (orange) and protected (green) DNA-origami barrel structures incubated over time (0–48 h) in cell medium with 10% FBS. (Right) Qualitative differences of bare and K_{10}-PEG_{5K} protected Cy5-labeled DNA-origami (5 nM, pink) incubated with Bone-marrow derived dendritic cells (BMDC) for 12 h, measured by confocal microscopy at 12 h post-washing. (Figure adapted with permission from Ref. [8])

to be compatible with cellular uptake assays, we will provide the protocols in detail. Protocols of other methods can be found in the respective publications [11].

1.2.4 Uptake Efficiency

Cellular uptake efficiency can differ largely between cell types as well as is subject to structural differences and influenced by surface charge. The aspect ratio and compactness of DNA origami has been shown to affect internalization, with the more compact and low aspect-ratio designs being favorized over wide and high aspect-ratio structures (*see* Fig. 2 (right panel)) [4]. However, even for the most favorable designs, generally a significant fraction remains associated with the outer cell surface. This membrane interaction is even more pronounced when a positive surface coating is present, as the electrostatic interaction with the negatively charged cell membrane causes a rapid, specific adsorption of DNA origami. It is therefore important to evaluate the fraction-internalized versus the membrane-adsorbed fraction when analyzing cellular uptake. Confocal microscopy can be used to qualitatively analyze the membrane adsorption, as the fluorescent signal from labeled origami will demonstrate a pronounced signal aligning at the membrane, but not in the cytoplasm. Contrary, DNA origami taken up into the cell will display either a punctate fluorescence that is present throughout the cell if the material is located in the endosomal vesicles, or a diffuse fluorescence when dispersed in the cytoplasm (*see* Fig. 5). The external fraction can be eliminated through incubation with a high dose of DNase I followed by several washing steps and a medium refresh [4]. Finally, the presence of dyes or targeting ligands can significantly alter the uptake efficiency as well (*see* **Notes 2** and **3**).

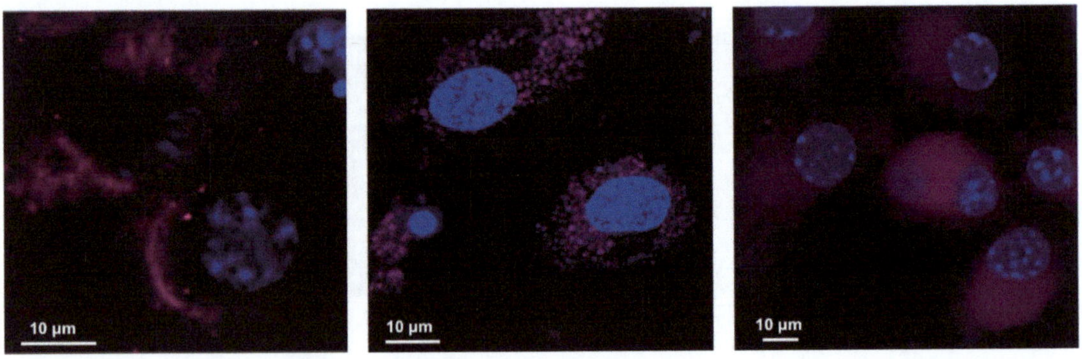

Fig. 5 Exemplary confocal images of DNA-origami at three different cellular locations: (left) attached to the cell membrane, (middle) punctate fluorescence resulting from DNA-origami in endosomal/lysosomal vesicles and (right) diffuse fluorescence resulting from distribution of particles in the cytoplasm (data not published)

1.2.5 Quantity and Purification

Cell uptake experiments require a large amount of material to be prepared, generally in the order of several milliliters of a 10 nM scaffold folding. This by itself poses demands on the purification strategy. To remove the excess of staples, the best method is PEG precipitation, as this can be easily scaled up to larger quantities, and all reagents can be sterilized/sterile filtered, to prevent contaminations that could negatively impact cell studies. Additionally, the origami structures can be dissolved in a high stock concentration necessary before dilution into cell media. Important, however, is to optimize the PEG precipitation conditions for each individual DNA origami shape, to make sure the excess of staples is adequately removed. Alternative purification using ultracentrifugation is discussed in **Note 4**.

To summarize, cellular uptake of DNA origami offers an exciting area of experimental exploration, since the DNA origami technique could provide us with many insights on cell function and decision making on the nanometer scale. However, the highest care should be given to ensure the assembled structures are adequately characterized, protected, and calibrated, in order to obtain a reliable readout of the experiments performed. Additionally, cell experiments should always be repeated with a newly split cell batch, as well as compared with internal positive and negative controls, that could rule out or shed light on intracellular degradation and structural integrity.

2 Materials

2.1 DNA Folding and Protection

1. *General DNA origami folding materials*: nanodrop analyzer, agarose gel electrophoresis equipment, TEM, grids and negative staining reagent for folding analysis, thermal cycler, M13 scaffold, staples dissolved in ultrapure-sterile water, fluorescent labeled staples, fluorescence plate reader.

2. *1× DNA origami folding buffer*: (typically) 5 mM Tris pH 8.5, *x* mM $MgCl_2$ (*x* specific per origami design), 1 mM EDTA, sterile filtered.
3. *Endotoxin free scaffold prep*: Triton X-114, Endosafe-PTS system and test cartridges or LAL assay kit, ultrapure sterile water, inversion mixer, Eppendorf tube centrifuge 16.8 xg max speed.
4. *DNA origami protection*: PEG5k-K10 for DNA origami protection, to be used in a 1:1 N:P ratio depending on the size of the DNA origami structure (amount of (+) from amines (N) versus (−) of phosphates (P)).
5. *PEG precipitation*: 8 k PEG, 0.25 M NaCl solution, 1× DNA origami folding buffer.

2.2 Cell Culture

1. *General cell culture materials*: T75 or T25 culture flasks, brightfield microscope to follow confluence, CO_2 incubator maintained at 37 °C, laminar flow cabinet, vacuum aspirator system, serological pipettes, pipettor, water bath at 37 °C, 15 mL and 50 mL sterile Falcon tubes, pipettes, sterile filter tips.
2. *Cell medium*: this will vary per cell type used. Please refer to the guidelines given by the supplier of your cell type of choice. In most cases, this will be either DMEM or RPMI-1640 medium, supplemented with 10% FBS and 1% penicillin/streptomycin solutions.
3. *Splitting*: Sterile PBS, trypsin, centrifuge.
4. *Cell counting*: Automated cell counter with disposable counting slides and trypan blue solution.

2.3 DNA Origami Uptake

1. *General*: plate reader for fluorescence calibration, nanodrop for concentration measurements, cell medium, 96/24 well plates (depending on assay size), sterile PBS.
2. *Microscopy*: confocal microscope with 63× objective, 1.5 glass bottom slides or plates, (optional but recommended) Hoechst 33324 (or Dapi).
3. *Flow cytometry*: BD LSRfortessa SORP flow cytometer with HTS sampler, cell scrapers, 96 wells plates, DNase I, PBS.

3 Methods

For all steps in this section, use sterile tubes, flasks, filter tips, ultrapure water, and autoclave and sterile filter all solutions. When working with cells, use a laminar flow cabinet and stay within a cell culture lab. Use appropriate personal protection equipment, and exchange gloves often. Protect samples from light when they contain fluorescent labels. Preheat solutions to 37 °C when interacting with cells.

3.1 Preparation of Cell-Compatible DNA Origami

3.1.1 Endotoxin Free Scaffold (Adapted from [10])

1. To a starting solution of 100 nM origami scaffold strand in ultra-pure, sterile water add the surfactant Triton X-114 to a final concentration of 2% (v/v) in a sterile Eppendorf tube.
2. Incubate the mixture at 4 °C on an inversion mixer for 30 min to solubilize endotoxin.
3. Mix the solution at 37 °C for 5 min; a phase separation will be visible.
4. Centrifuge the tube at 37 °C for 30 min at maximum speed (16.873 k × g). Transfer the top aqueous fraction to a new sterile tube; discard the viscous bottom fraction. Repeat this procedure at least four times to reduce endotoxin in the scaffold stock.
5. Quantify remaining endotoxin levels using the Endosafe-PTS system and test cartridges (Charles River) or following the guidelines of a LAL assay kit. An acceptable level of less than 0.5 EU/mL need to be achieved for cell experiments [7].
6. When the measured endotoxin level is too high (e.g. above FDA standards), repeat the endotoxin removal procedure described above.

3.1.2 Folding of the DNA Origami with Sterile Staples and Buffers

In order to keep the final product free of toxic contaminants that could interfere with cellular function, it is important to dissolve the staples in sterile ultrapure water and use only buffers that are autoclaved and sterile filtered. Additionally, to minimalize aerosol contaminants, sterile filter pipet tips are highly recommended.

1. Load 100 μL of the folding reaction (10 nM of scaffold and 10× excess of staples) to 200 μL (sterile) PCR tubes.
2. Run the annealing program in a thermal cycler to ensure proper heat distribution and high quality folding. As the structures contain fluorescent labels, it is important to work in a dark environment as much as possible.

3.1.3 Purification and Concentration

Given the large volume of the folding reactions, the preferred method to purify the DNA origami structures is PEG precipitation. However, the ultracentrifugation glycerol gradient method (*see* **Note 4**) can also be used, though requires more work, reagents, and specialized equipment.

1. For PEG precipitation, add x% 8 k PEG (*see* **Note 5**), 0.25 M NaCl in 1× folding buffer in a 1:1 ratio to the folding reaction.
2. Place the mixed solution on room temperature for 30 min.

3. Centrifuge at 16 k × g for 40 min at 25 °C. Now a clear small pellet should be visible at the bottom of the tube. Remove the supernatant from the pellet and set aside for agarose gel analysis.
4. Resuspend the pellet in the appropriate folding buffer according to the DNA origami, and let incubate for 30 min at room temperature to allow the purified DNA origami structures to unpack and dissolve (*see* **Note 6**).
5. Purification quality is then verified by agarose gel electrophoresis, by running an aliquot of the dissolved pellet (that contains the purified DNA origami structures) next to an aliquot of the supernatant (which should contain all the excess staples).
6. Run an agarose gel electrophoresis assay for the analysis of proper separation of staples and folded product. Separation is successful if the supernatant lane shows only staples, and the pellet (after dissolving in 1× folding buffer) only shows the pure DNA origami band. Typical percentages lay between 5% and 15%.

3.1.4 Calibration of Fluorescence (Optional, only if Multiple Structures Are to Be Compared)

Both the incorporation yield of staples in a DNA origami structure as well as the chemical labelling of DNA oligos with fluorescent tags are not quantitative in yield. Accepting a modest incomplete reaction yield and assuming a realistic less than 100% staple incorporation, this calibration step allows for a fair comparison of quantitative flow-cytometry fluorescent data between DNA origami batches or designs. When multiple designs are to be compared for their quantitative cellular uptake, it is crucial to internally calibrate the structures for their relative fluorescence. DNA origami for cell uptake studies is used at a 1 nM final concentration, and each structure should contain a minimum of three fluorescent labels. Only if all labels are present and the dye ligation on the staples was 100%, the structures can be directly compared.

1. To calculate the amount of dye per origami structure, measure the intensity of the fluorophore by plate reader using a known concentration of DNA origami, as measured by nanodrop.
2. Additionally, run an agarose gel to show the amount of fluorescent signal attached to the origami versus any unincorporated fluorescence.
3. Calculate a labeling correction factor by setting one batch or DNA origami structure as internal standard, and correct all other structures relative to the standard. In principle, all correction factors should be close to 1. When large deviations are measured, check the annealing reaction and verify that the dye functionalization of the staples was correct.

3.1.5 Coating of DNA Origami for Stabilization

As mentioned before, different protection methods are existing, and a proper decision based on the type of DNA origami as well as cellular experiment should be made [11]. However, the most generic strategy is coating with a PEG-oligolysine polymer [8]. As this strategy has been extensively tested and produced positive results in DNA origami uptake experiments [4], the protocol is provided in depth here. For other protection strategies, the reader is referred to detailed scientific publications elsewhere [11].

1. Dissolve oligolysine-PEG (K_{10}–PEG5K) to a stock concentration that is needed to achieve a 1-to-1 N:P ratio for 20 nM DNA origami design of interest.
2. Add the coating 1:1 (vol) to obtain a 10 nM DNA origami final concentration.
3. Incubate for 30 min at room temperature.

3.2 Cell Culture

3.2.1 Maintenance of Cells

Depending on the cell type used, culture methods might vary. Please refer to the instructions given for your cell type of choice to maintain and passage the cells before confluency. Maintenance of the parent cells is done in standard T75 or T25 culture flask, dependent on the amount of cells needed for the DNA origami uptake assays. Passaging should be performed regularly to keep cells actively growing. Standard passaging protocols can be used, as no special treatment is needed for DNA origami uptake later on (*see* **Note 7**).

1. Aspirate the growth medium.
2. Wash with sterile PBS (heated to 37 °C), aspirate PBS, add trypsin (1 mL for T25, 2 mL for T75), and incubate for 3 min in the CO_2 incubator at 37 °C.
3. Tap the flask gentle on the sides and verify under an inverted microscope if all cells are detached.
4. Add 5 mL fresh, heated culture medium to inhibit trypsin activity.
5. Pipette the cells up and down carefully to separate any aggregated cells and transfer the cell solution to a centrifuge tube (15 mL falcon tube).
6. Centrifuge for 5 min at 1200 rpm.
7. Aspirate the supernatant without touching the cell pellet, and add 2 mL cell medium to gently resuspend the pellet. If used for seeding in plates to continue with the DNA origami uptake assay, count the cells after resuspension in fresh culture medium after the centrifuge step.
8. Depending on the cell type, transfer 1:5 or 1:10 to a new culture flask (e.g., 400 μL to 4.6 mL fresh medium for T25

1:5, or 200 μL to 4.8 mL fresh medium for T25 1:10, and 400 μL to 14.6 mL fresh medium for T75 1:5, or 200 μL to 14.8 mL fresh medium for T75 1:10 splitting).

3.2.2 Counting of Cells

1. Pipette 10 μL of the resuspended mixture above in 90 μL culture medium (*see* **Note 8**).
2. Add 10 μL cell suspension to 10 μL tryphan blue in a fresh tube and gently mix together.
3. Transfer 10 μL into a clean cell counting slide and insert in the automatic cell counter. Follow the instructions of the equipment and make sure the cells are in focus.
4. Calculate the final cell concentration taking all previous dilutions into account.

3.2.3 Plating Cells of DNA Origami Uptake Assays

After counting the cell solution, a final dilution can be made to prepare the seeding of the correct number of cells in the culture plates used for the DNA origami uptake assays.

1. For microscopy analysis, seed 50,000 cells per well in a 1.5 glass bottom 96-well plate (in 100 μL per well) or 10,000 cells per well in tissue-treated 15-well ibidi slides (in 50 μL per well) (*see* **Note 9**).
2. For a full 96-well plate (60 wells of cells), 3 million cells are needed, present as 6 mL of 500,000 cells/mL. For an ibidi slide, only 150,000 cells are needed, in a total amount of 750 μL of 200,000 cells/mL.
3. For flow cytometry, seed 100,000 cells/well in 500 μL per well in a tissue culture-treated 24-well plate. For statistical significance, use at least 3 wells per tested condition, and leave a set of three untouched as isotype control.
4. For a 24-well plate, 7 uptake experiments can take place, with 3 wells remaining "cell-only" controls. To seed one full 24-well plate, 2.4 million cells are needed, which comes to 12 mL of a concentration of 200.000 cells/mL when seeding 500 μL per well.
5. For all cases, place the seeded plates back into the CO_2 incubator at 37 °C, and let the cells adhere overnight before adding any DNA origami.

3.3 Incubation with DNA Origami

3.3.1 Dilution in Cell Culture Buffer

To ensure an even distribution of DNA origami in the cell media as well as prevent the need to pipet small volumes into the wells when cells are already present, it is best to pre-dilute the DNA origami material in cell culture buffer. For triplicates, this ensures an equal amount is present per well. Perform this step just before you want to start the uptake experiments, to prevent long storage in cell medium which might lead to adhesion of medium components to the origami structures.

1. Use 1 nM final DNA origami concentration to accurately study uptake. With the 10 nM (protected) DNA origami stock, this means a 10× dilution in cell medium.
2. Calculate the amount of solution that needs to be prepared depending on the size of plate used (*see* **Note 10**).

3.3.2 Addition of DNA Origami to Cells

1. Take the cells that have been adhering to the 96, 24, or 15 wells plates/slides, out of the incubator, and verify the state of the cells by looking at some random wells under the inverted microscope. The cells should look nicely spread and not confluent so that they can be imaged individually. Be aware for rounded cells or lots of debris which might be a sign of cell death. It is best to plate out new cells in this case.
2. Remove the media by carefully aspirating the wells and replace with the heated media that contains the DNA origami.
3. Incubate for the desired time for your experiments, typically between 0 and 24 h.

3.4 Uptake Analysis

This section discusses two commonly used methods to analyze DNA origami uptake by cells, confocal microscopy, and flow cytometry. However, other methods could be used, and the reader is referred to the notes section for more discussion (*see* **Note 11**).

3.4.1 Confocal Microscopy

Via confocal microscopy, a clear qualitative view can be obtained on the internalization of DNA origami structures, versus external cell membrane association. Additionally, insights into the intracellular fate can be acquired based on the fluorescence signal inside the cell. As there are many different confocal microscopes, please refer to the operation manual of the machine for handling guidelines.

1. To obtain a sufficiently detailed view of the cells, a 63× objective is recommended.
2. Use a 1.5 glass bottom plate (96 wells or 15-well ibidi slide). Make sure the confocal stage is surrounded by a cell chamber that keeps the environment on 37 °C and 5% CO_2, so that the cells will stay comfortable during the imaging session.
3. Start the heat and CO_2 flow 15 min before the start of imaging to equilibrate the equipment.
4. If desired, add to the wells a nuclear localization stain (Hoechst 33324 or Dapi), while the microscope is equilibrating at 37 °C. This facilitates focusing and the identification of (live) cells, as well as helps to analyze the intracellular location of the DNA origami.
5. Add 1 μL to the cell medium in a well. Pre-dilution from the stock is required based on the manufacturer's guidelines.

6. Incubate for 15 min. It is recommended to not stain the nucleus too long before imaging, since it is known to be toxic and affect viability to the cells after several hours [12].
7. Replace the cell medium to remove all DNA origami that was not taken up or attached to the cells. If a kinetic evaluation of uptake over time is desired, the cell medium should be kept untouched (and in this case, also, no nuclear staining should be performed to prevent toxicity over the time course of the experiment).
8. Collect a z-stack of 0.2 μm intervals to analyze the presence of DNA origami inside the cell versus attached to the cell membrane, covering the full cell from top to bottom.
9. Deconvolute the stacks in the imaging software available to you (for instance, Imaris).
10. Obtain a quantitative analysis of intensities using Fiji. Make sure that the laser power settings remain constant during imaging to allow for correct comparison of intensities.

3.4.2 Flow Cytometry

Flow cytometry is a reliable means to quantitatively evaluate DNA origami particle uptake in cells and thus provides crucial details on the cellular uptake of DNA origami materials. It is preferably performed on living cells in order to not cause any artifacts by fixation solutions.

1. Seed 100.000 cells per well in 24 well plates and allow to adhere and proliferate overnight.
2. High amount of DNA origami sample is needed as the minimal well volume is 500 μL. For the evaluation of triplicates, this requires 1.5 mL of 1 nM DNA origami.
3. Use manual sample handling in tubes when only a few samples are evaluated (*see* **Note 12**).
4. It is advised to use a high-throughput sampler (HTS) that can handle a (round bottom) 96-well plate as input when full plates are being evaluated (*see* **Note 12**).
5. Carefully scrape the cells from the bottom of the wells using cell scrapers that fit in the 24 wells (larger scrapers can be cut to the correct size with a sterile knife or scissors) (*see* **Note 13**).
6. Split the volume over 2 wells in a round bottom 96-well plate, 200 μL in each well.
7. Add 20 U of DNase I to one of the wells, while leaving the other untouched.
8. Place the plate back in the cell culture incubator for 1 h. This dose will digest all external DNA origami structures; thus only those that were successfully internalized by the cells will

remain. One full 24-well plate will eventually thus yield 48 samples of 200 μL on a 96-well round bottom plate that fits in the HTS of a flow cytometer.

9. Make sure to include multiple (at least 3) cell-only wells to have a reliable isotype control. Treat these wells similarly with regard to the DNase I treatment, to rule out any artifacts or toxicity resulting from this procedure.
10. Additionally, use a "non-structured" DNA origami control which essentially is just the scaffold, stapled together randomly or linearly to generate a similar mass object but without any defined nanoshape. This type of control is relevant when the influence of a certain origami design on cellular uptake is investigated [4].
11. Run the filled 96-well plate on the flow cytometer and analyze for fluorescence per cell.
12. When comparing the DNase I treated and non-treated data, the fraction of internalized DNA origami is obtained. Often, only ~50% of cell associated fluorescent DNA origami material is actually internalized [4].
13. Analyze the data by the appropriate software, for instance, with FlowJo.
14. Use a gating for living cells and singlets only, to exclude debris and aggregates from the data. Dead cells and debris might be caused by violent cell scraping, hence the importance to be delicate when performing this procedure.

4 Notes

1. Labeling DNA origami through a handle-antihandle approach can lead to a loss in fluorescence caused by degradation during prolonged presence in cell medium [8]. While this is more pronounced for longer time points, it will eventually cause errors in the readout of cell uptake experiments. Besides a stability challenge, the presence of handles on the outside of a DNA origami structures also shows a negative influence on the actual cellular internalization quantity itself (*see* Fig. 6) [4, 8]. Even the presence of single-strand handles significantly reduces uptake efficiency. Therefore, the most reliable labeling strategy is to covalently functionalize a terminus of a staple strand that is part of the DNA origami structure. Although these results are measured in dendritic cells, if a handle-antihandle approach is used, the effect should be verified for the cell type of choice.
2. The choice of dye can significantly influence and be influenced by cellular uptake and intracellular localization [9]. The charge

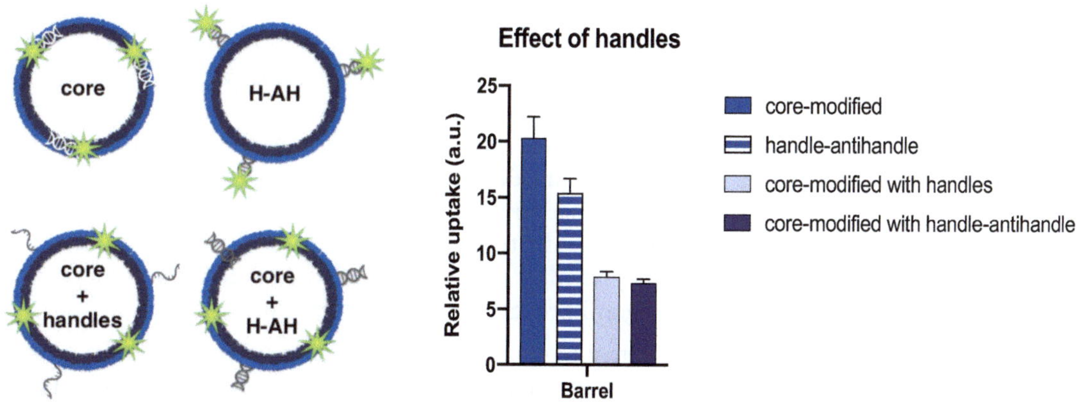

Fig. 6 (left) Cartoon of the 4 different labeling designs tested for uptake in dendritic cells. (right) Presence of ssDNA or dsDNA on the exterior of a DNA origami structure diminishes cellular uptake in dendritic cells. (Figure adapted with permission from Ref. [4])

of the dye can influence the binding of serum proteins, which can then alter fluorescence. Brightness of the dye can be altered by changes of pH in the different cellular organelles [13]. Therefore, to correctly relate fluorescence intensity to amount of DNA origami, precise colocalization and fluorescence analysis is required. The use of two chemically different dyes could be a strategy to confirm that results are not dye dependent as well as placing the dye functionalization inside the DNA origami structure, shielding it from potentially damaging environmental interactions. As a free dye control, a phosphorylated dye resembles the degradation product more closely than an unsubstituted dye molecule. For the ligation of the fluorescent label to a DNA oligo, the 3′ end has a preference over the 5′ end since in serum, 3′-exonucleases are the predominant danger for degradation [14]. Hence, the presence of a label on the 3′ end can help slow down degradation and release of the label and DNA origami structure. Furthermore, labeling of a strand that is buried in the core of a DNA origami structure can help the longevity of the fluorescent tag even further.

3. DNA origami structures have been found to enter cells on a structure and cell type-dependent manner without any targeting ligands; however, their uptake and selectivity can be enhanced by the introduction of ligands. Among many more, it has been demonstrated that cRGD can be used to triple the DNA origami uptake in HUVECs [8], and folate ligands have been used to achieve intracellular delivery of siRNA [15]. DNA aptamers binding to nucleolin could be used to target DNA nanostructures to tumor-associated endothelial cells [16]. The

iron transport protein transferrin was coupled to the exterior of DNA origami to facilitate its cellular uptake by 22-fold [17]. In all cases, the spatial presentation of ligands was found to play a key role in their effectivity. It needs to be mentioned that the presence of ligands will undoubtedly influence cellular uptake kinetics, total amounts, and intracellular localization. It goes beyond this chapter to provide protocols for all cases, as these are all very specific to the cell type used. However, the general protocols provided here can be used as a general guideline to DNA origami preparation and cell maintenance.

4. As alternative to the above-described PEG precipitation purification methods, a glycerol gradient-based ultracentrifugation approach can be used [18]. This method requires an ultracentrifuge capable of spinning at 50.000 rpm (~300.000 × *g* max) at 4 °C. Solutions of folded DNA origami need to be concentrated roughly tenfold using an appropriate MWCO ultracentrifugal filter device. Glycerol stocks of 15% and 45% should be prepared in 1× folding buffer containing same levels of $MgCl_2$ as required for folding, with ultra-pure water and autoclaved to make sure no contamination to the endotoxin free DNA origami occurs. After centrifugation for 1–3 h, depending on the structure, fractionation by hand or using an automated fractionator (which need to be thoroughly washed with EtOH to prevent contamination), fractions that contain the folded DNA origami structure, can be pooled, concentrated, and buffer exchanged to remove glycerol using a 30 k MWCO Amicon Ultra centrifugal filter device. Following purification, the stock solution is diluted appropriately for TEM imaging to verify quality. The stock concentration is determined by UV absorbance at 260 nm on a Nanodrop spectrophotometer, assuming that $A_{260} = 1$ for 50 μg/mL DNA nanostructures. Stock solutions can be stored at 4 °C until further use. The ultracentrifugation purification method provides an alternative when no suitable PEG precipitation conditions can be found. In other cases, the PEG precipitation method is preferred as it reduces materials and time and is better suited for large quantities.

5. Each structure requires optimization of the exact x, and thus an initial screening need to be performed.

6. Depending on the amount of buffer used to resuspend the pellet, the final concentration of the DNA origami can be adjusted. Note that for cell studies, it is best to use a high stock concentration of origami, since the sample will be diluted during protection and cell-medium addition. Best is to aim for a 20–50 nM stock concentration of folded, purified DNA origami.

7. Make sure that when keeping cells for long duration, the presence of mycoplasma is tested at least every 3 months using a PCR-based mycoplasma detection kit. The presence of mycoplasma in the cell medium can lead to the degradation of DNA particles outside the cell environment [19].
8. The dilution depends on the cell type; from the counting results, one will quickly notice if the dilution was too high or too low and adjust accordingly.
9. Depending on the amount of DNA origami in hand, a choice on recipient needs to be made. The 96-well plates offer better handling and a larger imaging field of view, as well as more conditions can be screened at the same time. It is recommended to fill the outer wells of the 96-well plate with buffer to maintain a humidified environment; hence 60 positions remain available. With experiments being performed in triplicate and 3 wells as isotype control, 19 conditions can be tested on 1 plate. On the ibidi slides, buffer can be placed in the reservoir around the wells; thus all 15 wells remain available, yet only four conditions can be tested with 3 wells as isotype control.
10. For a 96-well plate, 3 wells of 100 μL per well means you need to prepare 300 μL of 1 nM DNA origami in cell medium. Typically, some material is lost while pipetting and stays in the Eppendorf tube. Therefore, as common practice, make ˜10% more than needed to not encounter unpleasant surprises of running out of materials while adding the solution to the cells. Thus, for the example of 3 wells on a 96-well plate, prepare 330 μL of 1 nM DNA origami solution (33 μL 10 nM DNA origami to 297 μL of cell medium at 37 °C). For the 24-well plates, 500 μL per well is used, and for the ibidi slides, 50 μL; thus prepare for triplicates, respectively, 1650 μL and 165 μL of 1 nM solution.
11. As a label-free alternative to quantify the amount of DNA origami taken up by cells, the qPCR method developed by the Kjems lab can be followed [20]. It relies on the amplification of a part of the origami scaffold, since the M13 scaffold is a key component of most DNA origami structures and thus always present. The range of detection was demonstrated to be five orders of magnitude with the minimal detectable quantity as low as 1000 DNA origami structures. However, any (partly) degraded structures will be counted if the scaffold amplification site is still intact and thus could diminish the accuracy (the same is true for fluorescence based techniques). Additionally, loss of DNA during the preparation steps could lead to an underestimation of the quantity measured. Overall, this method provides a good alternative when label-free DNA origami is preferred.

12. It is important to realize that after the incubation of the cells with DNA origami (as described in Subheading 3.3.2), there will be a fraction of DNA origami attached to the outer cell membrane. With flow cytometry analysis, no difference is measured between a fluorescent signal resulting from intracellular DNA origami, and DNA origami extracellularly attached to the membrane. Therefore, an additional digestion step needs to be performed to compare the internalized versus the full cell-associated fraction of DNA origami.
13. Alternatively to scraping, cells can be fixed using a formaldehyde solution after detaching them from the wells. As cell detaching and fixation conditions might vary per cell type, please refer to external guidelines for these procedures. Note, however, that when using fixation strategies, DNase I treatment need to be performed before detaching the cells, and free-floating DNA origami structures need to be washed away to prevent any formaldehyde-related crosslinking artefacts. To have a control set where no DNase I treatment has been used, a double amount of wells is required, adding a large burden to the sample preparation. Besides these procedural remarks, formaldehyde fixation can wash away the actual fluorescent signal, as was observed for fluorescence resulting from mitochondrial staining [9]. Methanol fixation cannot be used as it causes DNA precipitation. Given all the challenges and potential artefacts resulting from fixation methods, keeping the cells alive and using the cell-scraping method is preferred.
14. While we have focused on solely the uptake of DNA origami into cells, their intactness and fate inside the cell is a totally different question. So far, no definite answer has been given by experiments performed in the field of DNA nanotechnology that fully reveals the cellular fate of DNA origami that has entered the cell. From experiments performed with protected, stabilized DNA origami, the data from a FRET (Förster resonance energy transfer) assay based on donor-fluorescence enhancement induced by acceptor photobleaching, demonstrated that 50% of DNA origami particles stay intact after 24 h of presence inside cells [8]. Of note is that proper controls should be designed, as it has been observed that FRET can occur due to the concentrated presence of individual components inside the endosomal/lysosomal compartments of the cell, yielding false positives [9]. A different strategy to measure the structural integrity of uptaken DNA origami particles is the use of TEM. Using a DNA-origami rod-shaped particle, decorated with an array of gold nanoparticles that could function as structural barcode, the internalization process in cancer cells

was followed [21]. DNA origami structures were observed to be intact directly after internalization in the early endosomes, but the structural integrity was lost in the late endosome and lysosomes. However, no distinction between a detachment of the AuNPs from the DNA origami or a global origami degradation can be made here. No AuNPs were observed to escape from endosomes or lysosomes into the cytosol, which might present a challenge in utilizing DNA origami particles as delivery vehicles for cargoes that require cytosol or nuclei transportation.

15. Precise colocalization and fluorescence analysis help to follow the cellular fate of DNA origami, and correlate this to potential variations in fluorescence based on environmental (pH, oxidative state) parameters. Fluorescent staining methods of various organelles (early/late endosome, lysosome, mitochondria, nucleus) have been employed to track the voyage of DNA origami. Typically, when a punctate fluorescent DNA origami signal is obtained, the structures are present in endosomes or lysosomes. A broad diffuse fluorescence indicates cytoplasm distribution, whereas filamentous fluorescent was seen for mitochondrial localizing. Keep in mind that time of addition and incubation plays an important role both in the labeling as well as the intracellular trafficking of DNA structures.

16. Depending on the cell type used, uptake kinetics will vary, and performing time-course study to follow the particle uptake is essential. Equally interesting is to study the retention of the DNA origami particles inside the cells. Both experiments together provide the window of opportunity to perform intracellular manipulation or measurements using DNA origami. All previous notes and mentions of choice of dye, shape and size of DNA origami, coating, charge, etc. hold true, and the experimental design is subject to change based on the cell type and methods used. Important is to wash away, e.g., refresh the cell medium, after a certain incubation time, and monitor the decrease of fluorescent signal over time. As example, Fig. 7 demonstrates the retention of a DNA origami barrel in dendritic cells, where a retention half time of ~8 h was measured [8]. In order to compare the fluorescence intensity signals of the cells (using Fiji or ImageJ), make sure that the laser settings of the confocal imaging session are kept constant during the measurements.

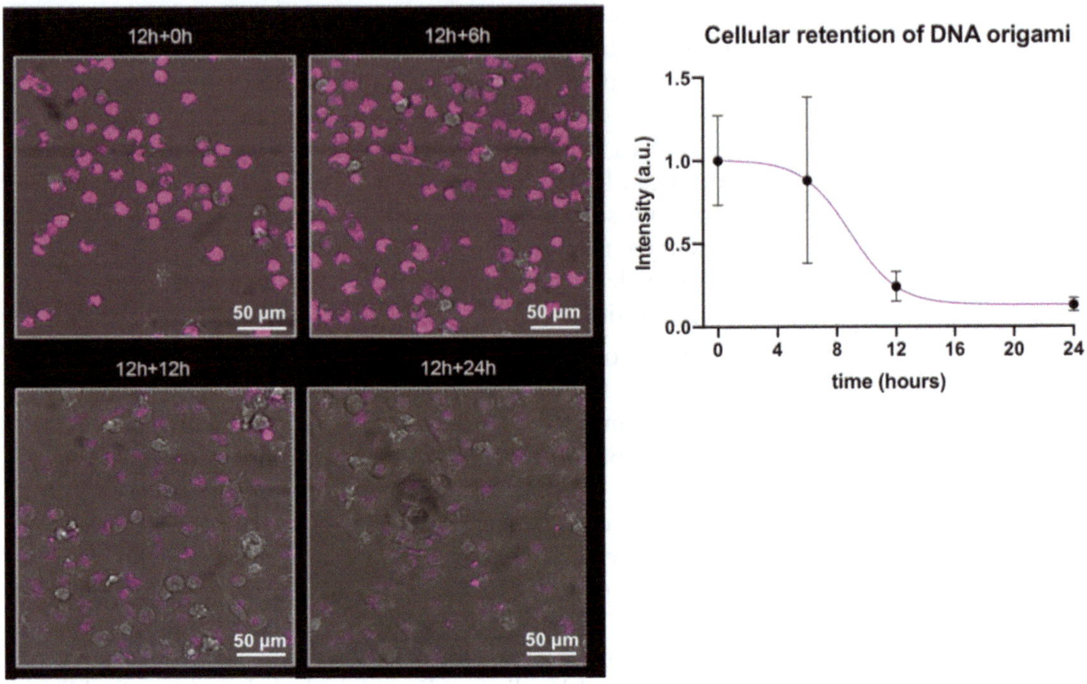

Fig. 7 Retention DNA origami (Pink) within dendritic cells at 37 °C, as monitored via confocal microscopy (left panels) at 0 h, 6 h, 12 h and 24 h after an initial incubation of 12 h. Fluorescence intensity was calculated using ImageJ and normalized to 0 h time point. The retention half-life of this particular DNA origami shape is ˜8 h post-washing. (Figure adapted with permission from Ref. [8])

References

1. (a) Rothemund PWK (2006) Folding DNA to create nanoscale shapes and patterns. Nature 440:297–302 (b) Douglas SM, Dietz H, Liedl T, Högberg B, Graf F, Shih WM (2009) Self-assembly of DNA into nanoscale three-dimensional shapes. Nature 459:414–418 (c) Dietz H, Douglas SM, Shih WM (2009) Folding DNA into twisted and curved nanoscale shapes. Science 325:725–730 (d) Han D, Pal S, Nangreave J, Deng Z, Liu Y, Yan HJ (2011) DNA origami with complex curvatures in three-dimensional space. Science 332:342–346
2. (a) Perrault SD, Shih WM (2014) Virus-inspired membrane encapsulation of DNA nanostructures to achieve *in vivo* stability. ACS Nano 8(5):5132–5140 (b) Robert Schreiber R, Do J, Roller E, Zhang T, Schüller VJ, Nickels PC, Feldmann J, Liedl T (2014) Hierarchical assembly of metal nanoparticles, quantum dots and organic dyes using DNA origami scaffolds. Nat Nanotech 9:74–78
3. Vinther M, Kjems J (2016) Interfacing DNA nanodevices with biology: challenges, solutions and perspectives. New J Phys 18:085005
4. Bastings MMC, Anastassacos FM, Ponnuswamy N, Leifer FG, Cuneo G, Lin CX, Ingber DE, Ryu JH, Shih WM (2018) Modulation of the Cellular Uptake of DNA Origami through Control over Mass and Shape. Nano Lett 18(6):3557
5. Koga M, Comberlato A, Rodríguez-Franco HJ, Bastings MMC (2022) Strategic Insights into Engineering Parameters Affecting Cell Type-Specific Uptake of DNA-Based Nanomaterials. Biomacromolecules 23(6):2586–2594
6. Schwarz H, Schmittner M, Duschl A, Horejs-Hoeck J (2014) Residual endotoxin contaminations in recombinant proteins are sufficient to activate human CD1c+ dendritic cells. PLoS One 9(12):e113840
7. U.S. Department of Health and Human Services Food and Drug Administration,

Guidance for Industry Pyrogen and Endotoxins Testing (2012) https://www.fda.gov/regulatory-information/search-fda-guidance-documents/guidance-industry-pyrogen-and-endotoxins-testing-questions-and-answers#_ftn27
8. Ponnuswamy N, Bastings MMC, Nathwani B, Ryu JH, Chou LYT, Vinther M, Li WA, Anastassacos FM, Mooney DJ, Shih WM (2017) Oligolysine-based coating protects DNA nanostructures from low-salt denaturation and nuclease degradation. Nat Commun 8: 15654
9. Lacroix A, Vengut-Climent E, Rochembeau D, Sleiman H (2019) Uptake and fate of fluorescently labeled DNA nanostructures in cellular environments: a Cautionary Tale. ACS Cent Sci 5:882–891
10. Hahn J, Wickham SFJ, Shih WM, Perrault SD (2014) Addressing the instability of DNA nanostructures in tissue culture. ACS Nano 8(9):8765–8775
11. Bila H, Kurisinkal EE, Bastings MMC (2019) Engineering a stable future for DNA as biomaterial. Biomater Sci 7:532–541
12. Siemann DW, Keng PC (1986) Cell cycle specific toxicity of the Hoechst 33342 stain in untreated or irradiated murine tumor cells. Cancer Res 46(7):3556–3559
13. Hilderbrand SA, Weissleder R (2007) Optimized pH-responsive cyanine fluorochromes for detection of acidic environments. Chem Commun 26:2747–2749
14. Eder PS, DeVine RJ, Dagle JM, Walder JA (1991) Substrate specificity and kinetics of degradation of antisense oligonucleotides by a 3′ exonuclease in plasma. Antisense Res Dev 1(2): 141–151
15. Lee H, Lytton-Jean AKR, Chen Y, Love KT, Park AI, Karagiannis ED, Sehgal A, Querbes W, Zurenko CS, Jayaraman M, Peng CG, Charisse K, Borodovsky A, Manoharan M, Donahoe JS, Truelove J, Nahrendorf M, Langer R, Anderson DG (2012) Molecularly self-assembled nucleic acid nanoparticles for targeted in vivo siRNA delivery. Nat Nanotechnol 7:389
16. Li S, Jiang Q, Liu S, Zhang Y, Tian Y, Song C, Wang J, Zou Y, Anderson GJ, Han J-Y, Chang Y, Liu Y, Zhang C, Chen L, Zhou G, Nie G, Yan H, Ding B, Zhao Y (2018) A DNA nanorobot functions as a cancer therapeutic in response to a molecular trigger in vivo. Nat Biotechnol 36:258–264
17. Schaffert DH, Okholm AH, Sørensen RS, Nielsen JS, Tørring T, Rosen CB, Kodal ALB, Mortensen MR, Gothelf KV, Kjems J (2016) Intracellular delivery of a planar DNA origami structure by the transferrin-receptor internalization pathway. Small 19:2634–2640
18. Lin C, Perrault SD, Kwak M, Graf F, Shih WM (2013) Purification of DNA-origami nanostructures by rate-zonal centrifugation. Nucleic Acids Res e40
19. Uphoff CC, Drexler HG (2011) Detecting mycoplasma contamination in cell cultures by polymerase chain reaction. Methods Mol Biol 731:93–103
20. Okholm AH, Nielsen JS, Vinther M, Sørensen RS, Schaffert D, Kjems J (2014) Quantification of cellular uptake of DNA nanostructures by qPCR. Methods 67:193
21. Wang P, Rahman MA, Zhao Z, Weiss K, Zhang C, Chen Z, Hurwitz SJ, Chen ZG, Shin DM, Ke Y (2018) Visualization of the cellular uptake and trafficking of DNA origami nanostructures in cancer cells. JACS 140(7): 2478–2484

Chapter 14

Binding and Characterization of DNA Origami Nanostructures on Lipid Membranes

Alena Khmelinskaia, Petra Schwille, and Henri G. Franquelim

Abstract

DNA origami is an extremely versatile nanoengineering tool with widespread applicability in various fields of research, including membrane physiology and biophysics. The possibility to easily modify DNA strands with lipophilic moieties enabled the recent development of a variety of membrane-active DNA origami devices. Biological membranes, as the core barriers of the cells, display vital structural and functional roles. Therefore, lipid bilayers are widely popular targets of DNA origami nanotechnology for synthetic biology and biomedical applications. In this chapter, we summarize the typical experimental methods used to investigate the interaction of DNA origami with synthetic membrane models. Herein, we present detailed protocols for the production of lipid model membranes and characterization of membrane-targeted DNA origami nanostructures using different microscopy approaches.

Key words DNA origami, Membrane interactions, Lipid vesicles, Supported lipid membranes, Lipid monolayers, Fluorescence microscopy, Atomic force microscopy, Transmission electron microscopy, Fluorescence correlation spectroscopy, Translational diffusion

1 Introduction

Biological membranes are barriers that separate the inner from the outer environment of cells and organelles. The prime building blocks of biomembranes are polar lipids (e.g., phospholipids), amphiphiles with hydrophilic headgroups, and hydrophobic tails that primarily organize into lamellar phases, i.e., bilayers, in aqueous solutions. In addition to their selective permeability to solutes, lipid bilayers serve as templates for a variety of integral and peripheral proteins that participate in fundamental phenomena, such as signal transduction, intracellular trafficking, cell division, and energy conversion. Interestingly, cellular membranes do not simply act as passive interfaces. On the one hand, lipid bilayers are highly dynamic entities, which can be controllably shaped and remodeled [1–3]. On the other hand, biological membranes are extremely heterogeneous in their lipid composition: not only between

Julián Valero (ed.), *DNA and RNA Origami: Methods and Protocols*, Methods in Molecular Biology, vol. 2639,
https://doi.org/10.1007/978-1-0716-3028-0_14,

organisms, but also within a single cell [4–6]. Hence, based on the pivotal role of biomembranes as functional interfacial barriers, research on membrane proteins and membrane-related processes are of particular biological significance and interest, with a wide range of applications in medicine and nanotechnology. Due to the complexity of cells, synthetic membrane models and in vitro reconstitution approaches revealed to be extremely attractive methods to investigate, in a minimalistic way, the interactions of proteins (and other biomolecules) with membranes [7–9].

Over the years, the usage of DNA nanostructures in the fields of synthetic biology and therapeutics has been increasingly popular [10–13]. The tight control over shape [14, 15] and biochemical functionality [16] simultaneously offered by DNA origami makes this material so unique and ideal for biomedical and biophysical applications. As lipid membranes are the first cellular barriers any external molecule will face, understanding how functional devices made of DNA origami can interact with these biological interfaces is of utmost significance.

Recently, various membrane-active DNA origami structures have been created and investigated using biophysical approaches [10, 17–21]. Hereof, seminal studies granted us a better overview on the minimal and underlying requirements for efficiently targeting DNA origami nanostructures to membranes. It was, for instance, demonstrated that the attachment of DNA origami to membranes can be primarily modulated by electrostatics and by the number, position, and accessibility of the chosen anchors [22–25] (Fig. 1a). All-in-all, a plethora of membrane-active DNA

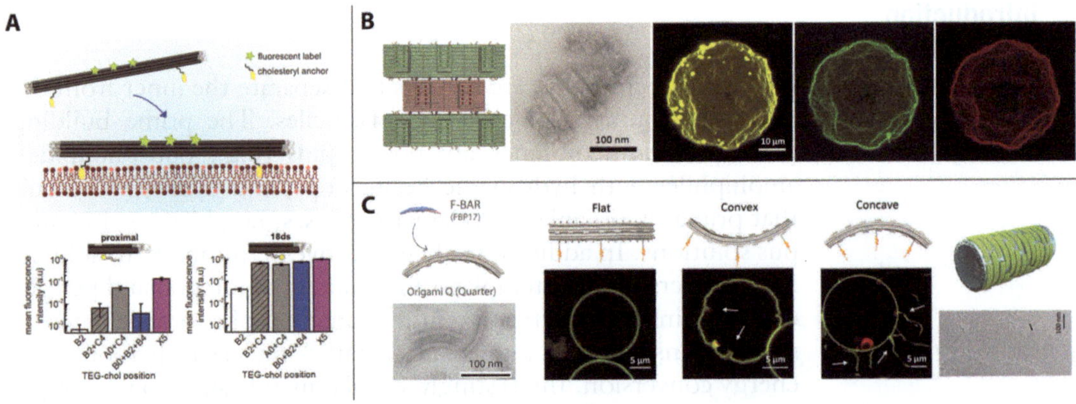

Fig. 1 Representative studies of DNA origami interacting with model lipid membranes. (**a**) Position, number, and accessibility of cholesterol anchors modulate binding of DNA origami to lipid membranes (adapted with permission from [23]); (**b**) DNA origami with lateral overhangs assemble into ordered lattices, leading to the macroscopic deformation of GUVs. (Adapted with permission from Ref. [33]). (**c**) Curved DNA origami can deform lipid membranes and type of deformation will depend on structure curvature. (Adapted with permission from Ref. [34])

origami structures have been successfully developed throughout the last decade, from nanoscale membrane channels [26–28] to membrane shaping coats [29–34] (Fig. 1b, c).

In order to attach DNA origami objects to lipid membranes, desired DNA strands are typically functionalized with membrane-anchoring moieties of hydrophobic nature. Cholesteryl-modified DNA oligonucleotides with tetraethylene glycol spacers (TEG-chol) are particularly attractive [10, 21], as these have been specially well characterized [24, 22, 23]; show a strong, mostly lipid composition independent, binding to membranes; and can be found in the catalogue of most DNA synthesizing companies. TEG-chol functionalized DNA staples have been implemented in distinct manners: these are either directly incorporated into the DNA origami structures during folding [22, 23, 34] (Fig. 2b, c) or can be added to the model lipid membranes for subsequent binding to complementary DNA handles on the DNA origami nanostructures [35, 34] (Fig. 2a). Other possible modifications of hydrophobic nature recurrently used to bind DNA to membranes are porphyrins [36] and tocopherol [26]. While the incorporation of multiple hydrophobic groups into DNA origami will guarantee a strong membrane insertion, increasing the total hydrophobicity will also give rise to additional solubility complications [37], which may require the addition of surfactants [38, 30] or single-stranded overhangs [39] in order to keep the functionalized DNA origami from forming aggregates. Usage of membrane-anchoring moieties with lower hydrophobicity, such as amphipathic peptides [40, 41], is an option to bypass such solubility issues. Alternatively, DNA origami objects can be further anchored to membranes via non-lipophilic moieties using, e.g., ligand-receptor types of binding. Herein, biotin-streptavidin-mediated interactions are extremely popular [34, 26] and of easy implementation (Fig. 3), as both biotinylated DNA oligonucleotides and lipids are commercially available. Direct covalent linkage of DNA staples to the headgroup of phospholipids is another viable option [30–32], ensuring a strong and permanent attachment to the membrane.

While all the abovementioned strategies assure an orientation-specific anchoring of DNA objects to membranes (important for membrane shaping applications, as depicted in Fig. 1c), binding of DNA nanostructures to lipid membranes may also happen unspecifically and in the absence of special functional moieties. Indeed, by taking advantage of the intrinsic properties of DNA as a highly negatively charged polymer, electrostatics can be used to promote the attachment of DNA origami to zwitterionic or cationic lipids. The inclusion of positively charged lipids, such as 1,2-dioleoyl-3-trimethylammonium-propane (DOTAP), will drastically increase the affinity of DNA origami to membranes [42]. Similarly, the amount of salt and valency of the cations in solution may greatly influence whether DNA origami is able to bind (or not)

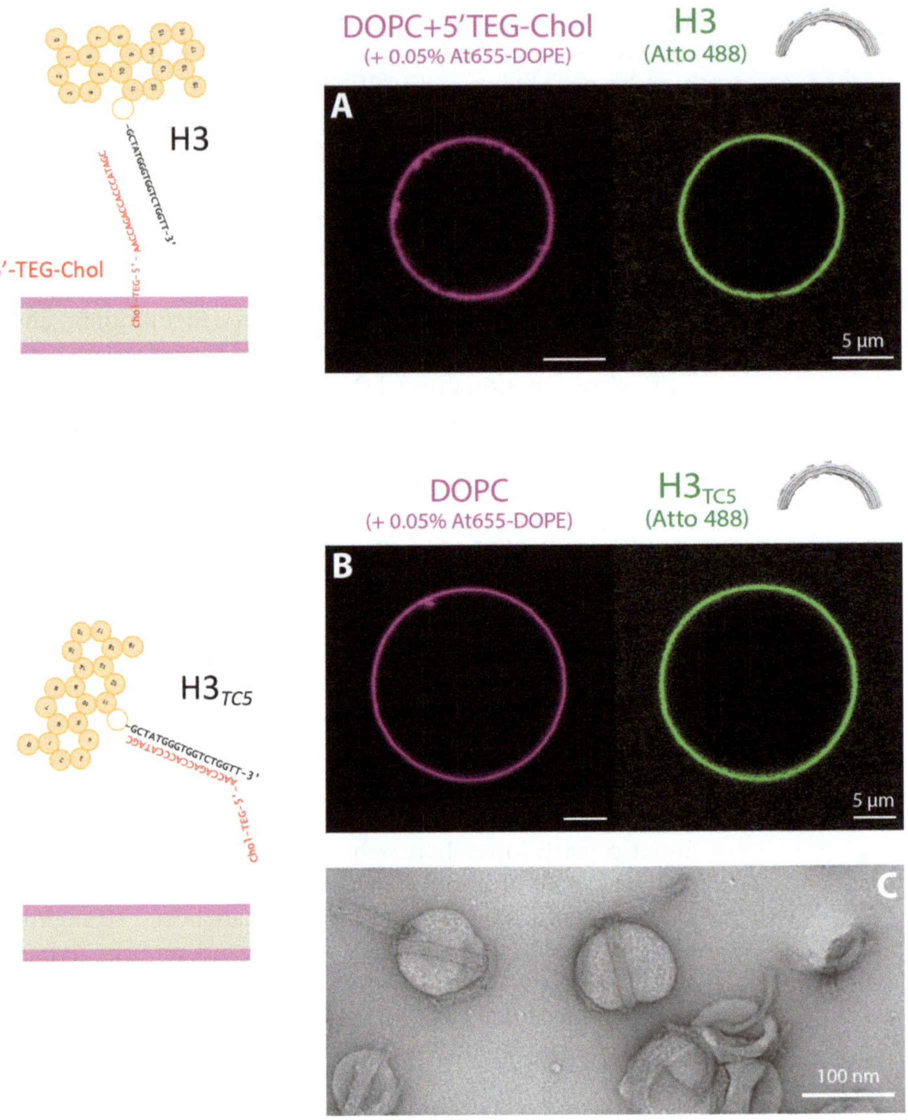

Fig. 2 Membrane binding of DNA origami nanostructure through TEG-chol modifications. (**a**) Confocal fluorescence image of DNA nanostructures, displaying single-stranded overhangs, binding to GUVs pre-incubated with a cholesteryl-modified complementary DNA strand (2 μM). Cholesteryl-modified DNA nanostructures directly binding to model lipid membranes, as observed by confocal fluorescence microscopy to GUVs (**b**) and by TEM to LUVs (**c**)

phospholipid membranes [43, 44]. While the attachment of DNA origami nanostructures to zwitterionic lipids (such as 1,2-dioleoyl-*sn*-glycero-3-phosphocholine; DOPC) can be promoted by divalent cations, like Mg^{2+}, that may act as a bridge between the negatively charged DNA and lipid headgroups (Fig. 4), monovalent cations, like Na^{+}, can easily screen such interactions.

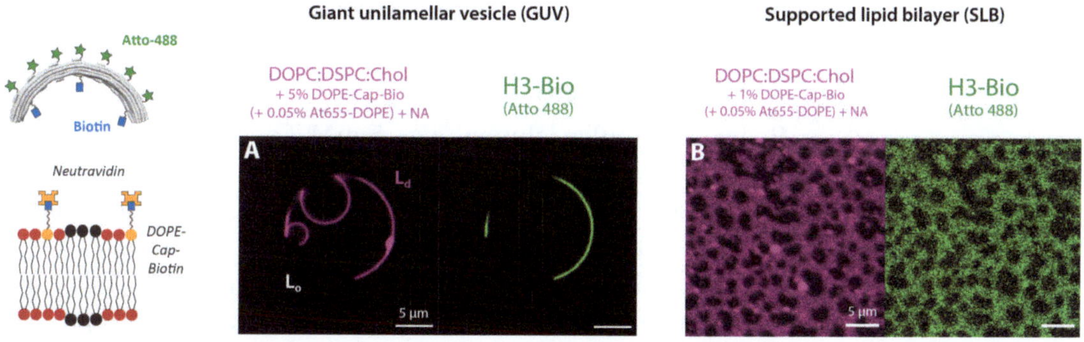

Fig. 3 Binding of DNA nanostructures to phase-separated model systems mediated by biotin-neutravidin interactions. Confocal fluorescence microscopy images of biotinylated curved DNA nanostructures binding to the liquid-disordered (L_d) phase of DOPC:DSPC:Chol (2:2:1 molar ratio) phase-separated GUVs (**a**) and SLBs (**b**) doped with biotinylated lipids (1–5%mol) and pre-incubated with neutravidin (1 μM)

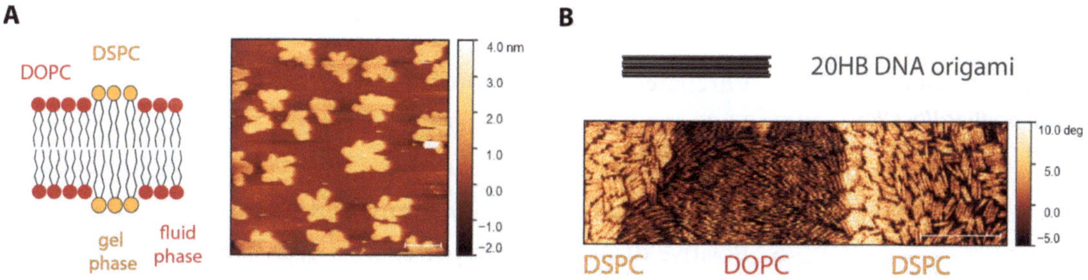

Fig. 4 Mg^{2+}-mediated binding of DNA origami nanostructures to phase-separated lipid bilayers deposited on mica. AFM images of a DOPC:DPSC (1:1) lipid mixture before (**a**) and after (**b**) incubation with a DNA 20-helix bundle in presence of 20 mM $MgCl_2$

In this context, the aim of this chapter is to describe step-by-step typical experimental methods used to investigate and characterize the interaction of DNA origami with synthetic membrane models.

2 Materials

2.1 Preparation of Giant Unilamellar Vesicles (GUV)

1. Lipids in chloroform or lyophilized (*see* **Note 1**), typically purchased from Avanti Polar Lipids, Alabaster, AL, USA.
2. Fluorescently labeled lipids (*see* **Note 2**).
3. Positive displacement pipettes with glass capillaries (Transferpettor, Brand, Wertheim, Germany).
4. Spectroscopic-grade chloroform.
5. Home-made polytetrafluoroethylene PTFE chambers with Pt wires for electroformation (for more details, see [45]).
6. If required, electric heating plate and home-made heating block for PTFE chambers.

7. Function generator, with connecting clamps, cables, and multimeter.
8. Micro-osmometer.
9. Imaging buffer solution (e.g., 5 mM Tris-HCl, 1 mM EDTA, 5 mM $MgCl_2$, 300 mM NaCl, pH 8.0).
10. Sucrose solution (matching the osmolarity of the buffer solution).
11. Passivation solution: bovine serum albumin, BSA (1–2 mg/mL aqueous solution) or PLL-PEG (e.g. PLL(20)-g[3.5]-PEG(2) from SuSoS AG, Dübendorf, Switzerland; 0.5 mg/mL aqueous solution).
12. Observation chambers with glass bottom (#1.5 thickness), e.g., SensoPlate multiwell plates (Greiner Bio-One, Kremsmünster, Austria).
13. Sealing film for well-plate chambers.

2.2 Preparation of Multilamellar Vesicles (MLV)

1. Purchased lipids in chloroform or lyophilized (*see* **Note 1**).
2. Fluorescently labeled lipids (*see* **Note 2**).
3. Glass vials or round flasks.
4. Spectroscopy-grade chloroform (and methanol, if required).
5. Positive displacement pipettes with glass capillaries.
6. N_2 stream (or bottled pressurized N_2).
7. Desiccator with vacuum pump.
8. Desired buffer solution (e.g., SLB buffer: 10 mM HEPES, 150 mM NaCl, pH 7.4).
9. Vortex mixer.
10. Electric heating plate.

2.3 Preparation of Large Unilamellar Vesicles (LUV)

1. MLV suspension (see Subheading 3.2).
2. Dewar with liquid N_2.
3. Beaker with ddH_2O.
4. Electric heating plate.
5. Desired buffer solution (e.g., SLB or origami buffer).
6. Lipid vesicle extruder (e.g., Avestin LiposoFast Basic or Avanti Mini Extruder) and respective accessories.
7. Polycarbonate track-etch membrane filters with the desired pore size (e.g., 50–100 nm).

2.4 Preparation of Supported Lipid Bilayers (SLB)

1. MLV suspension (see Subheading 3.2).
2. Glass vials.
3. Bath ultrasonicator.
4. Clamp stand.

5. Desired buffer solution (e.g., SLB or origami buffer).
6. 0.1 M aqueous solution of $CaCl_2$.
7. Electric heating plate.

2.4.1 Specific for Deposition on a Glass Substrate

1. Plasma cleaner.
2. High precision #1.5 thickness coverslips.

2.4.2 Specific for Deposition on Top of Freshly Cleaved Mica

1. Muscovite mica sheets.
2. Adhesive tape.
3. High precision #1 thickness coverslips.
4. UV Lamp.
5. UV-cured optical adhesive (e.g., #68 from Norland, Cranbury, NJ, USA).

2.5 Preparation of Lipid Monolayers

1. Purchased lipids in chloroform or lyophilized (*see* **Note 1**).
2. Fluorescently-labelled lipids (*see* **Note 2**).
3. High-precision scale for gravimetry.
4. Home-made PTFE chambers with a central aperture of 15 mm (for more details, see [46–48]).
5. Bath ultrasonicator.
6. High precision #1.5 thickness coverslips of approximately the size of the chamber.
7. Plasma cleaner.
8. Inert glue (e.g., twinsil 22, two component glue, picodent, Wipperfürth, Germany).
9. Imaging buffer (e.g., SLB buffer).
10. #1 thickness coverslips and grease.
11. Temperature-controlled chambers (*see* **Note 3**).

2.6 DNA Origami Folding and Purification

1. Commercial thermal cycler.
2. Single-stranded scaffold plasmid (e.g., M13 p7249).
3. Purchased DNA staple oligonucleotides for scaffold folding.
4. Functionalized DNA oligonucleotides for labeling and membrane binding as well as respective DNA handles, depending on the chosen binding strategy (*see* **Note 4**).
5. 10× origami buffer (e.g., 50 mM Tris-HCl, 10 mM EDTA, 200 mM $MgCl_2$).
6. Benchtop centrifuge.
7. Amicon Ultra 100 kDa MWCO filters.
8. Imaging buffer solution.

2.7 Fluorescence Confocal Imaging and Correlation Spectroscopy (FCS)

1. Commercial confocal laser scanning microscope equipped with a water immersion objective with numerical aperture >0.9 and correction ring; lasers of suitable wavelength; beamsplitter to separate the excitation from the emission; carefully chosen band-path filters, taking into account the used fluorophores; fiber-coupled avalanche photodiodes (APDs) for the detection; and a hardware correlator, for FCS measurements.

   Typically, we use an LSM 780 with a ConfoCor3 unit with C-Apochromat, 40x/1.2 W UV–VIS–IR objective (Zeiss, Jena, Germany). For different lipid model systems, small modifications are made to the setup to guarantee optimal acquisition settings (*see* **Note 3**).
2. High precision #1.5 thickness glass coverslips or chambers.
3. Nonfluorescent spectral grade immersion water.
4. Solutions of fluorophores (e.g., Alexa 488 or Atto 655) with known diffusion coefficient (10–100 nM in water) for FCS calibration measurements.
5. Analysis software (*see* **Note 5**).

2.8 Negative-Stain Transmission Electron Microscopy (TEM) Imaging

1. Commercial transmission electron microscope (TEM) equipped with a Lab6 filament and a camera.
2. Glow discharger.
3. Formvar-supported carbon coated copper grids, 400 mesh.
4. Negative action tweezers for grid manipulation.
5. 2% aqueous solution of uranyl formate with 25 mM NaOH.
6. Analysis software (*see* **Note 5**).

2.9 Atomic Force Microscopy (AFM) Imaging

1. Commercial atomic force microscope (AFM). We typically use a standard Nanowizard 3 or fast-scanning Nanowizard Ultra from JPK (Berlin, Germany).
2. Active vibration isolation table.
3. Soft cantilevers for imaging in aqueous solution. For applications involving DNA origami and supported lipid membranes, we typically use Ultra-Short Cantilevers USC-F0.3-k0.3 from Nanoworld (Neuchâtel, Switzerland) or BioLever mini BL-AC40-TS from Olympus (Tokyo, Japan).
4. Analysis software (*see* **Note 5**).

3 Methods

3.1 Preparation of GUVs by Electroformation (See Note 6)

1. Prepare the desired lipid mixture in chloroform (or chloroform/methanol 7:3, for certain lipid compositions, see **Note 1**) in a glass vial, typically at 2 mg/mL concentration. The stock solution can be stored at −20 °C for several months.

2. Check the osmolarity of the desired imaging buffer using, e.g., a freezing-point osmometer (for standard imaging buffer, the osmolarity is ~580 mOsm/kg).
3. Prepare a sucrose solution in Milli-Q ddH_2O water, iso-osmolar to the imaging buffer (ideally below 5% variation). The sucrose solution can be stored refrigerated for a week, but larger stocks should be stored in the −20 °C. The osmolarity of the sucrose solutions should be checked and adjusted prior usage (*see* **Note 7**).
4. Clean the Pt-wired PTFE chamber sequentially with chloroform and ethanol (bath sonication can be used). Blow-dry chamber with pressurized air or N_2.
5. Using solvent-resistant micropipette tips (or Hamilton syringe), homogeneously spread 3 μL of lipid stock solution on each Pt wire.
6. To completely evaporate solvent remnants, place the Pt-wired chamber under vacuum in a desiccator for 10–30 min.
7. Fill the electroformation chamber with 350 μL of sucrose solution and screw the cap with the Pt wires.
8. Use a multimeter to check the voltage of the function generator. Connect the cables from the function generator to the Pt wires, and apply a sine wave AC electric field of 2 V (RMS) at a frequency of 10 Hz for ~1–1.5 h, followed by 2 Hz for ~30–45 min. Electroformation should be performed above the melting temperature (T_m) of the lipids (*see* **Note 1**), on top of a heating block if necessary.
9. Disconnect the chamber from the function generator and allow PTFE chamber to slowly cool down to room temperature. The GUV suspension is now ready.

3.2 Preparation of MLVs

1. Prepare desired lipid mixture in chloroform (or chloroform/methanol 7:3, for certain lipid compositions, see **Note 1**) in a glass vial (or round flask), typically to a final lipid concentration of 10 mM.
2. Produce an even lipid film in the vial by drying the chloroform under a stream of N_2. To do so, gently rotate and tilt the vial in order to have a homogeneously distributed lipid film on the flask walls.
3. Dry remnants of chloroform under constant N_2 flow for at least 20 min, and then vacuum-dry in the desiccator, ideally for several hours (minimum 30–60 min).
4. Rehydrate lipid film by adding desired volume of buffer (typically SLB buffer).

5. Shake vial and vortex vigorously for several minutes, in order to completely solubilize the lipid film from the walls and produce a suspension of multilamellar vesicles (MLVs). Before this process, assure that the used buffer is above the T_m of the used lipid mixture (*see* **Note 1**).
6. For long storage (several months), aliquot the obtained MLVs into Eppendorf tubes and store them at −20 °C.

3.3 Preparation of LUVs by Extrusion

1. Transfer the previously prepared MLV suspension (see Subheading 3.2) into a glass vial or round flask. At this initial step, the MLV suspension should be opaque.
2. Perform freeze-thawing cycles to disrupt the multi-lamellarity of the vesicles and facilitate distribution of solutes. To do so, plunge the vial/flask, with the help of tweezers, between a small dewar containing liquid N_2 (to freeze sample) and a heated beaker of ddH_2O on a hot plate (to thaw sample) in a sequential manner. Repeat freeze-thawing cycles ~8 times. At the end of this step, the MLV suspension should appear clearer.
3. Assemble clean syringe-based extruder according to the manufacturer's instructions using polycarbonate track-etch membrane filters with the desired pore size (30–400 nm). For common applications, we use a membrane filter with a pore size of 100 nm.
4. Pre-rinse the extruder with the desired buffer to check for possible leakages and reduce loss of lipid vesicles.
5. Extrude the vesicles through the polycarbonate filter. Typically, we perform 21 passes through the filter, in order to ensure size monodispersity; however, a larger number of passes with a heated extruding system may be necessary for lipid mixture of high T_m (*see* **Note 1**). The number of passes should be odd in order to reduce the contamination with larger particles (i.e., do not take the final sample from the starting syringe).
6. The initially opaque lipid suspension (MLVs) should be clear after successful extrusion and production of LUVs. Extruded LUVs should ideally be used fresh, but can be kept at 4 °C for 2–3 days without significantly compromising monodispersity and lipid quality.

3.4 Preparation of SLBs by Vesicle Fusion

3.4.1 Prepare Small Unilamellar vesicles (SUVs) from MLVs (See Subheading 3.2)

1. Dilute previously prepared MLVs to the desired concentration. Typically, we dilute the MLV aliquots 1:10, to a final volume of 200 μL at a lipid concentration of 1 mM.
2. Transfer diluted MLV suspension into a glass vial. Immerse the bottom of the vial in the ultrasonic bath, and attach it at the required height using a clamp stand.
3. Sonicate the MLV suspension for 10–20 min (*see* **Note 8**).

4. The initially opaque lipid suspension (MLVs) should look clear after successful sonication and production of SUVs. SUVs should be used fresh, and extended storage is not recommended for SLB deposition.

3.4.2 Prepare Hydrophilic Substrate of Choice (See ***Note 9****)*

1. *For glass coverslip*: Pre-clean #1.5 thickness glass coverslips (e.g., using EtOH, standard glass-etching protocol [49], etc.). Subsequently, plasma-clean coverslips (or imaging chambers – *see* **Note 10**) prior to utilization. Mount chamber.
2. *For freshly cleaved mica*: Punch muscovite mica sheets into small discs (ø = 8–12 mm). Cleave mica disc using adhesive tape. Remove the resulting thin mica layer from the tape using tweezers and glue it (cleaved face up) on top of #1 thickness glass coverslip using a small drop of UV-curing optical adhesive. Let optical adhesive cure for 5–10 min under a UV lamp (365 nm). Mount chamber on the coverslip with freshly cleaved mica.

3.4.3 Form SLB Via Deposition of SUVs on Top of Hydrophilic Substrate

1. Deposit 100 μL of SUVs (1 mM lipid concentration) on top of substrate.
2. Add 2 μL of $CaCl_2$ (0.1 M stock); *see* **Note 1**.
3. Fill chamber with additional buffer to avoid evaporation (~200 μL buffer).
4. Incubate for 10–20 min above T_m of lipid mixture (*see* **Note 1**).
5. Wash the formed lipid bilayer 10× with buffer to remove excess of unfused vesicles without drying the SLB. If the sample was heated, let it slowly reach room temperature before usage.
6. The sample is now ready.

3.5 Preparation of Lipid Monolayers

1. Prepare desired lipid mixture in chloroform (or chloroform/methanol 7:3, for certain lipid compositions, *see* **Note 1**) in a glass vial, typically at a concentration of 0.1 mg/mL and determine concentration by gravimetry.
2. Sequentially clean PTFE spacer by sonication in acetone, chloroform, isopropanol, and ethanol.
3. Glue PTFE spacer to #1.5 thickness glass coverslip using inert glue and plasma-clean the assembled chamber.
4. Fill the chamber with buffer (~200 μL).
5. Deposit lipid mixture drop-by-drop at the air-water interface to the desired mean molecular area (MMA) (*see* **Note 11**). The lipid monolayer will form spontaneously upon chloroform evaporation.
6. The sample is ready for further manipulation and should be used right away, as the large surface area-volume ratio results in fast evaporation when the chamber is not sealed.

3.6 Folding, Purification, and Quantification of DNA Origami Nanostructures

3.6.1 DNA Origami Folding

Typically, we fold our DNA origami structures in a one-pot reaction using the protocol described in [14]. Briefly:

1. 20 nM p7249 plasmid and 200 nM staple oligonucleotides are mixed in 1× origami buffer.
2. Thermal annealing is performed over a cooling cycle scheme from 65 to 60 °C in 1 h and from 59 to 40 °C in 40 h on a thermal cycler.

3.6.2 DNA Origami Purification

Typically, we purify the folded DNA origami structures from the excess of staple strands using size-exclusion centrifugal filtration with Amicon Ultra 100 kDa MWCO filters. Herein, we use the standard manufacturer's protocol, with minor modifications:

1. Assemble Amicon filters and tubes according to the manufacturer's instructions.
2. Add the desired volume of folded DNA origami solution and fill the remaining volume (to 500 μL) with the desired experimental buffer (e.g., imaging buffer, see Subheading 2.1).
3. Centrifuge at 14000 *g* for 3 min.
4. Discard the flow-through and add 450 μL buffer into the Amicon filter.
5. Repeat **steps 3** and **4** two more times.
6. Centrifuge at 14000 *g* for 5 min.
7. Place the Amicon filter inverted into a clean tube.
8. Centrifuge at 1000 *g* for 2 min to collect the purified DNA origami.
9. Estimate the total concentration of DNA origami by determining its absorbance at 260 nm or putatively by measuring its fluorescence intensity, in case the structure is labeled with a known fluorescent dye (*see* **Note 14**).

3.7 Characterization of DNA Origami Binding to GUVs under Confocal Microscopy

1. Plasma-clean #1.5 glass coverslip/observation chamber (e.g., SensoPlate 384-multiwell plates with glass coverslip bottom from Greiner Bio-One, Kremsmünster, Austria).
2. Fill the observation chamber with the desired blocking agent for surface passivation. Incubate for 30 min and rinse the observation chamber with excess Milli-Q ddH_2O (~4×) and imaging buffer (~4×).
3. Pre-dilute (e.g., 1:10 or 1:50) the formed GUVs (see Subheading 3.1) into an Eppendorf tube using sucrose solution, if needed.
4. Add the imaging solution, with fluorescently labeled DNA origami at the desired concentration (e.g., 0.5–5 nM), into the observation chamber. Typically, we prepare an extra sample without DNA origami nanostructure to inspect the quality of the formed GUVs.

5. Cut the end of a micropipette tip (to widen the opening and reduce shear stress). Gently transfer the GUV suspension (e.g., 1:5–1:10 v/v) into the observation chamber, by slowly adding the suspension from the top.
6. Let the sample equilibrate for several minutes, in order for the GUVs to settle on the chamber surface. Cover the observation chamber accordingly to avoid evaporation (e.g., for SensoPlate, we use glueable DNase-free sealing films).
7. In principle, the sample is now ready for imaging. Nevertheless, depending on the selected strategy to bind DNA origami to membranes, we recommend a minimal 30 min incubation time (at room temperature), or even a maximal overnight incubation (at 4 °C), to guarantee a homogeneous coverage of the GUVs by DNA origami nanostructures.
8. Transfer the observation chamber to an inverted fluorescence confocal microscope. Choose an appropriate objective and light path, and acquire fluorescence images. Typically, we acquire images at the equatorial plane of individual GUVs (e.g., Figs. 1, 2, 3 and 5).
9. All additional sample manipulations, such as the addition of new components (e.g., DNA origami, functional DNA staples, glucose, $MgCl_2$, or any other desired buffer excipients) to the

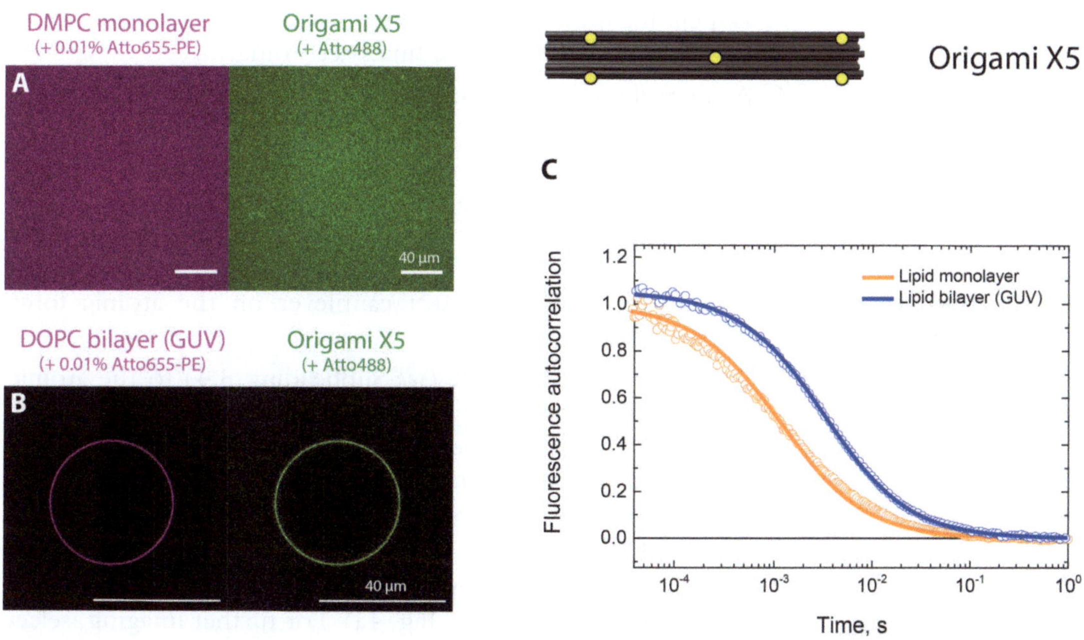

Fig. 5 Diffusion of DNA origami on lipid monolayer vs. lipid bilayer. Confocal fluorescence microscopy images of DNA origami binding to DMPC lipid monolayers (**a**) and DOPC GUVs (**b**). Comparison of the correlation curves obtained by FCS in each model membrane (**c**): diffusion in lipid monolayers is significantly faster in comparison with GUVs, due to its lower lipid packing. (Adapted with permission from Refs. [22, 46])

observation chamber containing GUVs, need to be done with a cut-end micropipette tip from the top, in order to reduce GUV perturbation, shear stress, and movement.

10. Quantitative comparison of membrane binding can be done by determining fluorescence intensity of DNA origami at the equatorial plane of GUVs using a previously home-developed script [50]. Through a titration of DNA origami nanostructure concentration, the membrane affinity K_D can be further determined.

3.8 Characterization of DNA Origami Binding to SLBs under Confocal Microscopy

1. Deposit SLBs in appropriate chambers according to Subheading 3.4 and cover the observation chamber to avoid evaporation.
2. The sample is transferred to an inverted fluorescence microscope as above (Subheading 3.7).
3. Check the quality of the formed lipid bilayers, by inspecting for defects, holes, and unfused vesicles by fluorescence imaging. Infer whether a continuous membrane was formed, by assessing the fluorescence recovery after photobleaching on a small area of the membrane. In case no recovery is observed, do not use the SLB.
4. Add desired amount of fluorescently labeled DNA origami (e.g., 0.5–5 nM) and pipette the solution up and down.
5. Depending on the selected strategy to anchor DNA origami to the SLB, adjust incubation times accordingly.
6. Acquire fluorescence images of DNA origami nanostructures on SLBs (e.g., Fig. 3).
7. Binding to the lipid membrane can be compared between structures by determining DNA origami fluorescence at the lipid bilayer (*see* **Note 5**).

3.9 Characterization of DNA Origami Binding to SLBs under AFM

1. Mount and calibrate a soft cantilever on the atomic force microscope.
2. Transfer the SLB sample (*see* Subheading 3.4) to the atomic force microscope. We typically perform imaging in aqueous solution under AC mode (with a selected frequency near resonance and oscillation amplitude below 10 nm), by continuously adjusting the setpoint, and gain to minimize the force applied on the sample.
3. Start by imaging an empty SLB, to check the quality of the formed lipid bilayer (e.g., Fig. 4a). For further imaging, select large regions without major defects, holes, and unfused vesicles. However, membrane holes can be used to determine the SLB height (~4–6 nm, depending on lipid mixture). When

combined with fluorescence microscopy, the initial quality check of the SLB can be performed using fluorescence instead (see Subheading 3.8).

4. Add the desired amount of DNA origami (e.g., 0.5–1 nM). Depending on the selected strategy to anchor DNA origami to the SLB, adjust incubation times accordingly. For optimal imaging of DNA origami on SLBs, a strong membrane attachment and slow DNA origami diffusion are required. Hence, we recommend the usage of high $MgCl_2$ concentrations (e.g., 20 mM) – Fig. 4.

3.10 Characterization of DNA Origami Binding to Lipid Monolayers under Confocal Microscopy

1. Cover the prepared lipid monolayer (see Subheading 3.5) with a glass coverslip and transfer to the inverted fluorescence microscope. If lipid monolayer is out of focus reach, pierce the monolayer with a pipette and remove a small amount of buffer to ensure that the interface is within the working distance of the objective.
2. Visually verify the quality of the formed monolayer according to the chosen lipid mixture (e.g., phase separated vs. homogeneous) and the respective MMA, taking into account its Langmuir isotherm.
3. Inject the desired amount of DNA origami into the buffer and incubate according to the used membrane anchor. After incubation, remove the same amount of buffer to guarantee correct positioning of the lipid monolayer relative to focus.
4. Seal the chamber with a greased coverslip and transfer to temperature chamber (for details, see [46]).
5. Proceed with fluorescence imaging (e.g., Fig. 5).
6. As for GUVs and SLBs (see Subheadings 3.7 and 3.8, respectively), membrane binding can be quantified considering the fluorescence intensity at the lipid monolayer (*see* **Note 2**).

3.11 Characterization of DNA Origami Binding to LUVs by Negative-Stain TEM

1. Incubate previously prepared LUVs (1:10 dilution) with DNA origami nanostructures of interest (e.g., at 0.5–1 nM) in an Eppendorf tube. Adjust incubating times according to the chosen membrane binding anchor.
2. Plasma-clean Formvar grids and prepare samples for negative-stain TEM. Typically, we follow the protocol described in [14]. To that end, first incubate 3 μL of diluted sample (1:10 into buffer of preference) on a Formvar grid (1–2 min). During this time, deposit on a clean parafilm sheet two 6 μL droplets of 2% uranyl formate aqueous solution (w/v) and 20 μL water.
3. Blot the grid with filter paper to remove excess liquid.
4. Quickly wash the grid with the first droplet of uranyl formate solution by simply approaching the grid to the droplet. If the

grid has been successfully plasma cleaned, the droplet will promptly transfer to the grid. Immediately blot the grid with filter paper to remove excess liquid.

5. Repeat the procedure above for grid staining by incubating the sample with uranyl formate solution for 30 s before removing excess liquid.
6. Use the deposited water droplet for washing the grid from excess uranyl formate, and let the grid dry before transferring to transmission electron microscope.
7. Samples are ready to be imaged with standard TEM procedures (e.g., Fig. 2c).

3.12 Characterization of Diffusion of DNA Origami on Membranes Using FCS

1. Switch on the lasers roughly 30 min prior to the measurement session to stabilize the system.
2. The fluorescence microscope should be calibrated on a daily basis with a dye solution of known diffusion coefficient (e.g., Alexa 488 or Atto 655 [51, 52]) with similar spectral properties to the dye's used for DNA origami labeling. For this, use a sample of 10–100 nM dye solution on a glass coverslip/chamber similar to the one later used in the experiment. Use the same optical settings as the ones used for data collection (*see* **Note 13**).
3. Transfer the calibration sample onto the objective and place the focal volume inside the sample, typically at least 15 μm above the surface of the coverslip.
4. Adjust the pinhole diameter to one Airy unit and align it in the x and y directions to maximize the fluorescence count rate on the detector and the counts per molecule (molecular brightness).
5. Adjust the correction ring of the objective in order to maximize the count rate per molecule and correct for the coverslip thickness.
6. Record FCS curves at sufficiently low irradiance (1–10 μW) in order to avoid triplet saturation and photobleaching [53–55]. The recording time should be long enough so that the noise level does not exceed 5–10% of the autocorrelation curve amplitude at short lag times.
7. The recorded correlation curves can be fitted by [55]

$$G(\tau) = N^{-1}\left(1 + \frac{T}{1-T}e^{-\tau/\tau_T}\right)\left(1 + \tau/\tau_D\right)^{-1}\left(1 + \tau/\left(S^2\tau_D\right)\right)^{-1/2} \tag{1}$$

Here, $\tau_D = w_0^2/4D$ is the diffusion time, which depends on the lateral e^{-2}-value of the confocal detection volume, S is the structure parameter which represents the ratio of axial/lateral extent of the detection volume, T is the triplet fraction, and τ_T is the triplet decay time. For proper setup calibration, the diffusion coefficient of the used dye should be corrected to the measurement temperature (i.e., temperature at the objective) according to $D(T)\ \alpha\ T/\eta(T)$ [56–59].

8. Prepare sample as for confocal microscopy and place it under the inverted fluorescence microscope (see Subheadings 3.7, 3.8 and 3.10). Verify the quality of the lipid model membrane as described above, and incubate with the DNA origami nanostructure of interest as previously described, however at a dilute regime (e.g., 50–500 pM) (*see* **Note 14**).
9. To perform point FCS measurements, first place the sample on the microscope and position the respective membrane in the middle of the confocal volume, i.e., upper pole of a GUV or center of a monolayer (to reduce the meniscus effect) or SLB containing chamber.
10. Focus the laser beam onto the membrane plane by finding the highest fluorescence intensity in z through a z-scan.
11. As for calibration, use low laser power to avoid photobleaching and high triplet fractions.
12. Several FCS measurements (10–20×) of 30–100 s should be performed. If drifting of the fluorescence signal is observed, stop acquisition, refocus the laser beam onto the membrane as in b), and proceed with the measurements.
13. Diffusion time τ_D and the number of particles in the confocal volume N can be determined by fitting a 2D diffusion model to the averaged autocorrelation curve, by setting $S = \infty$ (Fig. 5):

$$G(\tau) = N^{-1}\left(1 + \frac{T}{1-T} e^{-\tau/\tau_T}\right)\left(1 + \tau/\tau_D\right)^{-1} \tag{2}$$

14. The diffusion coefficient of DNA origami can then be calculated through

$$D_{\text{DNAorig.}} = D_{\text{cal.dye}} \frac{\tau D_{\text{DNAorig.}}}{\tau D_{\text{cal.dye}}} \tag{3}$$

4 Notes

1. Lipid handling and lipid model system particularities:

   For homogenous fluid lipid membranes, DOPC or 1-palmitoyl-2-oleoyl-*sn*-glycero-3-phosphocholine (POPC) lipids are typically used. For fluid-gel phase separation, we typically use DOPC/1,2-distearoyl-*sn*-glycero-3-phosphocholine

(DSPC), 1:1 molar ratio. For fluid liquid disordered (L_d)-liquid ordered (L_o) systems, DOPC/sphingomyelim (SM)/cholesterol (Chol), 2:2:1 molar ratio, or DOPC/DSPC/Chol, 2:2:1 molar ratio, mixtures are typically used.

All lipid model systems containing saturated lipids, including phase-separated lipid bilayers, should be prepared above T_m and slowly brought to room temperature after preparation.

Certain sphingolipids and charged phospholipids may not fully solubilize in chloroform. In those cases, stocks should be prepared in chloroform/methanol, 7:3 (volume ratio).

Chloroform lipid stock solutions should be stored at −20 °C in a glass vial filled with argon gas and sealed with PTFE tape, to avoid lipid oxidation, hydrolysis, and contamination by solubilized plastic contaminants. Prepared LUVs and SUVs can be stored at −20 °C for longer periods of time if necessary, but we advise to check vesicle size/polydispersity by dynamic light scattering (DLS) or to repeat extrusion/sonication prior to use in further experimental approaches.

SLBs containing positively charged lipids do not require the addition of $MgCl_2$ solution for deposition.

For lipid monolayers, fully saturated lipids such as 1,2-dimyristoyl-*sn*-glycero-3-phosphocholine (DMPC) are preferred, to avoid lipid oxidation at the air-buffer interface.

2. Typical membrane dyes are fluorescent lipid analogs, e.g., DiD, DiO, DiI, or fluorescent lipids, like BODIPY-cholesterol, Alexa Fluor 488-functionalized 1,2-dioleoyl-*sn*-glycero-3-phosphoethanolamine (Alexa Fluor 488-DOPE) or Atto 655-DOPE. The use of 0.01–0.1 mol% of fluorescently labeled lipids results in bright membrane staining for confocal microscopy.

3. Particularities of measurements in different membrane model systems:

   To obtain high quality GUV images by avoiding the effects of polarization section in excitation, a λ/4 waveplate can be installed in the excitation path.

   For lipid monolayers, due to the high surface area to volume ration of the used setup and the consequent fast subphase evaporation, a temperature control system compatible with the microscope stage should be used to stabilize the position of the interface in relation to the focus position.

4. Particularities of sample preparation according to the chosen membrane binding strategy of DNA origami nanostructures:

   When a large number of membrane anchors of hydrophobic nature are chosen/necessary, as an alternative to the direct folding DNA origami nanostructures with the respective modified oligonucleotides, we advise to fold DNA origami nanostructures without the anchors in order to avoid aggregation in solution. Instead, the modified oligonucleotides can be incubated with the prepared model lipid membranes (typically at 0.01–2 μM). Handles incorporated in the DNA origami nanostructure, complementary to the modified oligonucleotide, will assure a proficient membrane binding. To reduce the probability of aggregation or loss of membrane binding due to excess modified oligonucleotides in solution, the lipid membrane systems should be washed prior to the addition of DNA origami nanostructures.

A similar approach is used to membrane bind DNA origami nanostructures via biotin-streptavidin (or neutravidin) interactions. Shortly, lipid membranes prepared with 0.1–10 mol% biotinylated lipids (e.g., DOPE-cap-biotin) are incubated with excess streptavidin (e.g., 1–2 μM), washed with buffer and further incubated (at the desired concentration) with DNA origami nanostructures carrying biotinylated DNA handles.

Several protocols have been described, for covalent binding of DNA nanostructures to membranes, all with slight variations concerning the type and chemistry of the functional groups involved [30, 60]. Nevertheless, one simple approach, with commercially available components, is to prepare model lipid membranes (e.g., LUVs) doped with 0.1–5 mol% maleimide--functionalized lipids (e.g., DOPE-MPB) and incubate the vesicles with a 10–100× excess of thiol-functionalized DNA oligonucleotides (overnight at 4 °C or ~1–2 h at room temperature) in a standard buffer (pH < 7.5) with additional 1 mM TCEP. The vesicles can then be subsequently pelleted (or washed) out from the excess of unbound thiolated oligonucleotides. DNA origami nanostructures with complementary handles can then be incubated with the model membranes, allowing it to bind to the liposomes presenting DNA-functionalized lipids.

Due to the instability of GUVs towards shear stress, washing steps are discouraged, and the number of manipulation steps should be kept to a minimum. The incubation of GUVs with modified oligonucleotides (or streptavidin) should be done prior to the final GUV dilution (e.g. 1:10 or 1:50) for fluorescence confocal imaging. As such, the dilution step will not only guarantee a correct proportion of lipid membrane to DNA origami nanostructure, but also reduce the concentration of free modified oligonucleotides.

Concerning binding through electrostatic interactions, increasing the Mg^{2+} concentration to 20 mM, or incorporating a small percentage (1–10 mol%) of positively charged lipids (e.g., DOTAP) in the lipid model system will assure a strong membrane binding of DNA origami. Importantly, when Mg^{2+}-mediated membrane binding of DNA origami is applied on a phase-separated membrane system, the DNA origami nanostructures will preferentially bind to the ordered lipid domains [44].

5. Several open source software are available for image analysis. ImageJ (or Fiji) has been widely used for general image analysis [61, 62], independently of its experimental nature. For initial AFM image processing, typically each company provides the user with their commercial software version (e.g., JPK SPM Data Processing). However, one can use the open-source program Gwyddion [63]. For FCS analysis, besides the commercial software (e.g., Zen from Zeiss), users can either write their own fitting codes or resort to the previously developed user-friendly platform PyCorrFit [64].
6. Electroformed GUVs can be prepared not only on Pt wires but also ITO-coated coverslips [65]. In this case, 5 μL of lipid mixture at 10 mg/mL is spread in a snake-like pattern on the conductive side of an ITO-coated coverslip, to which a copper wire is attached. A second ITO-coated coverslip with a copper wire is used to assemble a flow chamber in which the GUVs are electroformed. Besides electroformation, other GUV-forming methods based on assisted swelling [66] or emulsion transfer [67] can be employed.
7. For FCS measurements, in order to guarantee a stable membrane with minimal fluctuations, GUVs should be prepared with a slightly hyper-osmolar sucrose solution (5% higher osmolarity) relatively to the buffer of choice.
8. Alternatively to bath ultrasonication, SUVs can be produced by tip ultrasonication or extrusion using a membrane filter with small pore size (e.g., 50 nm, as seen in Subheading 3.3).
9. SLBs can be prepared by other methods and then vesicle fusion, such as Langmuir-Blodgett [68] or spin coating [69]. The preferred SLB support should be chosen according to the used lipid mixture and the characterization method of choice. More specifically, mica is preferred over glass when using phase-separated lipid mixtures, as the lower roughness of mica favors the formation of larger lipid domains [70]. In this case, chambers should have freshly cleaved mica sheet glued on a 1# thickness glass coverslip as substrate. For AFM applications, the smoother mica surface is also the preferred support for imaging membranes.
10. Typical home-made chambers consist of an inverted bottom-cut Eppendorf tube of chosen volume glued to a coverslip with UV-curing or inert glue [49]. Alternatively, 2 mm thick silicon isolators of chosen diameter wells (Grace Bio-Laboratories, Oregon, U.S.A.) cleaned with tape to remove any dust are pressed on plasma-cleaned high precision #1.5 thickness glass coverslips. Chambers can be sealed by a second coverslip.

11. Traditionally, lipid monolayers are prepared in Langmuir troughs which require large volumes of buffer and membrane binding macromolecules. While in this method, the MMA of the lipid monolayer is controlled by physically compressing the lipids deposited at the air-water interface by movable barriers; at the small home-made PTFE chambers, the MMA of the lipid monolayer is controlled by the amount of lipids deposited at the interface [47]. As such, the exact concentration of the lipid mixture should be checked by gravimetry prior to sample preparation.
12. Other methods can be used for determining DNA origami diffusion on lipid membranes, such as single particle tracking (for more details, consult [35, 71]) or fluorescence recovery after photobleaching (for more details, consult [72]).
13. Typically, we use pseudo-cross-correlation settings to avoid after-pulsing effects. More specifically, the fluorescent signal was split in two equivalent channels, and the output of the two detectors was cross-correlated [55].
14. Particularities regarding fluorescence labeling of DNA origami for simplification of FCS measurements and interpretation:

    For FCS applications, while the incorporation of a large number of dyes on the DNA origami nanostructures is desirable as in standard confocal microscopy, additionally the chosen fluorophores should have high quantum yield, low photobleaching, and no or little blinking and triplet population, such as Alexa Fluor 647. When fluorescently labeled, the fluorescence intensity can be used to determine the DNA origami concentration when compared to a calibration curve of free dye-labeled oligonucleotide, corrected to the number of incorporated dyes.

    In order to reduce the contribution of the rotational diffusion on the collected FCS curves, we strongly recommend to place the fluorophores near the center of the DNA nanostructures. By doing so, only translation diffusion will contribute to the measured curves [24, 25], and Eq. 2 can be used for fitting.

    Using dyes with no triplet blinking (such as Atto 655) will further simplify the fitting model by setting $T = 0$.

    FCS measurements should be performed at low surface densities of membrane-bound DNA origami in order for the simple one component 2D diffusion model (Eq. 2) to hold [25].

    When GUVs are used as model membranes, FCS measurements should be performed on the pole of vesicles with diameters larger than 20 μm, to assure the correct positioning of the focal volume, and thus the applicability of the fitting model.

Acknowledgments

The authors thank the Deutsche Forschungsgemeinschaft (DFG, German Research Foundation) – SFB-863 – Project ID 111166240 for funding. Additional support from the Graduate School of Quantitative Biosciences Munich to A.K., Max Planck Society to P.S., and Alexander von Humboldt Foundation to H.G. F (Humboldt Research Fellowship PTG/1152511/STP) is also acknowledged.

References

1. McMahon HT, Gallop JL (2005) Membrane curvature and mechanisms of dynamic cell membrane remodelling. Nature 438(7068): 590–596. https://doi.org/10.1038/nature04396
2. Zimmerberg J, Kozlov MM (2006) How proteins produce cellular membrane curvature. Nat Rev Mol Cell Biol 7(1):9–19. https://doi.org/10.1038/nrm1784
3. Baumgart T, Capraro BR, Zhu C, Das SL (2011) Thermodynamics and mechanics of membrane curvature generation and sensing by proteins and lipids. Annu Rev Phys Chem 62:483–506. https://doi.org/10.1146/annurev.physchem.012809.103450
4. Lombard J, Lopez-Garcia P, Moreira D (2012) The early evolution of lipid membranes and the three domains of life. Nat Rev Microbiol 10(7): 507–515. https://doi.org/10.1038/nrmicro2815
5. van Meer G, Voelker DR, Feigenson GW (2008) Membrane lipids: where they are and how they behave. Nat Rev Mol Cell Biol 9(2): 112–124. https://doi.org/10.1038/nrm2330
6. Harayama T, Riezman H (2018) Understanding the diversity of membrane lipid composition. Nat Rev Mol Cell Biol 19(5):281–296. https://doi.org/10.1038/nrm.2017.138
7. Schwille P, Diez S (2009) Synthetic biology of minimal systems. Crit Rev Biochem Mol 44(4): 223–242. https://doi.org/10.1080/10409230903074549
8. Sezgin E, Schwille P (2012) Model membrane platforms to study protein-membrane interactions. Mol Membr Biol 29(5):144–154. https://doi.org/10.3109/09687688.2012.700490
9. Kretschmer S, Ganzinger KA, Franquelim HG, Schwille P (2019) Synthetic cell division via membrane-transforming molecular assemblies. BMC Biol 17(1):43. https://doi.org/10.1186/s12915-019-0665-1
10. Czogalla A, Franquelim HG, Schwille P (2016) DNA nanostructures on membranes as tools for synthetic biology. Biophys J 110(8): 1698–1707. https://doi.org/10.1016/j.bpj.2016.03.015
11. Ke Y, Castro C, Choi JH (2018) Structural DNA nanotechnology: artificial nanostructures for biomedical research. Annu Rev Biomed Eng 20:375–401. https://doi.org/10.1146/annurev-bioeng-062117-120904
12. Mathur D, Medintz IL (2019) The growing development of DNA nanostructures for potential healthcare-related applications. Adv Healthc Mater 8(9):e1801546. https://doi.org/10.1002/adhm.201801546
13. Suo ZG, Chen JQ, Hou XL, Hu ZH, Xing FF, Feng LY (2019) Growing prospects of DNA nanomaterials in novel biomedical applications. RSC Adv 9(29):16479–16491. https://doi.org/10.1039/c9ra01261c
14. Castro CE, Kilchherr F, Kim DN, Shiao EL, Wauer T, Wortmann P, Bathe M, Dietz H (2011) A primer to scaffolded DNA origami. Nat Methods 8(3):221–229. https://doi.org/10.1038/nmeth.1570
15. Wagenbauer KF, Engelhardt FAS, Stahl E, Hechtl VK, Stommer P, Seebacher F, Meregalli L, Ketterer P, Gerling T, Dietz H (2017) How we make DNA origami. Chembiochem 18(19):1873–1885. https://doi.org/10.1002/cbic.201700377
16. Hong F, Zhang F, Liu Y, Yan H (2017) DNA origami: scaffolds for creating higher order structures. Chem Rev 117(20):12584–12640. https://doi.org/10.1021/acs.chemrev.6b00825
17. Mishra S, Feng Y, Endo M, Sugiyama H (2019) Advances in DNA origami-cell interfaces. Chembiochem 20:1–13. https://doi.org/10.1002/cbic.201900481
18. Dong Y, Mao Y (2019) DNA origami as scaffolds for self-assembly of lipids and proteins.

Chembiochem 20(19):2422–2431. https://doi.org/10.1002/cbic.201900073

19. Langecker M, Arnaut V, List J, Simmel FC (2014) DNA nanostructures interacting with lipid bilayer membranes. Acc Chem Res 47(6):1807–1815. https://doi.org/10.1021/ar500051r
20. Bae W, Kocabey S, Liedl T (2019) DNA nanostructures in vitro, in vivo and on membranes. Nano Today 26:98–107. https://doi.org/10.1016/j.nantod.2019.03.001
21. Shen Q, Grome MW, Yang Y, Lin CX (2019) Engineering lipid membranes with programmable DNA nanostructures. Adv Biosyst 1900215. https://doi.org/10.1002/adbi.201900215
22. Khmelinskaia A, Franquelim HG, Petrov EP, Schwille P (2016) Effect of anchor positioning on binding and diffusion of elongated 3D DNA nanostructures on lipid membranes. J Phys D Appl Phys 49:194001. https://doi.org/10.1088/0022-3727/49/19/194001
23. Khmelinskaia A, Mücksch J, Petrov EP, Franquelim HG, Schwille P (2018) Control of membrane binding and diffusion of cholesteryl-modified DNA origami nanostructures by DNA spacers. Langmuir 34(49): 14921–14931. https://doi.org/10.1021/acs.langmuir.8b01850
24. Czogalla A, Petrov EP, Kauert DJ, Uzunova V, Zhang Y, Seidel R, Schwille P (2013) Switchable domain partitioning and diffusion of DNA origami rods on membranes. Faraday Discuss 161:31–43; discussion 113–150. doi:https://doi.org/10.1039/c2fd20109g
25. Czogalla A, Kauert DJ, Seidel R, Schwille P, Petrov EP (2015) DNA origami nanoneedles on freestanding lipid membranes as a tool to observe isotropic-nematic transition in two dimensions. Nano Lett 15(1):649–655. https://doi.org/10.1021/nl504158h
26. Krishnan S, Ziegler D, Arnaut V, Martin TG, Kapsner K, Henneberg K, Bausch AR, Dietz H, Simmel FC (2016) Molecular transport through large-diameter DNA nanopores. Nat Commun 7:12787. https://doi.org/10.1038/ncomms12787
27. Langecker M, Arnaut V, Martin TG, List J, Renner S, Mayer M, Dietz H, Simmel FC (2012) Synthetic lipid membrane channels formed by designed DNA nanostructures. Science 338(6109):932–936. https://doi.org/10.1126/science.1225624
28. Göpfrich K, Li CY, Ricci M, Bhamidimarri SP, Yoo J, Gyenes B, Ohmann A, Winterhalter M, Aksimentiev A, Keyser UF (2016) Large-conductance transmembrane Porin made from DNA origami. ACS Nano 10(9): 8207–8214. https://doi.org/10.1021/acsnano.6b03759
29. Dong Y, Yang YR, Zhang Y, Wang D, Wei X, Banerjee S, Liu Y, Yang Z, Yan H, Liu D (2017) Cuboid vesicles formed by frame-guided assembly on DNA origami scaffolds. Angew Chem Int Ed Engl 56(6):1586–1589. https://doi.org/10.1002/anie.201610133
30. Perrault SD, Shih WM (2014) Virus-inspired membrane encapsulation of DNA nanostructures to achieve in vivo stability. ACS Nano 8(5):5132–5140. https://doi.org/10.1021/nn5011914
31. Yang Y, Wang J, Shigematsu H, Xu W, Shih WM, Rothman JE, Lin C (2016) Self-assembly of size-controlled liposomes on DNA nanotemplates. Nat Chem 8(5):476–483. https://doi.org/10.1038/nchem.2472
32. Zhang Z, Yang Y, Pincet F, Llaguno MC, Lin C (2017) Placing and shaping liposomes with reconfigurable DNA nanocages. Nat Chem 9(7):653–659. https://doi.org/10.1038/nchem.2802
33. Czogalla A, Kauert DJ, Franquelim HG, Uzunova V, Zhang Y, Seidel R, Schwille P (2015) Amphipathic DNA origami nanoparticles to scaffold and deform lipid membrane vesicles. Angew Chem Int Ed Engl 54(22): 6501–6505. https://doi.org/10.1002/anie.201501173
34. Franquelim HG, Khmelinskaia A, Sobczak JP, Dietz H, Schwille P (2018) Membrane sculpting by curved DNA origami scaffolds. Nat Commun 9(1):811. https://doi.org/10.1038/s41467-018-03198-9
35. Kocabey S, Kempter S, List J, Xing Y, Bae W, Schiffels D, Shih WM, Simmel FC, Liedl T (2015) Membrane-assisted growth of DNA origami nanostructure arrays. ACS Nano 9(4): 3530–3539. https://doi.org/10.1021/acsnano.5b00161
36. Burns JR, Göpfrich K, Wood JW, Thacker VV, Stulz E, Keyser UF, Howorka S (2013) Lipid-bilayer-spanning DNA nanopores with a bifunctional porphyrin anchor. Angew Chem Int Ed Engl 52(46):12069–12072. https://doi.org/10.1002/anie.201305765
37. Johnson-Buck A, Jiang S, Yan H, Walter NG (2014) DNA-cholesterol barges as programmable membrane-exploring agents. ACS Nano 8(6):5641–5649. https://doi.org/10.1021/nn500108k
38. Burns JR, Howorka S (2019) Structural and functional stability of DNA nanopores in biological media. Nanomaterials (Basel) 9(4). https://doi.org/10.3390/nano9040490

39. Ohmann A, Göpfrich K, Joshi H, Thompson RF, Sobota D, Ranson NA, Aksimentiev A, Keyser UF (2019) Controlling aggregation of cholesterol-modified DNA nanostructures. Nucleic Acids Res 47(21):11441–11451. https://doi.org/10.1093/nar/gkz914
40. Grome MW, Zhang Z, Lin C (2019) Stiffness and membrane anchor density modulate DNA-nanospring-induced vesicle tubulation. ACS Appl Mater Interfaces 11(26):22987–22992. https://doi.org/10.1021/acsami.9b05401
41. Grome MW, Zhang Z, Pincet F, Lin C (2018) Vesicle tubulation with self-assembling DNA nanosprings. Angew Chem Int Ed Engl 57(19):5330–5334. https://doi.org/10.1002/anie.201800141
42. Kurokawa C, Fujiwara K, Morita M, Kawamata I, Kawagishi Y, Sakai A, Murayama Y, Nomura SM, Murata S, Takinoue M, Yanagisawa M (2017) DNA cytoskeleton for stabilizing artificial cells. Proc Natl Acad Sci U S A 114(28):7228–7233. https://doi.org/10.1073/pnas.1702208114
43. Suzuki Y, Endo M, Sugiyama H (2015) Lipid-bilayer-assisted two-dimensional self-assembly of DNA origami nanostructures. Nat Commun 6:8052. https://doi.org/10.1038/ncomms9052
44. Sato Y, Endo M, Morita M, Takinoue M, Sugiyama H, Murata S, Nomura SIM, Suzuki Y (2018) Environment-dependent self-assembly of DNA origami lattices on phase-separated lipid membranes. Adv Mater Interfaces 5(14):1800437
45. Garcia-Saez AJ, Carrer DC, Schwille P (2010) Fluorescence correlation spectroscopy for the study of membrane dynamics and organization in giant unilamellar vesicles. Methods Mol Biol 606:493–508. https://doi.org/10.1007/978-1-60761-447-0_33
46. Khmelinskaia A, Mücksch J, Conci F, Chwastek G, Schwille P (2018) FCS analysis of protein mobility on lipid monolayers. Biophys J 114(10):2444–2454. https://doi.org/10.1016/j.bpj.2018.02.031
47. Chwastek G, Schwille P (2013) A monolayer assay tailored to investigate lipid-protein systems. ChemPhysChem 14(9):1877–1881. https://doi.org/10.1002/cphc.201300035
48. Vogel SK, Greiss F, Khmelinskaia A, Schwille P (2017) Control of lipid domain organization by a biomimetic contractile actomyosin cortex. elife 6. https://doi.org/10.7554/eLife.24350
49. Ramm B, Glock P, Schwille P (2018) In vitro reconstitution of self-organizing protein patterns on supported lipid bilayers. J Vis Exp 137. https://doi.org/10.3791/58139
50. Thomas FA, Visco I, Petrasek Z, Heinemann F, Schwille P (2015) Introducing a fluorescence-based standard to quantify protein partitioning into membranes. Biochim Biophys Acta 1848(11 Pt A):2932–2941. https://doi.org/10.1016/j.bbamem.2015.09.001
51. Petrov EP, Ohrt T, Winkler RG, Schwille P (2006) Diffusion and segmental dynamics of double-stranded DNA. Phys Rev Lett 97(25):258101. https://doi.org/10.1103/PhysRevLett.97.258101
52. Dertinger T, Pacheco V, von der Hocht I, Hartmann R, Gregor I, Enderlein J (2007) Two-focus fluorescence correlation spectroscopy: a new tool for accurate and absolute diffusion measurements. ChemPhysChem 8(3):433–443. https://doi.org/10.1002/cphc.200600638
53. Widengren J, Mets U, Rigler R (1995) Fluorescence correlation spectroscopy of triplet-states in solution – a theoretical and experimental-study. J Phys Chem 99(36):13368–13379. https://doi.org/10.1021/j100036a009
54. Gregor I, Patra D, Enderlein J (2005) Optical saturation in fluorescence correlation spectroscopy under continuous-wave and pulsed excitation. ChemPhysChem 6(1):164–170. https://doi.org/10.1002/cphc.200400319
55. Petrov EP, Schwille P (2008) State of the art and novel trends in fluorescence correlation spectroscopy. In: Resch-Genger U (ed) Standardization and quality Assurance in Fluorescence Measurements II: bioanalytical and biomedical applications. Springer, Berlin Heidelberg, Berlin, Heidelberg, pp 145–197. https://doi.org/10.1007/4243_2008_032
56. Sutherland W (1905) LXXV. A dynamical theory of diffusion for non-electrolytes and the molecular mass of albumin. Lond Edinb Dublin Philos Mag J Sci 9(54):781–785. https://doi.org/10.1080/14786440509463331
57. Einstein A (1905) Über die von der molekularkinetischen Theorie der Wärme geforderte Bewegung von in ruhenden Flüssigkeiten suspendierten Teilchen. Ann Phys 322(8):549–560. https://doi.org/10.1002/andp.19053220806
58. Kestin J, Sokolov M, Wakeham WA (1978) Viscosity of liquid water in the range −8 °C to 150 °C. J Phys Chem Ref Data 7(3):941–948. https://doi.org/10.1063/1.555581
59. Korson L, Drost-Hansen W, Millero FJ (1969) Viscosity of water at various temperatures. J Phys Chem 73(1):34–39. https://doi.org/10.1021/j100721a006

60. Hoffecker IT, Takemoto N, Arima Y, Iwata H (2015) Sequence-specific nuclease-mediated release of cells tethered by oligonucleotide phospholipids. Biomaterials 53:318–329. https://doi.org/10.1016/j.biomaterials.2015.02.059
61. Schneider CA, Rasband WS, Eliceiri KW (2012) NIH image to ImageJ: 25 years of image analysis. Nat Methods 9(7):671–675. https://doi.org/10.1038/nmeth.2089
62. Schindelin J, Arganda-Carreras I, Frise E, Kaynig V, Longair M, Pietzsch T, Preibisch S, Rueden C, Saalfeld S, Schmid B, Tinevez JY, White DJ, Hartenstein V, Eliceiri K, Tomancak P, Cardona A (2012) Fiji: an open-source platform for biological-image analysis. Nat Methods 9(7):676–682. https://doi.org/10.1038/nmeth.2019
63. Nečas D, Klapetek P (2012) Gwyddion: an open-source software for SPM data analysis. Centr Eur JPhys 10(1):181–188. https://doi.org/10.2478/s11534-011-0096-2
64. Müller P, Schwille P, Weidemann T (2014) PyCorrFit-generic data evaluation for fluorescence correlation spectroscopy. Bioinformatics 30(17):2532–2533. https://doi.org/10.1093/bioinformatics/btu328
65. Betaneli V, Mücksch J, Schwille P (2019) Fluorescence correlation spectroscopy to examine protein-lipid interactions in membranes. Methods Mol Biol 2003:415–447. https://doi.org/10.1007/978-1-4939-9512-7_18
66. Weinberger A, Tsai FC, Koenderink GH, Schmidt TF, Itri R, Meier W, Schmatko T, Schroder A, Marques C (2013) Gel-assisted formation of giant unilamellar vesicles. Biophys J 105(1):154–164. https://doi.org/10.1016/j.bpj.2013.05.024
67. Litschel T, Ganzinger KA, Movinkel T, Heymann M, Robinson T, Mutschler H, Schwille P (2018) Freeze-thaw cycles induce content exchange between cell-sized lipid vesicles. New J Phys 20. https://doi.org/10.1088/1367-2630/aabb96
68. Kurniawan J, Ventrici de Souza JF, Dang AT, Liu GY, Kuhl TL (2018) Preparation and characterization of solid-supported lipid bilayers formed by Langmuir-Blodgett deposition: a tutorial. Langmuir 34(51):15622–15639. https://doi.org/10.1021/acs.langmuir.8b03504
69. Thormann E, Simonsen AC, Nielsen LK, Mouritsen OG (2007) Ligand-receptor interactions and membrane structure investigated by AFM and time-resolved fluorescence microscopy. J Mol Recognit 20(6):554–560. https://doi.org/10.1002/jmr.850
70. Goodchild JA, Walsh DL, Connell SD (2019) Nanoscale substrate roughness hinders domain formation in supported lipid bilayers. Langmuir 35(47):15352–15363. https://doi.org/10.1021/acs.langmuir.9b01990
71. Kempter S, Khmelinskaia A, Strauss MT, Schwille P, Jungmann R, Liedl T, Bae W (2019) Single particle tracking and super-resolution imaging of membrane-assisted stop-and-go diffusion and lattice assembly of DNA origami. ACS Nano 13(2):996–1002. https://doi.org/10.1021/acsnano.8b04631
72. Journot CMA, Ramakrishna V, Wallace MI, Turberfield AJ (2019) Modifying membrane morphology and interactions with DNA origami Clathrin-mimic networks. ACS Nano 13(9):9973–9979. https://doi.org/10.1021/acsnano.8b07734

Chapter 15

Electrical Actuation of DNA-Based Nanomechanical Systems

Jonathan List, Enzo Kopperger, and Friedrich C. Simmel

Abstract

DNA nanotechnology provides efficient methods for the sequence-programmable construction of mechanical devices with nanoscale dimensions. The resulting nanomachines could serve as tools for the manipulation of macromolecules with similar functionalities as mechanical tools and machinery in the macroscopic world. In order to drive and control these machines and to perform specific tasks, a fast, reliable, and repeatable actuation mechanism is required that can work against external loads. Here we describe a highly effective method for actuating DNA structures using externally applied electric fields. To this end, electric fields are generated with controllable direction and amplitude inside a miniature electrophoresis device integrated with an epifluorescence microscope. With this setup, DNA-based nanoelectromechanical devices can be precisely controlled. As an example, we demonstrate how a DNA-based nanorobotic system can be used to dynamically position molecules on a molecular platform with high speeds and accuracy. The microscopy setup also described here allows simultaneous monitoring of a large number of nanorobotic arms in real time and at the single nanomachine level.

Key words DNA nanotechnology, DNA origami, Molecular programming, Nanorobotics, NEMS, Nanopositioning, Electrophoresis, Single molecule studies, Force spectroscopy

1 Introduction

Nanotechnology ultimately aims at the precise control of matter at the finest possible scale. Within the realm of biomolecules, DNA nanotechnology has been shown to provide powerful methods for the programmable self-assembly of large macromolecular structures that are precisely addressable at the nanoscale [1, 2]. In this context, the DNA origami method has been established as a particularly user-friendly and robust assembly technique [3–5]. While the complexity of static assemblies has grown steadily over the past decades, strategies for the creation and actuation of dynamic DNA-based nanoassemblies have remained comparatively limited. One actuation strategy relies on DNA strand displacement reactions [6–11], which is, in principle, also capable of creating

Julián Valero (ed.), *DNA and RNA Origami: Methods and Protocols*, Methods in Molecular Biology, vol. 2639,
https://doi.org/10.1007/978-1-0716-3028-0_15,

autonomous motion [12]. Alternatively, changes in buffer conditions [13], light-switchable moieties [14], or enzyme reactions [8, 15] were used to drive DNA-based nanomachines. These methods usually resulted in relatively low movement speeds, however. Furthermore, the devices often suffered from low operation yields (e.g., in terms of fraction of devices actuated per cycle), and they have not been shown to work against significant external loads. Biophysical single molecule manipulation setups such as optical tweezers [16], magnetic tweezers [17], and atomic force microscopes [18] can achieve much higher actuation speeds, but are limited to the control of only single nanodevices at a time. A similar level of control for larger numbers of structures in parallel would therefore be highly desirable. One possibility to achieve this is to utilize the high electrical charge of the sugar-phosphate backbone of DNA, which facilitates its manipulation by electric fields [19]. In this context, previous work has demonstrated the electrical actuation of DNA levers in z-direction [20, 21], the continuous electrical rotation of double-stranded DNA [22], and also the dielectrophoretic attraction and capture of DNA nanostructures with micro-electrodes [23].

While previous attempts to control the motion of DNA nanostructures focused on either translation or rotation of the structures as a whole, we recently demonstrated the use of electric fields as an effective strategy to drive the relative motion of the movable components of DNA-based nanomachines with respect to each other with a high degree of control and at very high speeds [24].

To this end, we built a dedicated miniature electrophoresis setup that allowed the generation of electric fields within a microscopy sample chamber with precisely defined planar orientations that were realized by superposition of two perpendicular electric fields (Fig. 1). Driven by computer controlled high voltage amplifiers, our system also establishes an interface for software control of DNA nanomachines. In our specific embodiment, a nanorobotic arm mounted on a DNA plate can be positioned rapidly with high angular resolution. This configuration allows us to move parts of the structure relative to each other and to addressably set different configurations or machine states. In principle, the actuated DNA origami structures can be customized to implement more sophisticated mechanisms, and they can be easily interfaced with other biomolecules or inorganic nanostructures.

In this MiMB chapter, we will provide detailed guidelines for the experimental implementation of electrically actuated DNA nanodevices. We describe critical design considerations for mechanical nanodevices that facilitate actuation with this technique and also their microscopic observation (a general guide for DNA origami design can be found in Refs. [25, 26]). We further describe the minimal single molecule microscopy setup required for tracking the movement of the devices and the design of microscopy-compatible chambers for electric actuation. One

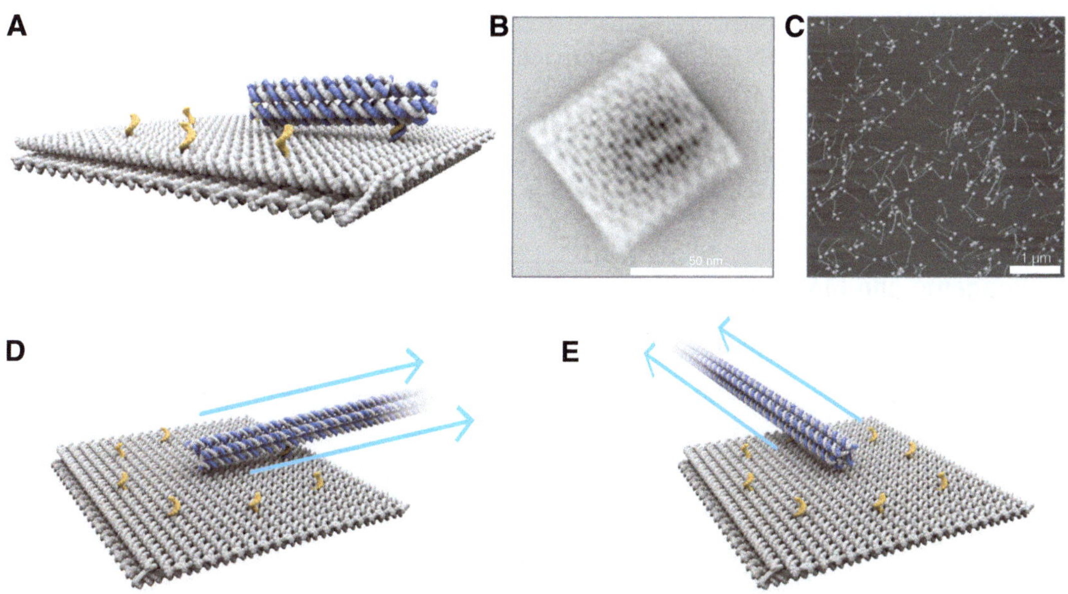

Fig. 1 (**a**) An electrically controllable DNA nanostructure consisting of a movable robotic arm and a base plate, which is fixed to the substrate. The base plate acts as a molecular pegboard for placing functional moieties and sticky DNA ends (shown in yellow) used to temporarily fix the arm. A flexible joint carries the robotic arm. (**b**) TEM class average picture of platform and arm. (**c**) To simplify observation, the arm can be extended by a 6-helix (6H) DNA origami tube. AFM image shows plates with extended robot arms. Z-scale: 10 nm. (**d**) Schematic illustration of the platform including the robot arm and the 6H extension, aligned along the electric field. (**e**) Alignment of the arm along an orthogonal direction. Superposition of both field directions allows selection of arbitrary angles

version of these chambers enables angular control of the electric field, while a simpler version for basic control over nanostructures via binary polarity switching is easier to implement. We also describe the required electrical setup, the preparation of the glass substrate, and the experimental protocol for actuation. Further, we introduce data analysis strategies for tracking the motion of the nanomachines, and we finally suggest possible applications for this actuation method by demonstrating several proof of concept experiments.

To conclude, in this chapter we describe an experimental platform that allows interfacing externally applicable, freely programmable electric signals with nanoscale movements generated on a DNA-based nanomechanical platform. The resulting hybrid nanorobotic system enables fast, repeatable, and highly parallel actuation of biomolecular assemblies with the experimental benefit of direct observation in real time. Already with commonly available laboratory equipment, a simple actuation and observation setup for electrical switching in one dimension can be realized. Superposition of two orthogonal electric fields generated by software-controlled high voltage amplifiers allows circular actuation with high angular precision and the performance of arbitrary voltage-controlled protocols. Fast and precise positioning of molecular components by DNA-based nanomachines is expected to open up a wide range of new experimental possibilities, which may prove useful, e.g., in the context of reconfigurable sensor devices, single molecule studies of biomolecular interactions, actuation of nanoplasmonic structures, or electromechanically controlled chemical reactions.

2 Reagents and Materials

2.1 Microscope Setup (See Fig. 2)

1. Microscope body (Olympus IX71).
2. 100× TIRF objective (UAPON 100xO TIRF, NA 1.49 oil, Olympus).
3. 640 nm diode laser (Toptica iBeam smart, 150 mW).
4. Dichroic mirror.
5. Translational alignment stage that parallel shifts the beam.
6. Camera: Andor iXon 897 emCCD (Andor Technologies, Oxford Instruments) for high sensitivity experiments or Andor NEO sCMOS camera (for measurements that require frame rates faster than 100 fps).

   A detailed instruction for building low-cost TIRF microscopes can be found in [27] (*see* **Note 1**).

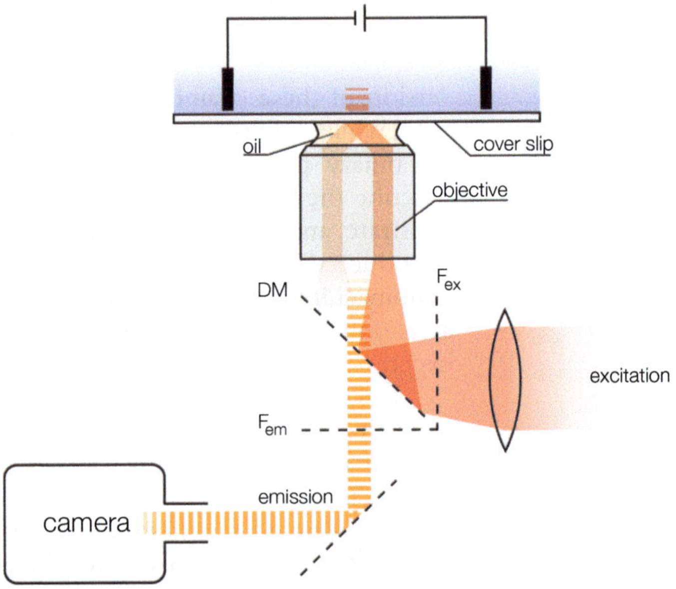

Fig. 2 Schematic fluorescence microscopy setup for the observation of nanostructures during actuation. The sample chamber is mounted on an inverted microscope. Accessible space on top of the sample chamber simplifies liquid handling and electrode placement. Objective type total internal reflection illumination reduces the fluorescence background by only exciting dyes close to the glass-water interface

2.2 Setup for Binary Electrical Switching

1. Driller (4 mm diameter).
2. 5 mm thick piece of PMMA.
3. Double-sided adhesive tape 3M 467MP (i.e., 3M Company, Maplewood, Minnesota, USA).
4. Plastic syringes (e.g., Braun Injekt Solo 5 mL, B. Braun Melsungen AG, Germany).
5. Standard electrophoresis or other high voltage laboratory power supplies.

2.3 Preparation of Glass Slides

1. #1.5H (180 μm thick) microscopy cover glass slides.
2. Biotinylated bovine serum albumin (Biotin-BSA) (e.g., Pierce, Thermo Scientific).
3. Tween-20.
4. Optional: Biotin-PEG-silane (*see* **Note 1**).
5. NeutrAvidin.

2.4 Preparation of Origami Structures and Actuation

Scaffold strand (e.g., M13mp18).

Staple oligonucleotides.

Folding buffer: 5 mM Tris-HCl pH 7.5, 1 mM EDTA, 5 mM NaCl, 20 mM $MgCl_2$.

Thermocycler.

Precipitation solution: 5 mM Tris-HCl pH 7.5, 1 mM EDTA, 250 mM NaCl, 20 mM $MgCl_2$, 15% w/v PEG-8000.

Bench centrifuge.

Actuation buffer: e.g., 0.5× TBE buffer, 6 mM $MgCl_2$.

3 Methods

3.1 Preparation of Origami Structures

1. Base plate-arm structures with integrated arms and pointer structures are designed following certain considerations (for details *see* Subheading 4) prepared in a one pot folding reaction. For that, mix the scaffold strand (33 nM) and the staple strands (each at 165 nM) in folding buffer.
2. Put the mixture in the thermocycler using a ramp from 70 °C to 20 °C over 12 h followed by a 40 °C holding period of 3 h.
3. Remove excess staple strands by polyethylene glycol (PEG) precipitation, adapted from [28]. To this end, mix the sample with an equal volume of a precipitation solution.
4. Centrifuge the mixture at 20,000 *g* for 20 min.
5. Pointer structures are produced and purified following the same protocol (**steps 1–4**).

6. Assemble the pointers to the base plate-arm structure by mixing a 1.25-fold excess of pointers to the plate-arm structures and incubate the solution for 1 h at 37 °C.

3.2 Preparation of Glass Slides

1. Use #1.5H (180 μm thick) microscopy cover glass slides and apply a coating for passivation (*see* **Note 2**).
2. A quick-and-easy protein coating can be achieved with biotinylated bovine serum albumin (Biotin-BSA).
3. The readily assembled sample chamber is incubated for 1 min with 0.5 mg/mL of a Biotin-BSA solution and subsequently rinsed with water (*see* **Notes 1** and **3**).

3.3 A Simple Setup for Binary Electrical Switching (see Fig. 2)

The following method for electric control in one dimension can be considered an easy route to rapidly realize a simple type of electric actuation of nanostructures and gain experience before developing a more sophisticated setup:

Positioning the nanorobotic arm at an arbitrary angle requires several pieces of custom-built equipment such as sample chambers, plugs for electrode fixation, and high voltage electronics.

Some experimental tasks might not even require rotational angle control but merely application of an electric field in a single direction. Binary switching of nanomachines between two states with electric fields can be realized more easily using only commercial components found in most biochemistry labs. The following method for electric control in one dimension can be considered an easy route to rapidly realize a simple type of electric actuation of nanostructures and gain experience before developing a more sophisticated setup. A single linear channel is sufficient as a sample chamber for one-dimensional actuation (*see* **Note 4**). To create such channel:

1. Drill two holes of 4 mm diameter (roughly compatible with standard luer syringe fittings) into a 5 mm thick piece of PMMA at a distance of 22 mm (*see* **Note 5**).
2. Cut a piece of double-sided ahesive tape i.e. 3M 467MP (3M Company, Maplewood, Minnesota, USA) with a scalpel or laser cutter into the shape of a channel, connecting the two holes. Bind a microscopy coverslip to the PMMA chip via the double sided tape piece.
3. Attached plastic syringes (e.g., Braun Injekt Solo 5 mL, B. Braun Melsungen AG, Germany) to the luer ports act as buffer reservoirs. The syringe pistons hold the platinum wire electrodes in place (*see* Fig. 3a) and keep the cable and soldering joint separated from the buffer solution.
4. Make holes in the piston that allow insertion of the electrodes without pressurizing the sample chamber.

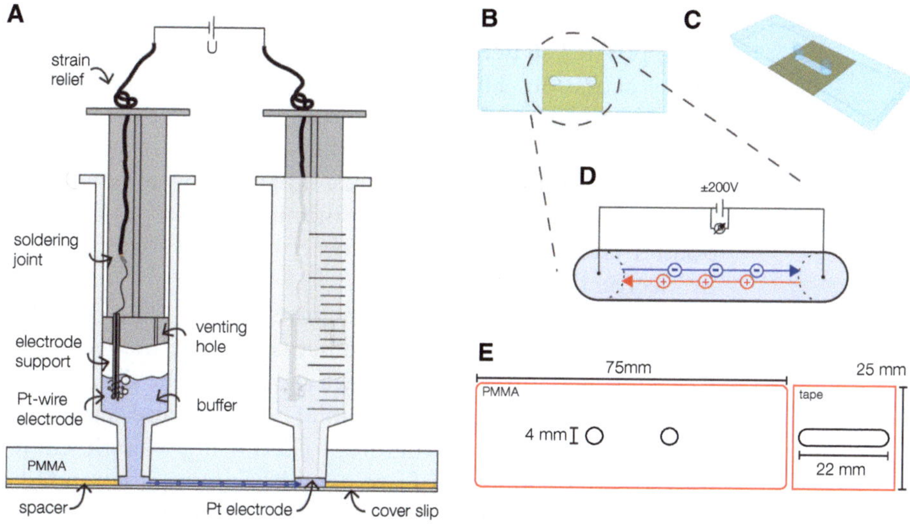

Fig. 3 Easy-to-build linear actuation setup. (**a**) Cross-sectional schematic view. The measurement channel is formed between a microscopy coverslip and a PMMA sheet bonded with a thin tape layer. Plastic syringes serve as buffer reservoirs with each piston holding a platinum electrode. (**b**) Top view of the assembled microscopy chip. (**c**) Isometric view. (**d**) Schematic electric connection of the microscopy channel. (**e**) Contours of PMMA and adhesive tape for laser cutting

5. Use standard electrophoresis or other high voltage laboratory power supplies are well suited as voltage sources, potentially controlled by a mechanical switch or a relay to reverse the direction of the electric field.
6. These microscopy chambers allow alignment of structures like the robotic arm not only by electric fields but also by fluid flow (*see* **Note 6**). To that end, attach the necessary tubing to the access ports and connect them to a peristaltic or syringe pump.

4 Design Considerations for Electrically Driven DNA Nanostructures

Electrically actuated DNA nanosystems can be regarded as non-autonomous nanomachines that perform mechanical movements controlled by externally applied electric fields, which enables the performance of specific tasks at the nanoscale.

A simple embodiment of such a machine is composed of an immobilized component that acts as a "stator" that is combined with a mobile component that acts as a "rotor." In our case, the stator is a DNA origami plate with addressable anchor positions for the attachment of various functionalities. Multiple biotinylated staples at the bottom of the base plate connect the stator element to the biotin-functionalized substrate via NeutrAvidin. A rod-like arm – essentially a DNA six-helix (6H) bundle – serves as the mobile rotor element, which also contains addressable attachment positions that can be designed to interact with positions on the plate. We can extend the arm with an additional, roughly 400 nm long, pointer structure. This pointer structure aids observation as well as actuation. The tip of the pointer can be easily followed when labeled with multiple fluorophores. The pointer places the fluorescent dyes far enough from the center of rotation to enable diffraction-limited high contrast imaging over extended periods of time. In addition, the large number of charges along the long pointer creates a strong lever force when the structure is actuated by electric fields.

A key element of the structure is the joint, which flexibly connects the two structural elements that are moved with respect to each other and which confines motion to the desired degrees of freedom. In the described example, both arm and platform are part of one DNA origami structure, folded using a single circular scaffold strand. Composed of two neighboring single-stranded scaffold segments with a length of 3–4 bases, the joint allows the arm to perform a complete rotation within the plate plane. By adjusting the position of the joint on the arm, the maximum accessible elevation of the arm above the platform can be tuned. The two scaffold connections ensure that the pulling force acting on the staples close to the joint is not applied in a mechanically unfavorable "unzipping" geometry for at least one of the two connecting strands, which prevents tearing the arm out of the structure [29]. Besides the strategy chosen here, various other bearings, hinges [30–32], and joints [33, 34] implemented with DNA nanostructures have been demonstrated and could find application in electrically driven DNA nanorobotics [35]. Examples are connections via staple strands which link two origami structures, biotin-streptavidin or other protein-mediated connections, organic linker elements [36], or mechanically interlocked motifs such as in rotaxanes or catenanes [11, 18, 37, 38].

In our roboarm design, application of larger forces (for instance, when using the nanorobotic arm to unzip DNA duplexes) was not possible when connecting the pointer structure head-on (co-linearly) with the arm on the platform, as this connection turned out not to be sufficiently stable. We therefore employed a mechanically more robust connection strategy, in which the pointer structure was "plugged" onto the arm using shape complementary "host-guest" interactions.

5 Observation and Microscopy

As the nanorobotic arm was specifically designed for relatively straightforward single molecule observation, typical inverted fluorescence microscopy setups are sufficient for tracking its movements in real time. By using an inverted microscope body, the sample chamber which contains the electrodes and buffer reservoirs can be placed on top of the sample (Fig. 3). Such a configuration avoids optical components above the sample chamber which would hinder handling of the sample. In the following, we describe a minimal configuration required for such measurements.

A 640 nm diode laser (Toptica iBeam smart, 150 mW) is used as excitation light source. The excitation beam is coupled into the microscope body (Olympus IX71) which holds a 100× TIRF objective (UAPON 100xO TIRF, NA 1.49 oil, Olympus). Total internal reflection fluorescence (TIRF) illumination reduces the background signal. The TIRF illumination angle is adjusted by a translational alignment stage that parallel shifts the beam. A dichroic mirror in the microscope body separates emission from the excitation light. For experiments that mostly require high sensitivity, the images are recorded with an Andor iXon 897 emCCD (Andor Technologies, Oxford Instruments). Measurements that require frame rates faster than 100 fps were performed with an Andor NEO sCMOS camera. A detailed instruction for building low-cost TIRF microscopes can be found in [27].

Depending on the particular type of experiment, additional microscopy techniques can be integrated. If more detailed information on small structure features is of interest, the super-resolution microscopy technique DNA-PAINT provides image data beyond the diffraction limit. This allows, for example, determination of the orientation of the base plate on the substrate. DNA-PAINT requires relatively strong excitation light and an autofocus system to maintain the focus position over extended observation periods [39].

Another way to study motion of the nanorobot and determine its positional state is multi-color FRET measurements. For instance, strategically positioned donor and acceptor dyes on plate and arm allow to detect DNA hybridization events that are induced by positioning the arm in close proximity. FRET requires an additional, short wavelength excitation light source and simultaneous observation of two different emission channels.

Dark field microscopy techniques can observe the movement of plasmonic nanoparticles attached to the structures. A dark field method compatible with TIRF illumination requiring no condenser can be implemented with a custom-built dot mirror [40].

6 A Sample Chamber for 2D Rotational Actuation

Compared to simple switching along one direction, control of the electric field in two dimensions requires a more elaborate electro-actuation chamber. A design that proved to be reliable, easy to handle, and reproducible is shown in Fig. 4. The chamber is constructed with the same sandwich principle described in the previous section and is based on a glass slide, tape, and a PMMA top part. However, as the design is quite delicate, PMMA chip and tape layer

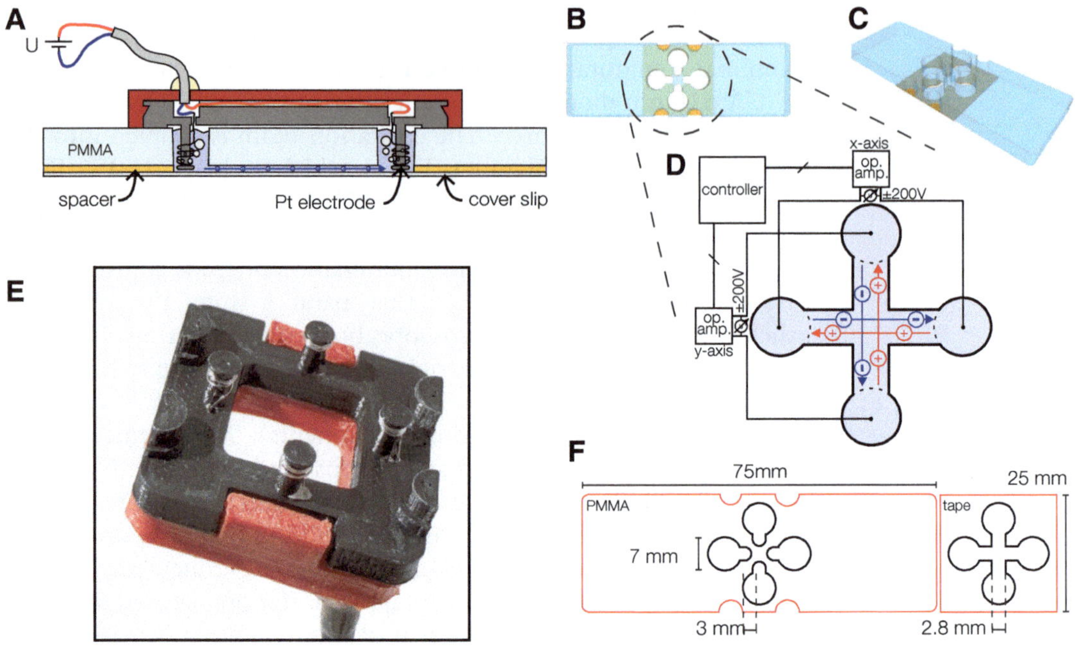

Fig. 4 Compact chamber for circular actuation. (**a**) Cross-section of the setup. Similar to the linear actuation setup, the sample chamber is formed with a PMMA chip, adhesive tape, and a cover slide. Here the buffer reservoirs are directly integrated into the PMMA chamber. The plug placed on top of the chip carries the four platinum electrodes and the connecting wires. (**b**) Top view. (**c**) Isometric view. (**d**) Diagram of the microscopy channel and the connections of the two electrode pairs. (**e**) Image of the 3D-printed plug. (**f**) Cut lines for the production of PMMA chip and adhesive tape

have to be fabricated with a more precise manufacturing technique, for example, by using a laser cutting machine. These easy-to-use instruments are available, e.g., in most publicly accessible "Maker-space" workshops. The chamber design is optimized for this production method and allows fast sample chip production in larger quantities. With a standard commercial laser cutter, batches of more than 50 PMMA chips and adhesive tape spacers at a time are possible. 5 mm thick PMMA is used as it is a transparent, low cost, electrically insulating material that can be easily machined. The chamber design is also compatible with fabrication by milling, allowing to use different materials, e.g., polymer materials or quartz glass. For the tape layer, an acrylic adhesive tape without carrier substrate (3M 467MP) is well suited as it seals tightly and can be easily processed via laser cutting. The particularly low thickness of the 50 μm tape reduces channel height and thereby focuses the voltage drop to the channel and reduces the current.

The two crossed channels each connect one pair of electrodes, enabling independent control over x- and y-component. Field superposition in the central observation area allows to produce arbitrary field directions in the x–y plane. The large reservoirs, containing about 200 μL of buffer, ensure reliable escape of gas bubbles, reduce rapid electrochemically induced changes in buffer conditions, and minimize electric access resistance. The geometry of the chamber is designed to reduce leak currents through the corners for diagonally applied electric fields while maximizing the central observation area, in which the deviation from the desired electric field direction is negligible. To identify this area, position-dependent electric field directions were determined with finite element simulations.

7 Custom Electrode Plugs for 2D Rotational Actuation

Sample chambers for angular actuation require a special plug to hold four high voltage electrodes safely in place inside the buffer reservoirs during the course of the experiment (Fig. 4a, e). Electrodes are connected to the high voltage amplifier with a 4-wire cable using a 4 pin LEMO connector. This plug also allows buffer exchange without removal of the plug or moving the chamber on the microscopy stage during the course of the experiment. The plug is fabricated from two parts, a base and top part that can snap into each other.

The base contains posts to fix the electrodes, channels for the connecting cables, and protrusions that allow to click the plug firmly into indentations in the PMMA chambers. The bottom part of the plug is 3D-printed with an Ultimaker 3 Extended fused filament fabrication machine with black polylactic acid (PLA) filament (Innofil3D, Emmen, Netherlands). Holes with 0.8 mm diameter are drilled next to the electrode posts, and platinum wire (2 cm, ø 0.2 mm) is threaded through these holes. The wire is wrapped around the posts, soldered to connecting the cables, and the holes are sealed with UHU Schnellfest (UHU GmbH, Bühl, Germany) epoxy glue.

We print the top part with flexible thermoplastic copolyester (TPC) filament (InnoFlex 45, Innofil3D, Emmen, Netherlands). Its flexibility allows a better handling and tighter fit of the lid for increased water resistance. The 4-wire cable is threaded to a 2 mm hole in one corner of the top part and fixed with glue to provide a strain relief connection after clicking it onto base part.

8 High Voltage Amplifier for Computer-Controlled Actuation (*See* Note 7)

In order to create programmed routines for variable electric field orientations and to achieve 2D control over the nanorobots, we initially generate control signals ±3.8 V for both x- and y-component of the electric field. These signals are produced by the control computer's data acquisition (DAQ) card

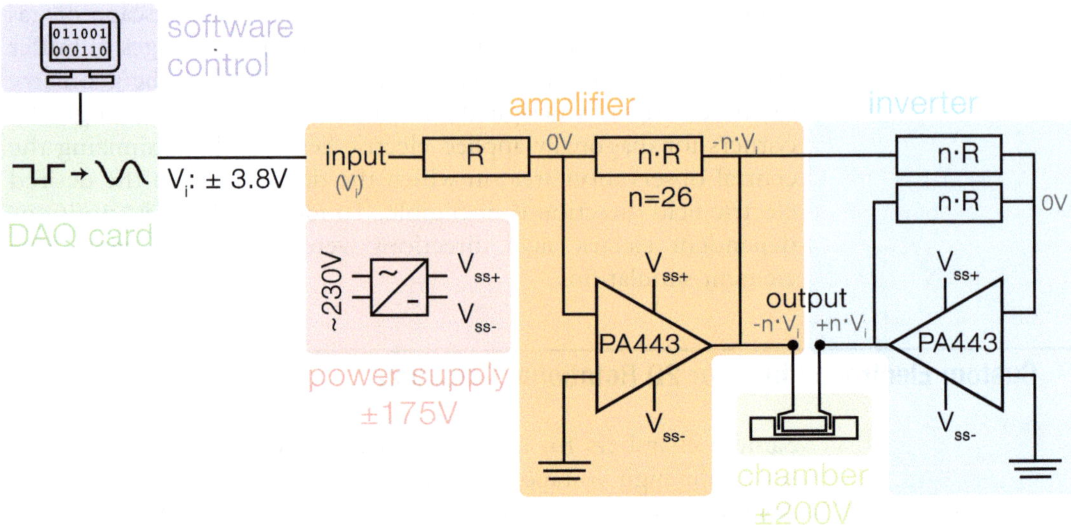

Fig. 5 Schematic electronic setup for one pair of electrodes. For programmed actuation routines, a computer provides the analog control voltages of ±3.85 V via a DAQ card. These inputs are amplified up to ±200 V by a home-built bridge amplifier. In this schematic circuit diagram, one operational amplifier drives one electrode with half the final gain, while a second operational amplifier, controlled by the output of the first amplifier, acts as an inverter (with gain = −1) to drive the corresponding second electrode. The amplifiers are supplied by switching power supplies

(NI PCI-6036E, National Instruments, USA) with a home-built LabView software. The signals are amplified to voltages of ±200 V, which unfortunately requires custom-built electronics as variable high DC voltages are uncommon for commercial electronics, especially if bipolar operation is necessary. Each channel is built as a bridge amplifier, containing two high voltage operational amplifiers (OpAmp) (Apex PA443, Apex Microtechnology), each driving one of the platinum electrodes (Fig. 5). While the gain of the first OpAmp is adjusted to half of the desired final gain, the second OpAmp acts as an inverter of the first signal creating the desired final gain. A switching power supply provides the required bipolar DC power supply of 175 V. Although limited by software to 200 V, the system can provide voltages of up to 350 V.

9 Overall Experimental Procedure

9.1 Setting Up an Experiment

In a typical experiment, an assembled chip is filled with buffer medium. Structures are added at low, nanomolar concentrations into the channel and flushed after a defined time period. Thereby, excess pointer structures that did not bind to arms are removed. The buffer medium is exchanged to the desired buffer for actuation (*see* **Note 8**).

The chip is placed on the microscope stage and the electrode containing plug is inserted. In the case of a 1D syringe system, the syringes are inserted and filled with buffer before placing the electrode holder pistons and starting actuation experiments. Avoiding air bubbles is critical to guarantee an appropriate electrical connection. Attachment of structures can also be performed on the microscope in situ, allowing to flush the chamber once the desired surface density of structures is reached. Liquid handling is usually performed by pipetting, but for charged species like DNA strands or nanostructures, these components can also be moved within the chip by electrophoresis using the actuation electronics. Best positioning results were obtained with a 0.5× TBE buffer containing 6 mM $MgCl_2$. Note that applying voltages to the chip can lead to changes in the z-position and require refocusing.

9.2 Tracking of Actuated Arms and Data Processing

Tracking of the pointer positions is necessary for a quantitative analysis of the microscope video data. The necessary tasks of fluorescent spot detection, 2D Gauss fitting of these spots, and drift correction are tasks that are routinely solved by software for super-resolution microscopy analysis. We borrowed these functionalities from the DNA-PAINT image reconstruction software package Picasso [39] that creates an event list of all pointer positions over all frames. Picasso is user-friendly and allows extremely fast event fitting when utilizing its GPU computing capabilities. The event list can be further processed with MATLAB analysis routines to determine the angular position of individual nanorobot arms over time.

10 Performance and Application Potential

We performed several proof of concept experiments to study the performance of our nanorobotic system and to explore potential future applications of electrically actuated DNA nanostructures [24].

The nano-positioning capabilities of our system were initially tested with various movement patterns and speeds. Figure 6a shows binary actuation and rotation of the arm without additional programmed interactions between arm and plate. By applying electrical fields, a positioning accuracy of 0.1 rad can be reached. Fast actuation routines were applied to examine the speed limits of the system as shown in Fig. 6b. The arm can follow rotating electrical fields of up to 25 actuation cycles per second, which also enables a rapid control of distances between molecules attached to the arm and to the platform.

Extension of staple strands on arm and plate with complementary single-stranded domains allows the arm to latch into defined positions on the platform. This facilitates switching of the nanorobotic arm between multiple, well-defined mechanical states (Fig. 6c).

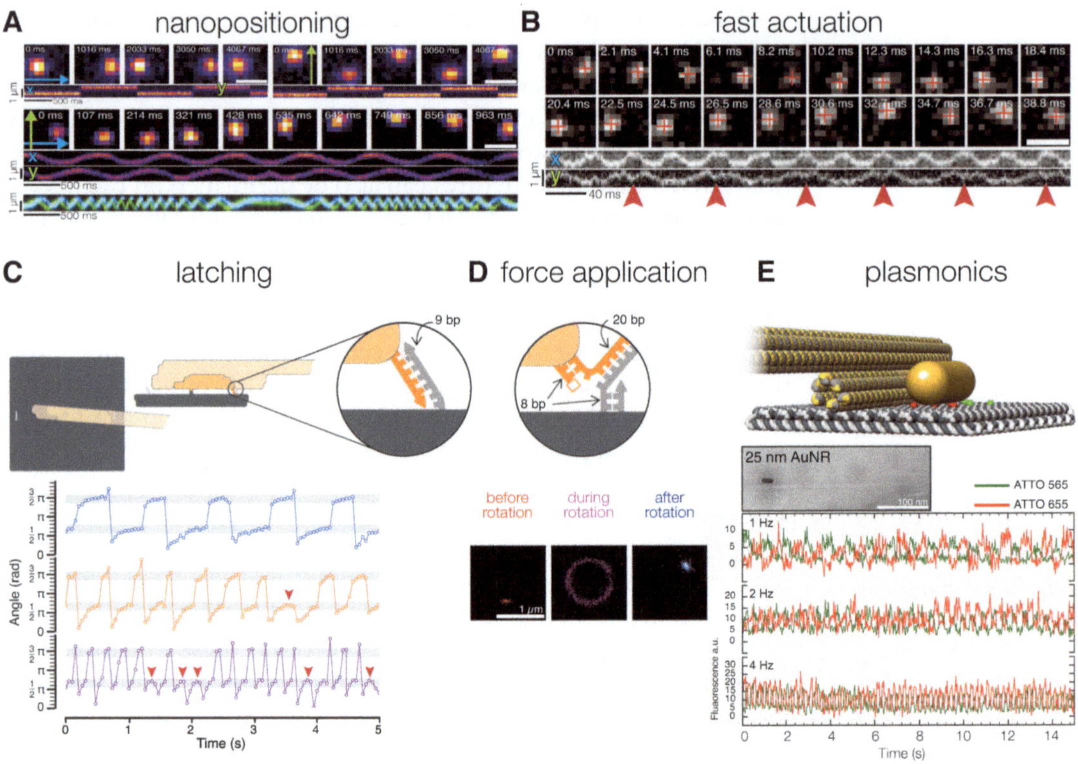

Fig. 6 Experiments realized with various actuation routines. (**a**) TIRFM images of accurate nano-positioning. Kymographs show the time-dependent position of the tip of the pointer extension. (**b**) Video snapshots and kymographs of fast actuation. The structures were rotated back and forth at 25 Hz. (**c**) Schematic configuration of the nanodevice for latching experiments. Metastable interaction sites on the base plate serve as positions for temporary latching as the arm passes. (**d**) Schematic configuration for force-induced unzipping of DNA duplexes. (**e**) Demonstration of actuated plasmonics. Two different species of fluorophores were immobilized on the platform and periodically quenched by a gold nanorod modified arm

The DNA domains defining these machine states can be constructed to be thermally stable, which causes the structures to remain in the defined state after the external control was switched off. For switching to another state, electrical actuation can be used to separate these domains via force-induced DNA melting (Fig. 6d).

In principle, these forces could be also used to probe other biomolecular interactions, indicating the potential application of our nanorobotic system as a test bench for single molecule force experiments.

We furthermore utilized the nanopositioning capabilities of the system to demonstrate its potential for the realization of switchable plasmonic devices [41] as shown in Fig. 6e. To this end, we attached a gold nanorod to the arm and placed two different species of fluorescent dyes on opposite sides of the base plate. Upon rotation, the fluorophores were quenched by the periodically passing gold nanorod. This can be observed as alternating fluorescent signals in two distinct emission channels. Apart from plasmonic metal particles, the system is capable of transporting any other inorganic or organic molecular cargo that can be attached to DNA.

11 Notes

1. A better coating can be provided by a Biotin-PEG-silane coating [24, 42]. Covalently coupled PEG-silane is more laborious to apply but provides a better protection against nonspecific binding of fluorescent impurities. Furthermore, this coating prevents sticking of the DNA arms to the substrate even without the addition of the surfactant. Covalently attached biotin moieties on the coating material allow the specific attachment of biotinylated DNA-structures via NeutrAvidin.
2. The coating has to fulfil the following requirements: (i) the coating must allow specific attachment of the static component; (ii) it has to passivate against sticking of the rotating or moving parts; and (iii) it has to prevent nonspecific binding of background contaminants.
3. While providing basic passivation against nonspecific binding of impurities and simultaneously facilitating specific binding of structures, unintended sticking of the arm to the substrate occurs frequently. Addition of a surfactant-like TWEEN 20 to the buffer medium to reduce this effect is necessary. A drawback of this surfactant is the stabilization of gas bubbles that occur on the electrodes during electrophoresis which can deteriorate the electric contact.
4. A similar geometry can be easily realized with commercial microscopy chambers (e.g., sticky-Slide VI 0.4, ibidi GmbH) which offer well-defined channel geometry and leak-free luer fittings and can be used with in-house functionalized coverslips.
5. We chose the outer dimensions of these chips to match the size of standard microscopy slides.
6. This actuation strategy is very simple to implement, but only allows slower actuation frequencies and requires comparatively large mechanical components.
7. Due to the dangers associated with the use of high voltages involved in these experiments, we here list precautions for constructing and handling high voltage equipment. These somewhat obvious measures ensure a safe operation and furthermore prevent laborious troubleshooting during the experiments. Besides the following suggestions, always follow the general and local safety guidelines for working with high voltages:

It is advisable to use only high-quality plugs, for example, LEMO plugs or safety type banana plugs and jacks with an insulating sheath covering the voltage carrying pins at all times. Cables should be well insulated, and connections should be protected by heat-shrink tubing. Shortening of circuit components – for example, with spilled buffer medium – must be avoided, especially for the platinum electrodes that cannot be sealed. Electrical components should be enclosed in appropriate housing to avoid electric shocks and mechanical damage. These housings should be properly insulated or grounded. Mechanical switches and software control further aid safe operation. Fuses and current limits for applied outputs can reduce the risks. In our studies, we used a 2.8 mA current limit for the amplifier stage. Note, however, that upstream components like the driving high voltage power supplies might operate at much higher maximum output currents. Besides electric shock, high power devices can also create fire hazards caused by extensive heating.

8. Typical experiments which observe the redundantly labeled pointers do not require optimized buffer solutions that reduce photobleaching. If the lifetime of the dyes needs to be prolonged, e.g., for single dye studies, typical photo-cocktails consisting of oxygen scavenging systems and triplet state quenchers are compatible with electric actuation. Two established oxygen scavenging systems are glucose/glucose oxidase/catalase [43] and glucose/protocatechuate 3,4-dioxygenase (PCD)/protocatechuic acid (PCA), both in combination with the triplet state quencher Trolox [44].

Acknowledgments

We gratefully acknowledge support by the Deutsche Forschungsgemeinschaft through SFB 1032 "Nanoagents" (TPA2).
Jonathan List gratefully acknowledges support by a generous stipend of the "Peter und Traudl Engelhorn-Stiftung".

References

1. Tikhomirov G, Petersen P, Qian L (2017) Fractal assembly of micrometre-scale DNA origami arrays with arbitrary patterns. Nature 552(7683):67–71. https://doi.org/10.1038/nature24655
2. Ong LL, Hanikel N, Yaghi OK, Grun C, Strauss MT, Bron P, Lai-Kee-Him J, Schueder F, Wang B, Wang P, Kishi JY, Myhrvold C, Zhu A, Jungmann R, Bellot G, Ke Y, Yin P (2017) Programmable self-assembly of three-dimensional nanostructures from 10,000 unique components. Nature 552(7683):72–77. https://doi.org/10.1038/nature24648
3. Rothemund PW (2006) Folding DNA to create nanoscale shapes and patterns. Nature 440(7082):297–302. https://doi.org/10.1038/nature04586
4. Douglas SM, Dietz H, Liedl T, Hogberg B, Graf F, Shih WM (2009) Self-assembly of

DNA into nanoscale three-dimensional shapes. Nature 459(7245):414–418. https://doi.org/10.1038/nature08016

5. Douglas SM, Marblestone AH, Teerapittayanon S, Vazquez A, Church GM, Shih WM (2009) Rapid prototyping of 3D DNA-origami shapes with caDNAno. Nucleic Acids Res 37(15):5001–5006. https://doi.org/10.1093/nar/gkp436
6. Zhang DY, Seelig G (2011) Dynamic DNA nanotechnology using strand-displacement reactions. Nat Chem 3(2):103–113. https://doi.org/10.1038/nchem.957
7. Yurke B, Turberfield AJ, Mills AP, Simmel FC, Neumann JL (2000) A DNA-fuelled molecular machine made of DNA. Nature 406(6796): 605–608
8. Wickham SFJ, Endo M, Katsuda Y, Hidaka K, Bath J, Sugiyama H, Turberfield AJ (2011) Direct observation of stepwise movement of a synthetic molecular transporter. Nat Nanotechnol 6(3):166–169
9. Omabegho T, Sha R, Seeman NC (2009) A bipedal DNA Brownian motor with coordinated legs. Science 324(5923):67–71. https://doi.org/10.1126/science.1170336
10. Gu H, Chao J, Xiao S-J, Seeman NC (2010) A proximity-based programmable DNA nanoscale assembly line. Nature 465(7295): 202–205. https://doi.org/10.1038/nature09026
11. Marras AE, Zhou L, Su H-J, Castro CE (2015) Programmable motion of DNA origami mechanisms. Proc Natl Acad Sci U S A 112(3):713–718. https://doi.org/10.1073/pnas.1408869112
12. Green S, Bath J, Turberfield A (2008) Coordinated chemomechanical cycles: a mechanism for autonomous molecular motion. Phys Rev Lett 101(23):art. no. 238101. https://doi.org/10.1103/PhysRevLett.101.238101
13. Gerling T, Wagenbauer KF, Neuner AM, Dietz H (2015) Dynamic DNA devices and assemblies formed by shape-complementary, non--base pairing 3D components. Science 347(6229):1446–1452. https://doi.org/10.1126/science.aaa5372
14. Yang Y, Tashiro R, Suzuki Y, Emura T, Hidaka K, Sugiyama H, Endo M (2017) A photoregulated DNA-based rotary system and direct observation of its rotational movement. Chemistry (Weinheim an der Bergstrasse, Germany) 23(16):3979–3985. https://doi.org/10.1002/chem.201605616
15. Lund K, Manzo AJ, Dabby N, Michelotti N, Johnson-Buck A, Nangreave J, Taylor S, Pei R, Stojanovic MN, Walter NG, Winfree E, Yan H (2010) Molecular robots guided by prescriptive landscapes. Nature 465(7295):206–210. https://doi.org/10.1038/nature09012
16. Pfitzner E, Wachauf C, Kilchherr F, Pelz B, Shih WM, Rief M, Dietz H (2013) Rigid DNA beams for high-resolution single-molecule mechanics. Angew Chem Int Ed Eng 52(30):7766–7771. https://doi.org/10.1002/anie.201302727
17. Lauback S, Mattioli KR, Marras AE, Armstrong M, Rudibaugh TP, Sooryakumar R, Castro CE (2018) Real-time magnetic actuation of DNA nanodevices via modular integration with stiff micro-levers. Nat Commun 9(1):1446. https://doi.org/10.1038/s41467-018-03601-5
18. List J, Falgenhauer E, Kopperger E, Pardatscher G, Simmel FC (2016) Long-range movement of large mechanically interlocked DNA nanostructures. Nat Commun 7: 12414. https://doi.org/10.1038/ncomms12414
19. Viovy JL (2000) Electrophoresis of DNA and other polyelectrolytes: physical mechanisms. Rev Mod Phys 72(3):813–872
20. Rant U, Arinaga K, Fujita S, Yokoyama N, Abstreiter G, Tornow M (2004) Dynamic electrical switching of DNA layers on a metal surface. Nano Lett 4(12):2441–2445. https://doi.org/10.1021/nl0484494
21. Kroener F, Traxler L, Heerwig A, Rant U, Mertig M (2018) Magnesium-dependent electrical actuation and stability of DNA origami rods. ACS Appl Mater Interfaces 11(2):2295–2301. https://doi.org/10.1021/acsami.8b18611
22. Klapper Y, Sinha N, Ng TWS, Lubrich D (2010) A rotational DNA nanomotor driven by an externally controlled electric field. Small 6(1):44–47. https://doi.org/10.1002/smll.200901106
23. Kuzyk A, Yurke B, Toppari JJ, Linko V, Torma P (2008) Dielectrophoretic trapping of DNA origami. Small 4(4):447–450. https://doi.org/10.1002/smll.200701320
24. Kopperger E, List J, Madhira S, Rothfischer F, Lamb DC, Simmel FC (2018) A self-assembled nanoscale robotic arm controlled by electric fields. Science 359(6373):296–301. https://doi.org/10.1126/science.aao4284
25. Castro CE, Kilchherr F, Kim DN, Shiao EL, Wauer T, Wortmann P, Bathe M, Dietz H (2011) A primer to scaffolded DNA origami. Nat Methods 8(3):221–229. https://doi.org/10.1038/nmeth.1570
26. Wagenbauer KF, Engelhardt FAS, Stahl E, Hechtl VK, Stommer P, Seebacher F, Meregalli L, Ketterer P, Gerling T, Dietz H

(2017) How we make DNA origami. Chembiochem 18(19):1873–1885. https://doi.org/10.1002/cbic.201700377

27. Auer A, Schlichthaerle T, Woehrstein JB, Schueder F, Strauss MT, Grabmayr H, Jungmann R (2018) Nanometer-scale multiplexed super-resolution imaging with an economic 3D-DNA-PAINT microscope. ChemPhysChem 19(22):3024–3034. https://doi.org/10.1002/cphc.201800630
28. Stahl E, Martin TG, Praetorius F, Dietz H (2014) Facile and scalable preparation of pure and dense DNA origami solutions. Angew Chem Int Edit 53(47):12735–12740. https://doi.org/10.1002/anie.201405991
29. Kufer SK, Puchner EM, Gumpp H, Liedl T, Gaub HE (2008) Single-molecule cut-and-paste surface assembly. Science (New York, NY) 319(5863):594–596. https://doi.org/10.1126/science.1151424
30. Douglas SM, Bachelet I, Church GM (2012) A logic-gated nanorobot for targeted transport of molecular payloads. Science 335(6070): 831–834. https://doi.org/10.1126/science.1214081
31. List J, Weber M, Simmel FC (2014) Hydrophobic actuation of a DNA origami bilayer structure. Angew Chem Int Ed Eng 53(16): 4236–4239. https://doi.org/10.1002/anie.201310259
32. Funke JJ, Dietz H (2016) Placing molecules with Bohr radius resolution using DNA origami. Nat Nanotechnol 11(1):47–52. https://doi.org/10.1038/nnano.2015.240
33. Kopperger E, Pirzer T, Simmel FC (2015) Diffusive transport of molecular cargo tethered to a DNA origami platform. Nano Lett 15(4): 2693–2699. https://doi.org/10.1021/acs.nanolett.5b00351
34. Tomaru T, Suzuki Y, Kawamata I, Nomura S-M, Murata S (2017) Stepping operation of a rotary DNA origami device. Chem Commun. https://doi.org/10.1039/C7CC03214E
35. Ijas H, Nummelin S, Shen B, Kostiainen MA, Linko V (2018) Dynamic DNA origami devices: from strand-displacement reactions to external-stimuli responsive systems. Int J Mol Sci 19(7). https://doi.org/10.3390/ijms19072114
36. Chen H, Zhang H, Pan J, Cha TG, Li S, Andreasson J, Choi JH (2016) Dynamic and progressive control of DNA origami conformation by modulating DNA helicity with chemical adducts. ACS Nano 10(5):4989–4996. https://doi.org/10.1021/acsnano.6b01339
37. Ketterer P, Willner EM, Dietz H (2016) Nanoscale rotary apparatus formed from tight-fitting 3D DNA components. Sci Adv 2(2): e1501209. https://doi.org/10.1126/sciadv.1501209
38. Powell JT, Akhuetie-Oni BO, Zhang Z, Lin C (2016) DNA origami rotaxanes: tailored synthesis and controlled structure switching. Angew Chem Int Ed Eng 55(38): 11412–11416. https://doi.org/10.1002/anie.201604621
39. Schnitzbauer J, Strauss MT, Schlichthaerle T, Schueder F, Jungmann R (2017) Super-resolution microscopy with DNA-PAINT. Nat Protoc 12(6):1198–1228. https://doi.org/10.1038/nprot.2017.024
40. Ueno H, Nishikawa S, Iino R, Tabata KV, Sakakihara S, Yanagida T, Noji H (2010) Simple dark-field microscopy with nanometer spatial precision and microsecond temporal resolution. Biophys J 98(9):2014–2023. https://doi.org/10.1016/j.bpj.2010.01.011
41. Kuzyk A, Urban MJ, Idili A, Ricci F, Liu N (2017) Selective control of reconfigurable chiral plasmonic metamolecules. Sci Adv 3(4): e1602803. https://doi.org/10.1126/sciadv.1602803
42. Jain A, Liu R, Xiang YK, Ha T (2012) Single-molecule pull-down for studying protein interactions. Nat Protoc 7(3):445–452. https://doi.org/10.1038/nprot.2011.452
43. Stein IH, Capone S, Smit JH, Baumann F, Cordes T, Tinnefeld P (2012) Linking single-molecule blinking to chromophore structure and redox potentials. ChemPhysChem 13(4): 931–937. https://doi.org/10.1002/cphc.201100820
44. Aitken CE, Marshall RA, Puglisi JD (2008) An oxygen scavenging system for improvement of dye stability in single-molecule fluorescence experiments. Biophys J 94(5):1826–1835. https://doi.org/10.1529/biophysj.107.117689

Chapter 16

Enzyme Cascade Reactions on DNA Origami Scaffold

Eiji Nakata, Huyen Dinh, Peng Lin, and Takashi Morii

Abstract

The protocols for constructing, characterizing, and analyzing enzyme cascade reaction systems on the DNA scaffold are described. Two-step and three-step enzyme cascade reactions were adapted from the xylose metabolic pathway as the example of natural metabolic pathway and were assembled on the DNA scaffold by using the DNA binding adaptors.

Key words Enzyme cascade reaction, DNA origami, DNA scaffold, DNA binding adaptor, Xylose metabolic pathway, AFM, Volume analysis, Assembling yield

1 Introduction

Biological systems to produce molecules, to convert energy, and to utilize them as vital activity are very complex and heterogeneous, but well-organized. These compounds are involved in biological processes, such as metabolism, which are mainly aided by cellular enzyme cascades. In these cascades, efficient transport of an intermediate is very important and is often driven by confining free diffusion in a compartment of spatially organized enzymes. When the enzymes are in close proximity to each other upon compartmentalization in the cell, the formation of byproducts is substantially reduced, leading to high turnover and reduction of unfavorable kinetics [1, 2]. When the enzymes are positioned near each other such that the intermediate produced by the first enzyme is processed efficiently by the second enzyme, before it diffuses in the bulk solution, a proximity effect is expected to enhance the sequential reaction [3, 4]. In order to understand the role of the spatial organization of enzymes, enzyme cascade reactions have been studied in vitro or inside the cell by immobilizing the enzymes on different scaffolds. Among the various macromolecules available for using as scaffold to create self-assembled artificial nanostructures, DNA-based nanostructures [5–10] are the promising biomacromolecules for construction of this complex,

Julián Valero (ed.), *DNA and RNA Origami: Methods and Protocols*, Methods in Molecular Biology, vol. 2639, https://doi.org/10.1007/978-1-0716-3028-0_16,

precisely designed, and refined nanostructure, in the field of bottom-up nanotechnology. DNA nanostructures prepared by the DNA origami technique have been applied as scaffolds for the spatial organization of enzymes to form artificial enzyme cascades, with which the efficient transportation of reaction intermediates is modeled [11–14]. The programmable nature of DNA nanostructures allows for the construction of a variety of spatially constrained enzyme assemblies. Analysis of a given sequential enzymatic reaction system in vitro enhances our understanding of the spatial factors and chemistry behind the highly efficient cascade reactions and provides a principle for further application of the current enzyme technology. There are several reviews focusing on the application of DNA nanostructures as templates for assembling enzyme cascades [15–17]. Furthermore, several methods have been developed for the conjugation of proteins of interest (POIs) such as enzymes with DNA. Each of these methods has its pros and cons; thus, the conjugation method should be carefully chosen depending on the purpose of experiment. In particular, both the efficiency of POI loading and their activity once assembled on the DNA scaffold are very important issues for the construction of the enzyme cascade reaction system and are significantly dependent on the choice of the method. Among them, the DNA binding adaptor-based method developed by us, which is classified into two categories, the reversible type DNA binding adaptor (*see* **Note 1**) and the irreversible type modular adaptor (*see* **Note 2**), successfully assemble recombinant enzymes in high loading yields with reasonable control on the number of enzyme molecules and the maintenance of their catalytic activities [18–21]. Although detailed characteristics of our developed DNA binding adaptor method are not described in this chapter (briefly described in *see* **Notes 1** and **2**), please refer to the original research papers [18–21] and reviews that include the comparison with other conjugation methods [22, 23].

Here, we focus on the protocol for constructing, characterizing, and analyzing the two-step and three-step enzyme cascade reaction systems on the DNA origami scaffold, adapted from the natural metabolic pathway (xylose metabolic pathway), by using the DNA binding adaptor-based method (*see* Fig. 1) [21, 24].

2 Materials

2.1 Buffers and Reagents

All the buffer solutions were prepared using ultrapure water (18.2 MΩ·cm at 25 °C).

1. Scaffold strand: M13mp18 bacteriophage single-stranded DNA (p7249) used as a scaffold strand was obtained from New England BioLabs (250 μg/mL, cat. No. N4040S) or Guild BioSciences (100 nM, cat. No. D441-010).

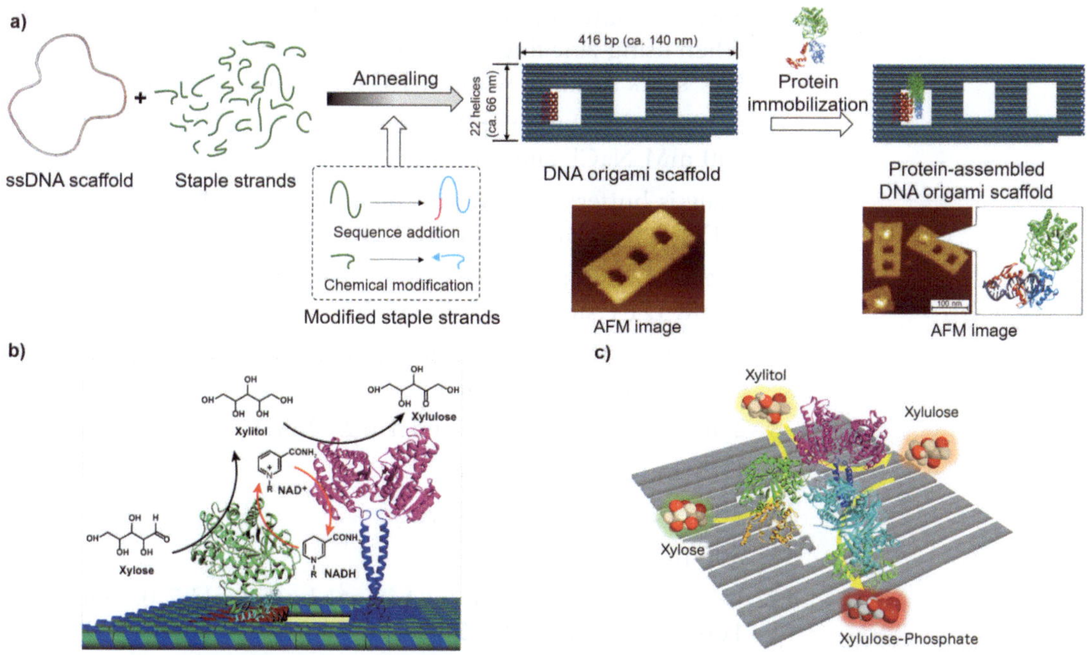

Fig. 1 (**a**) The general schematic illustration for the construction of a DNA origami scaffold containing the protein binding sites and the protein-assembled DNA origami scaffold. (**b**) Two-step cascade reaction [24] or (**c**) three-step cascade reaction [21] adapted from the xylose metabolic pathway assembled on the DNA scaffold

2. Prepare DNA staple strands (unmodified oligo DNA) (Thermo Fisher) at a 100 μM concentration, dissolved in TE buffer. Oligo DNA with amino-C6-dT at the defined position (Gene-Design, Inc. (Japan) or JBioS (Japan)). The tag-substrates, BG-GLA-NHS (New England Biolabs Japan, cat. No. S9151) for SNAP-tag [25] and BC-GLA-NHS (New England Biolabs Japan, cat. No. S9237) for CLIP-tag [26], and HaloTag Succinimidyl Ester (O2) Ligand (Promega, cat. No. P1691) for Halo-tag [27], were purchased from the respective companies or synthesized according to the previous report (*see* **Note 3**). The tag-substrate modified oligo DNA was purified by reversed-phase HPLC on a Cosmosil 5C18-MS II column (4.6 mm × 150 mm), elution with 100 mM triethylammonium acetate (TEAA) buffer (pH 7.0), linear gradient over 30 min from 2.5% to 30% acetonitrile at flow rate of 1.0 mL·min^{-1}) and characterized by MALDI-TOF-MS spectrometry (AXIMA-LNR, Shimadzu, HPA matrix).
3. 10× TAE-Mg^{2+} stock solution: 0.4 M Tris-base, 0.2 M acetic acid, 10 mM EDTA, and 125 mM $MgCl_2$. Adjust the pH of the solution to 8.3. This 10× stock solution was diluted to a 1× concentration. The pH of the buffer was adjusted from 7 to

9 without affecting the formation of the DNA nanostructure. When using zinc finger protein as the DNA binding adaptor, a 10× stock solution without EDTA should be prepared.

4. Equilibration buffer: 50 mM phosphate buffer, pH 7.0 with 200 mM NaCl, and 1 mM DTT.
5. Dialysis buffer: 50 mM phosphate buffer, pH 7.0 with 1 mM DTT, 50 μM $ZnCl_2$, and 50% glycerol.
6. Reaction buffer: 1× TAE-Mg^{2+}, pH 8.0 with 1 μM $ZnCl_2$, 1 mM DTT, and 0.02% Tween20.
7. Staining buffer: 1× TAE-Mg^{2+} buffer containing 0.5 μg/mL ethidium bromide (EtBr).
8. 2 and 0.15 mM NADH and 1 mM NAD^+.
9. 12.5 and 200 mM xylose.
10. 200 nM [1-^{3}H]-labeled xylose.
11. 300 mM xylitol.
12. Enzyme assay buffer: 1× TA-Mg^{2+} buffer, pH 7.0 containing 100 mM NaCl, 1 μM $ZnCl_2$, and 0.02% Tween20.
13. XK activity assay buffer: 1× TA-Mg^{2+} buffer (pH 7.6) containing 1 mM DTT, 1 μM $ZnCl_2$, 0.02% Tween20, 100 mM NaCl, 1.1 mM ATP, 2.3 mM phosphoenolpyruvate (PEP), 0.2 mM NADH, 4.8 units/mL PK, 4.5 units/mL LDH.
14. 3-enzyme cascade reaction buffer: 1× TA-Mg^{2+} buffer (pH 7.0) containing 1 mM DTT, 1 μM $ZnCl_2$, 0.02% Tween20, 100 mM NaCl, 2 mM NADH, and 1 mM ATP.
15. Pre-equilibrate Sephacryl S-400 HR resin thoroughly with 20 mM phosphate buffer followed by 1× TAE-Mg^{2+} buffer solution.

2.2 Equipment

1. Atomic force microscopy (AFM) imaging system (fast-scanning AFM system, Nano Live Vision, RIBM) with silicon nitride cantilever (Olympus BL-AC10DS-A2).
2. HPLC (high-performance liquid chromatograph) system.
3. MALDI-TOF Mass Spectrometer.
4. Agarose gel electrophoresis system with a buffer circulator connected to a cooling system.
5. Spectrophotometer (or microplate spectrophotometer).
6. Software for designing the DNA nanostructures: caDNAno 2.2.0 (https://cadnano.org/), Python 2.7.2 (https://www.python.org/). Microscope image analysis software: SPIP (Scanning Probe Image Processor, ver. 6.2.8, by Image Metrology), Gwyddion (GNU General Public License). Data analysis software: Origin Pro 9 (OriginLab Corporation).

3 Methods

3.1 Preparation of DNA Origami with the Modification Sites for Protein Assembly

3.1.1 Design of DNA Origami with Modification Sites for Protein Assembly

1. For DNA origami scaffold design, use Python 2.7.2 and caDNAno 2.2.0 (free, MIT license) software. Preferably use Autodesk Maya 2015 (free for students and educators, registration required) for visualizing DNA structures in the 3D space. For the details of the design of DNA origami from the beginning using caDNAno, please refer to the references [28, 29] and the tutorials.
2. Select staple strands at the desire locations on DNA origami scaffold. First, identify the position(s) for modification; they can be at the end point(s) (either 3′ or 5′) of the staple strand (in the case of DNA origami surface's modification) or at the crossover (when the modification sites are at the outer edge of DNA scaffold or inner sidewall of the cavity) (*see* Fig. 2a).
3. Insertion of a designed DNA sequence into the selected staple strand. As the specific binding site for the modular adaptor, the hairpin sequence, which includes the binding sequence for DNA-binding protein and amino-modified DNA for introducing the substrate with protein tags, is inserted (*see* Fig. 2a). The position of the modifier for DNA is decided based on the modeling data with the crystal structure of the complex formation, if available. Several modifiers of DNA can be applied as a modification site bearing the substrate of protein tags.

3.1.2 Preparation of Tag-Substrate Modified ODNs (See Fig. 2b–d)

1. Dissolve the succinimidyl derivative of protein tag-substrates (BG-GLA-NHS for SNAP-tag [25], BC-GLA-NHS for CLIP-tag [26], and HaloTag Succinimidyl Ester (O2) Ligand for Halo-tag [27]) in anhydrous DMSO (or DMF).
2. Perform coupling reaction between 100 μM amino-modified ODNs and 1 mM succinimidyl derivatives of the protein tag-substrates in a 100 mM phosphate buffer (pH 8.0) for 8 h at room temperature.
3. Purify the substrate-modified ODNs by reversed-phase HPLC using a Cosmosil $5C_{18}$-MS-II column.
4. Characterize the substrate modified ODNs by MALDI-TOF mass spectrometry (HPA matrix). The purity of the substrate modified ODN can be checked using reversed-phase HPLC.

3.1.3 Preparation and Purification of DNA Origami Scaffold with the Modification Sites (See Fig. 3a, b)

1. Prepare a stock solution of all the staple strands (master mix) mixing all the unmodified strands together with a final concentration of 500 nM for each staple strand. Stock solutions of all the modified strands should be prepared in the same manner and labeled as modified solutions (*see* **Note 4**).

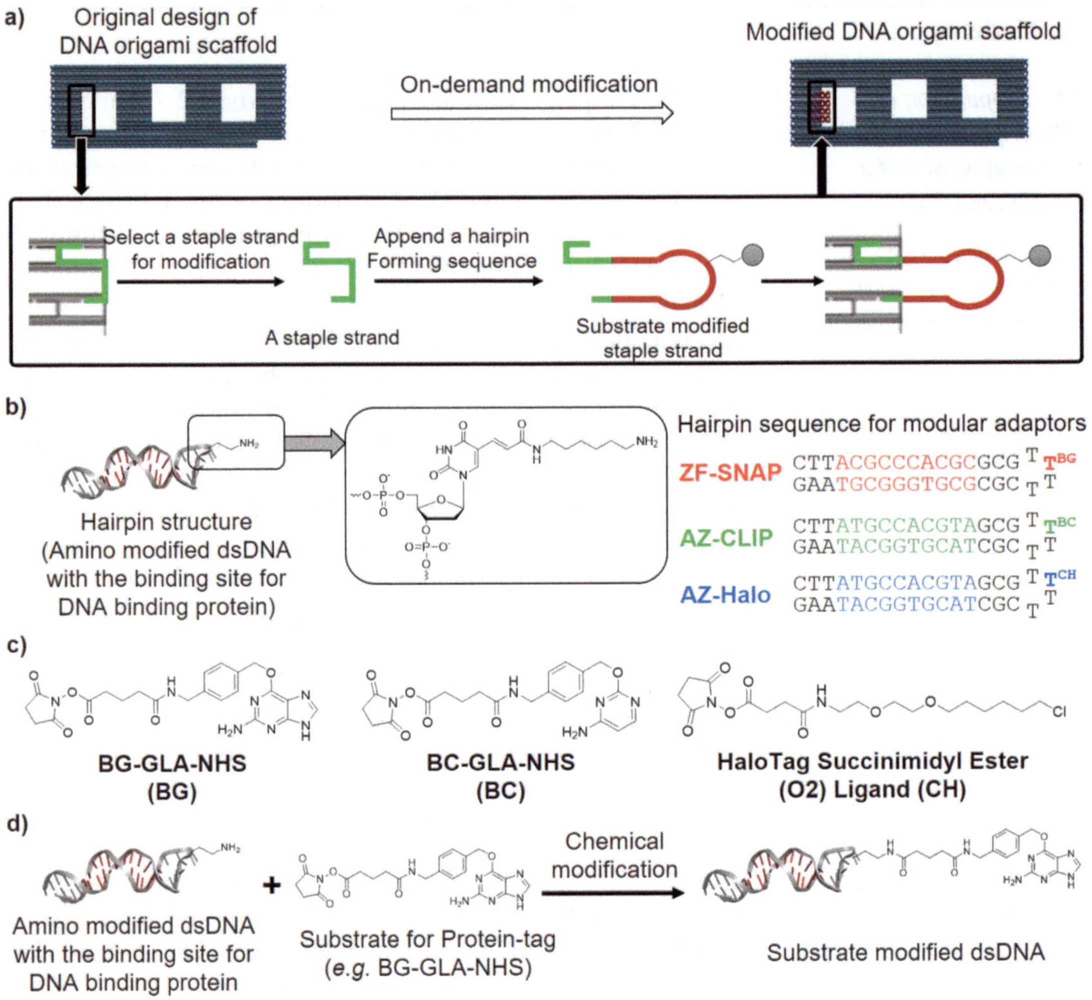

Fig. 2 (**a**) Scheme representing the modification at the desired position of staple strands to append the additional DNA sequence, including the chemically tethered functional molecule, such as the substrate of self-ligating protein tag. (**b**) An example of the DNA sequence of amino-modified ODN for forming the hairpin structure. (**c**) Chemical structure of the succinimidyl derivatives of protein tag substrates. (**d**) A scheme illustrates the modification of the target sequence of zif268 with the substrate of SNAP-tag (BG)

2. Place 50 μL of folding mix in a PCR tube. A typical reaction mixture contains 20 nM of scaffold strand (p7249) with a fivefold molar excess of each staple strand (unmodified and modified ones) in 1× TAE-Mg^{2+} buffer (pH 8.3) (*see* **Note 5** and Table 1).
3. Anneal the folding mix using a thermal cycler with the following thermal folding protocols: at 53 °C for 30 min (for single-layer DNA structures) or gradually decrease the temperature from 80 to 15 °C using a longer annealing program (for multiple layers or 3D DNA structures) [30, 31]. The folded DNA structures (without further purification) can be kept at 4 °C for several days to several weeks.

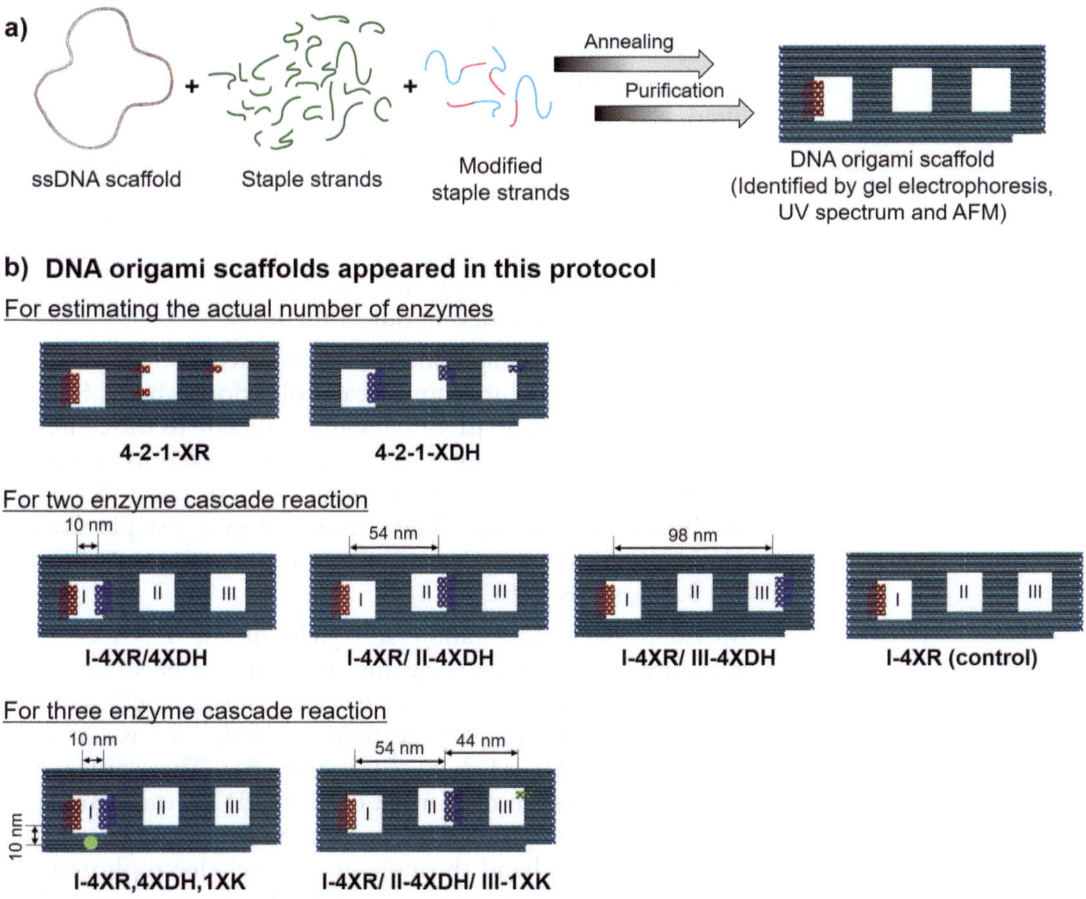

Fig. 3 (**a**) Preparation of a DNA origami scaffold with modification sites. (**b**) Different types of DNA origami scaffolds used in this protocol

Table 1
The folding mix for the DNA nanostructure

Components	Amount (μL)	Final concentration
Scaffold strand, 100 nM	10	20 nM
Master mix solution, 500 nM of each strand	10	100 nM of each strand
Modified solution, 500 nM of each strand	10	100 nM of each strand
Folding buffer, 10×	5	1×
H_2O	15	
Total	50	

4. For preparation of the gel-filtration column for the size-exclusion chromatography, add 500 μL of pre-equilibrated Sephacryl S-400 HR resin with 1× TAE-Mg^{2+} buffer to the filtering device, centrifuge for 4 min at 1000 × *g*, and discard the filtrate buffer. Wash the resin with 1× TAE-Mg^{2+} buffer three times (each time adding 500 μL of buffer) and centrifuge for 4 min at 1000 × *g*. The final centrifugation should be carried out without adding buffer to remove any remaining packing buffer. The columns should be used immediately after preparation to avoid drying out of the resin.
5. Remove the excess staple strands by size-exclusion chromatography. Place the column in a clean collection tube (1.5 mL), and carefully apply 50 μL of the sample directly to the center of the gel-filtration column. Centrifuge the column for 4 min at 1000 × *g*. The purified DNA origami structure is collected at the bottom of the tube. The removal of excess staple strands can be confirmed by agarose gel electrophoresis (*see* Subheading 3.3.2, **steps 1–6**).
6. Measure the concentration and yield of the DNA origami structure. The DNA origami structure concentration is measured using the estimated extinction coefficient at 260 nm (for p7249 scaffold strand, it is around $\varepsilon_{260} = 11.7 \times 10^7\ M^{-1}\cdot cm^{-1}$ (*see* **Note 6**)). There are several methods to confirm the formation of the DNA origami structure [6, 30]. Samples can be verified by the visualization of a single and sharp band through agarose gel electrophoresis or using an atomic force microscope (AFM) or transmission electron microscope (TEM) for visualization of the structure.

3.2 Design and Preparation of the DNA Binding Adaptor-Fused POI

Adaptor-fused POIs are prepared using two types of adaptors for assembly in a reversible [18, 19] or irreversible manner [20, 21] on the DNA scaffold depending on the purpose of experiment (*see* **Note 7**). In both the cases, the POI is genetically fused to the N- or C-terminal of the adaptor through an appropriate linker (*see* **Note 8**). Choosing the type of adaptor is very important when the POI forms an oligomer (*see* **Note 9**). As a typical example, the design and construction of ZF-SNAP-fused XR (ZS-XR) is described in the following procedure (*see* Fig. 4a). Other modular adaptor-fused POIs (AZ-CLIP-fused XK (AC-XK) and AZ-Halo-fused XK (AH-XK)) or adaptor-fused POI (G-XDH) used in this study are designed and prepared in the same manner.

1. Construction of vectors for ZS-XR (pET-30a-ZS-XR) (*see* Fig. 4b).

   Amplify by PCR the NADH-selective mutant of xylose reductase (XR) in the YEpM4 vector and ZF-SNAP in the pET-30a-ZF-SNAP vector, and purify the PCR products. The

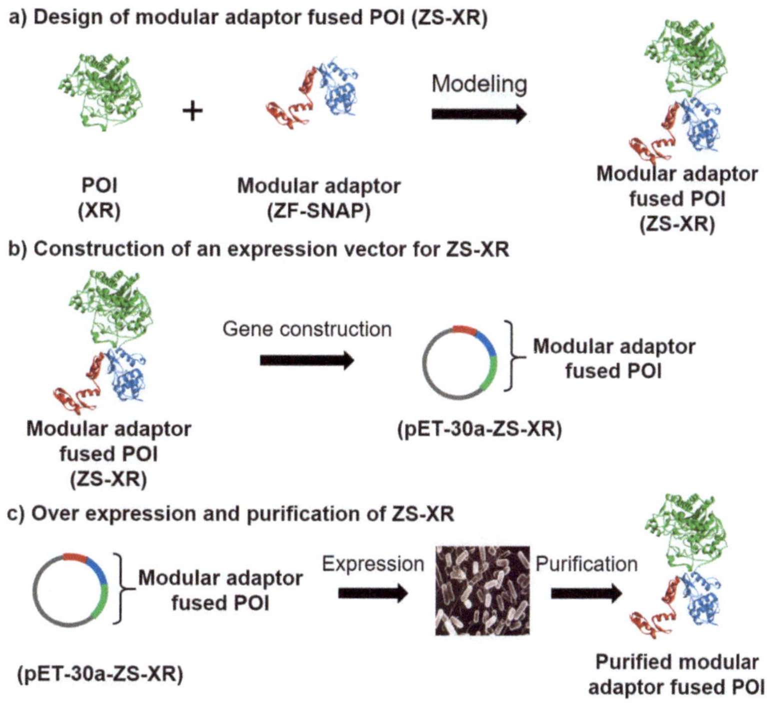

Fig. 4 (**a**) Design and preparation of modular adaptor-fused POI (ZS-XR). (**b**) Construction of vectors for ZS-XR. (**c**) Overexpression by *E. coli* and purification of ZS-XR

PCR products and pET-30a plasmid were digested with *Nde*I and *Hind*III and were purified, separately. Incubate the products with T4 DNA ligase. Transform the mixture into competent *E. coli* DH5α cells for amplification. Check the purity and sequence of the vector encoding ZS-XR (termed as pET-30a-ZS-XR), and transform into *E. coli* BL21(DE3) competent cells for overexpression.

2. Overexpression and purification of ZS-XR (*see* Fig. 4c).

   Grow the transformed cells at 37 °C until reaching an OD_{600} of 0.5 and induce protein expression with 1 mM IPTG for 24 h at 25 °C. Load the soluble fraction of the cell lysate containing recombinant protein to an HisTrap HP column equilibrated with equilibration buffer and elute using an imidazole gradient. Collect the main fractions containing the target protein, and load them to HiTrap SP HP column equilibrated with a 50 mM phosphate buffer (pH 7.0) containing 1 mM DTT and eluted by NaCl gradient. Dialyze the purified protein using dialysis buffer and stocked at −20 °C (*see* **Note 10**). Check the purity of the target protein by SDS-PAGE. For ZS-XR, the weight of the major band in SDS-PAGE corresponds to the calculated molecular weight with an estimated purity of over 95%.

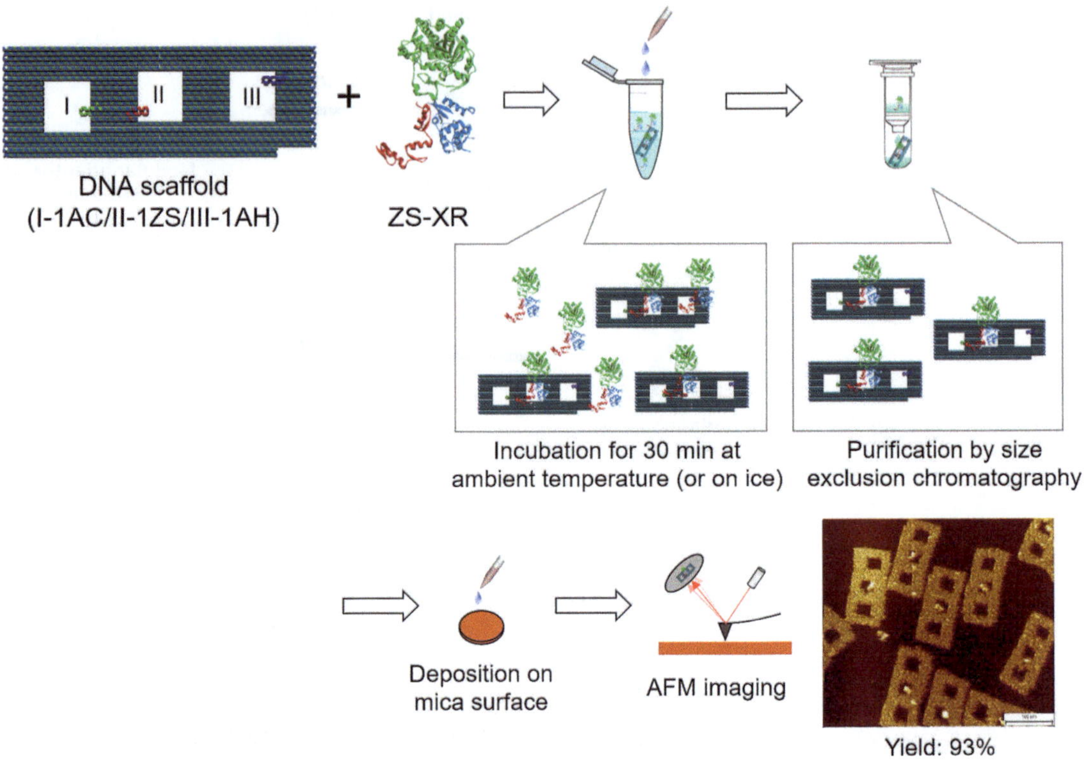

Fig. 5 Scheme shows typical procedures to assemble ZS-XR on a 3-well DNA scaffold. The yields of assembly at the expected or unexpected positions were estimated as described (*see* Subheading 3.3.2)

3. MALDI-TOF mass spectrometry confirmation.

   The modular adaptors or enzyme-fused modular adaptor should be characterized by MALDI-TOF mass spectrometry (SA matrix).

3.3 Preparation and Characterization of Adaptor-Fused POI Assembled DNA Origami Scaffold

As a typical example, the ZS-XR assembled on a 3-well DNA scaffold with an ZF-SNAP binding site is described (*see* Fig. 5). Other modular adaptor-fused POIs (AZ-CLIP-fused XK (AC-XK) and AZ-Halo-fused XK (AH-XK)) and/or adaptor-fused POI (G-XDH) can be assembled on the DNA scaffold in the same manner.

3.3.1 Preparation of POI Assembled DNA Origami Scaffold

1. Incubate 5 nM of purified DNA origami scaffold with 100 nM ZS-XR in reaction buffer for 30 min on ice.
2. Purify the mixture using size-exclusion chromatography (500 μL volume of Sephacryl S-400 in Ultrafree-MC-DV) to remove the unbound ZS-XR. The collected fractions containing DNA origami scaffold are identified as follows (Subheading 3.3.2).

3.3.2 Characterization of the POI Assembled DNA Origami Scaffold

1. For the identification of DNA origami scaffolds by agarose gel electrophoresis, first prepare 1% agarose gel. Fill the electrophoresis chamber with 1× TAE-Mg^{2+} buffer to cover the gel. To prevent heating up of buffer during electrophoresis, the chamber is connected to a circulation buffer cooling system.
2. Mix gently 10 μL of sample with 2 μL of 6× loading buffer and carefully load 10 μL of the mix to each well of the gel. A typical set of samples contains 1 kbp DNA ladders as the marker, the scaffold strand (M13mp18), the folded DNA origami scaffold before and after purification, and the POI-assembled DNA scaffold before and after purification.
3. Run the gel at constant voltage (50 V) for 2 h (for the mini gel of 6 cm in length). The running time is subjected to change according to the gel concentration and the length of gel.
4. Stain the gel with staining buffer for 20 min. Wash the gel with distilled water to remove excess of EtBr (*see* **Note 11**).
5. Analyze the DNA bands using UV transilluminator (when stained with EtBr) or other gel imaging system. The expected result should visualize a clear, single band corresponding to the well-folded DNA origami scaffold and the fuzzy, faster migrating bands of staple strand mixture. The absence of staple strand mixture in the purified DNA origami sample indicates the success of the purification step. A clear band shift with slower migration is often observed for the POI assembled DNA scaffold due to the mass increase upon protein loading.
6. Measure the band intensity using gel-analysis software: ImageJ (NIH image) or Quantity One (Bio-Rad). The band intensities of the sample before and after purification indicate the recovery yield after the purification process.
7. For the determination of the concentration of POI assembled DNA origami scaffold, measure the absorbance of the sample at 260 nm by an spectrophotometer such as Nanodrop.
8. Determine the molar absorbance coefficient of DNA origami at 260 nm (e.g., for p7249 scaffold strand, $\varepsilon_{260} = 11.7 \times 10^7\ M^{-1} \cdot cm^{-1}$) (*see* **Note 6**). The molar absorbance coefficients of modular adaptor-fused POI at 260 nm (e.g., ZS-XR and G-XDH (monomer) extinction coefficients were determined to be $(5.4 \pm 0.1) \times 10^4\ M^{-1}\ cm^{-1}$ and $(1.1 \pm 0.1) \times 10^4\ M^{-1}\ cm^{-1}$, respectively) were negligible to the value of DNA origami alone.
9. Calculate the concentration of POI-assembled DNA origami after purification by using the molar absorbance coefficient of DNA origami.

   Atomic force microscopy (AFM) measurements (**steps 10–12**) are conducted to determine the assembling yield, the actual number, and the inter-enzyme distance of ZS-XR and G-XDH molecules bound to the DNA origami scaffold (**steps 13–19**).

10. A sample is diluted to 1 nM concentration in 1× TAE-Mg^{2+} buffer. Apply 2 μL of the sample to the freshly cleaved, flat mica surface. Incubate for 5 min at ambient temperature in a sealed cover to prevent the sample to be dried up. Rinse the mica surface to remove unbound samples with 20–50 μL of 1× TAE-Mg^{2+} buffer solution. Avoid to totally dry up the mica surface.
11. Measure the sample with a fast-scanning AFM system with a silicon nitride cantilever (Olympus BL-AC10DS-A2) under tapping mode in 1× TAE-Mg^{2+} buffer solution. For high-resolution imaging, such small cantilever custom-made by Olympus is recommended. Take a clear image of DNA structures with POI while maintaining a low imaging force (*see* **Note 12**).
12. To analyze AFM data, export the raw file in AFM software to microscopy file (.bcrf format) for further analysis or general file (.jpg format) or video files according to the purpose. The POIs assembled on DNA origami scaffold are shown in a brighter color dot because of the different height compared to the surface of DNA origami structure. It is easy to detect and count the number of POI molecules on the DNA origami scaffold based on the color image. The height and distance analyses are carried out by the built-in software of AFM system or using the popular microscope image analysis software, such as Gwyddion (free and open source software) or a commercial software SPIP (by Image Metrology). Those software provide basic functions, such as cleaning and enhancing data, analyzing measurements (height or distance analysis), visualizing, and reporting analysis results.
13. To determine the assembling yield of ZS-XR or G-XDH on the DNA origami scaffold, the following procedures are conducted.

    The yield of ZS-XR on DNA scaffold (P_{specific}) is calculated as the percentage of the number of DNA scaffolds bearing ZS-XR at the expected position ($N_{\text{expected posi}}$) over the total number of well-formed DNA scaffold (N_{total}):

$$P_{\text{specific}} = \left(N_{\text{expected posi}}/N_{\text{total}}\right) \times 100$$

    The yield of ZS-XR located at unexpected positions ($P_{\text{non-specific}}$) is calculated as the percentage of cavities modified non-specifically by ZS-XR ($N_{\text{unexpected posi}}$) over the total number of cavities without ZS-XR binding sites on well-formed DNA scaffold ($2N_{\text{total}}$):

$$P_{\text{nonspecific}} = \left(N_{\text{unexpected posi}}/2N_{\text{total}}\right) \times 100$$

Typical examples are found in our previous reports [18–21, 24, 32].

14. To determine the co-assembling yield of ZS-XR and G-XDH on the DNA origami scaffold, the following procedure is conducted.

    Prior to determining the co-assembling yield of ZS-XR and G-XDH on the DNA scaffold, the yield of assembled ZS-XR on the DNA scaffold is calculated as described above. The yield of co-assembled ZS-XR and G-XDH on DNA scaffold ($P_{\text{coassembly}}$) is calculated as the percentage of the number of modified DNA scaffolds bearing the two enzymes (ZS-XR and G-XDH) at the expected position on the DNA scaffold ($N_{\text{expected posi}}$) over the total number of well-formed DNA scaffold (N_{total}) (*see* Fig. 6):

$$P_{\text{coassembly}} = \left(N_{\text{expected posi}}/N_{\text{total}}\right) \times 100$$

    The yield of ZS-XR and/or G-XDH found in nonspecific positions ($P_{\text{nonspecific}}$) is calculated as the percentage of the number of cavities modified nonspecifically by ZS-XR and/or G-XDH ($N_{\text{unexpected posi}}$) over the total number of cavities (without binding sites) of well-formed DNA scaffold (N_{total}):

$$\text{I-4XR/4XDH}: P_{\text{nonspecific}} = \left(N_{\text{unexpected posi}}/2N_{\text{total}}\right) \times 100$$

$$\text{I-4XR/II-4XDH or I-4XR/III-4XDH}: P_{\text{nonspecific}} = \left(N_{\text{unexpected posi}}/N_{\text{total}}\right) \times 100$$

15. The observed inter-enzyme distance (center-to-center) between ZS-XR and G-XDH is estimated from the AFM images by using SPIP software, "Cross section profile" function.

    To estimate the actual number of ZS-XR and G-XDH molecules on the DNA scaffold, volume analyses of AFM images (*see* **Note 13**) are conducted.

16. To evaluate the actual number of ZS-XR molecules (*see* Fig. 7a) bound to the predesigned binding sites on the DNA scaffold, the volumes of ZS-XR in each cavity of DNA scaffold (4-2-1-XR, *see* Fig. 4b), which is designed to have four binding sites for ZS-XR inside cavity I, two in cavity II, and one in cavity III (*see* Fig. 7b), are determined using volume analyses of AFM images (*see* Fig. 7c) and listed as frequency distribution of molecular volumes of ZS-XR (*see* Fig. 7d). The average volume of ZS-XR in each cavity of 4-2-1-XR is determined as 204 ± 40 nm^3 (cavity III), 363 ± 98 nm^3 (cavity II), and 778 ± 115 nm^3 (cavity I), which corresponds to the volumes

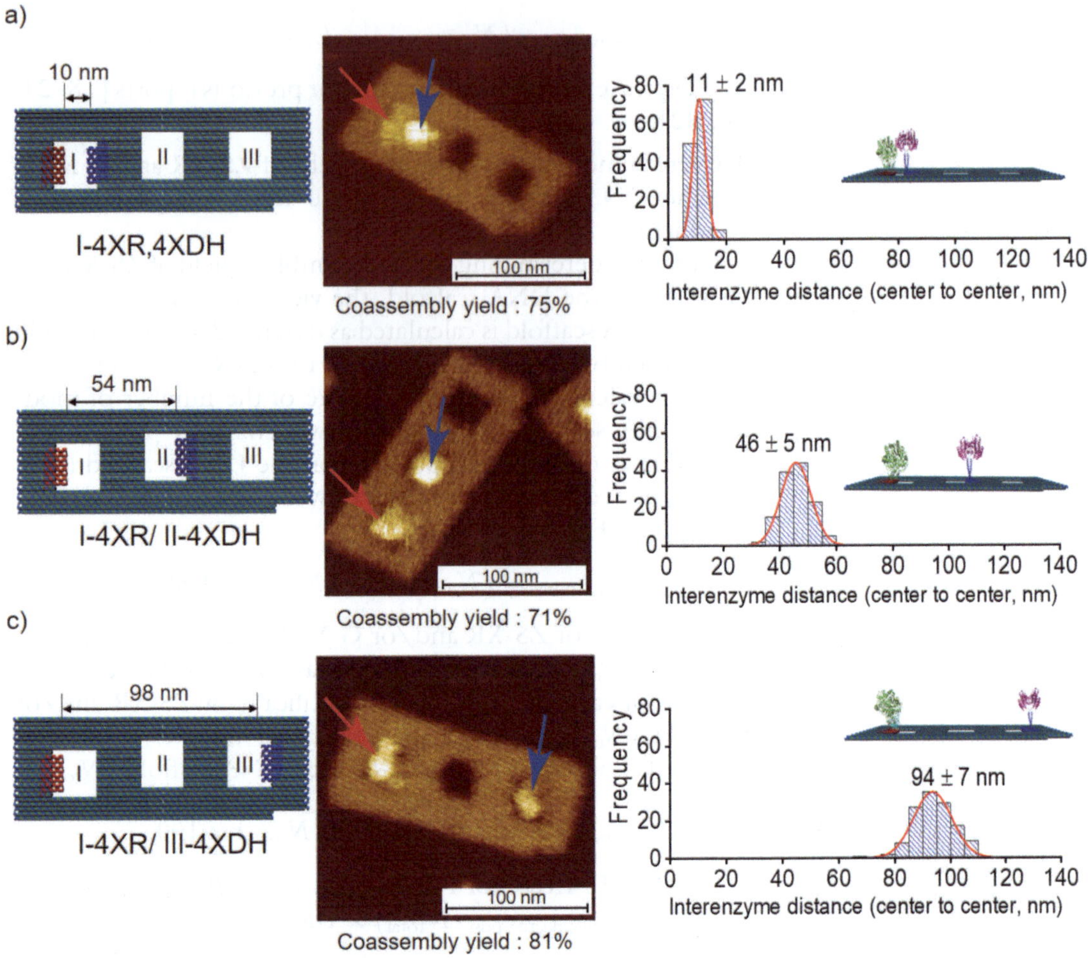

Fig. 6 Distance dependent co-assembly of enzymes inside the cavity of the DNA scaffold and analysis of the inter-enzyme distance (**a**) 10 nm, **b**) 54 nm, **c**) 98 nm). (**left**) Illustrations of the DNA scaffold containing a set of four binding sites for ZS-XR (red) and G-XDH (blue). (**middle**) AFM images of the DNA scaffold with bound enzymes. The red and blue arrows indicate ZS-XR and G-XDH, respectively. (**right**) Statistical analyses of the observed inter-enzyme distances (center-to-center) between ZS-XR and G-XDH

occupied by one, two, and four molecules of ZS-XR at the specific binding sites on the DNA scaffold (*see* Fig. 7d). Standard curve for the number of ZS-XR molecules versus molecular volumes is determined as shown in Fig. 7e.

17. In the case of G-XDH (*see* Fig. 7f), the volume of G-XDH in each cavity of the DNA scaffold (4-2-1-XDH, *see* Fig. 4b), which is designed to have four binding sites for G-XDH inside cavity I, two in cavity II, and one in cavity III (*see* Fig. 7g), is determined using volume analyses of AFM images (*see* Fig. 7h) and listed as frequency distribution of molecular volumes of - G-XDH (*see* Fig. 7i). In cavity III, unique distribution corresponding to one G-XDH dimer is observed

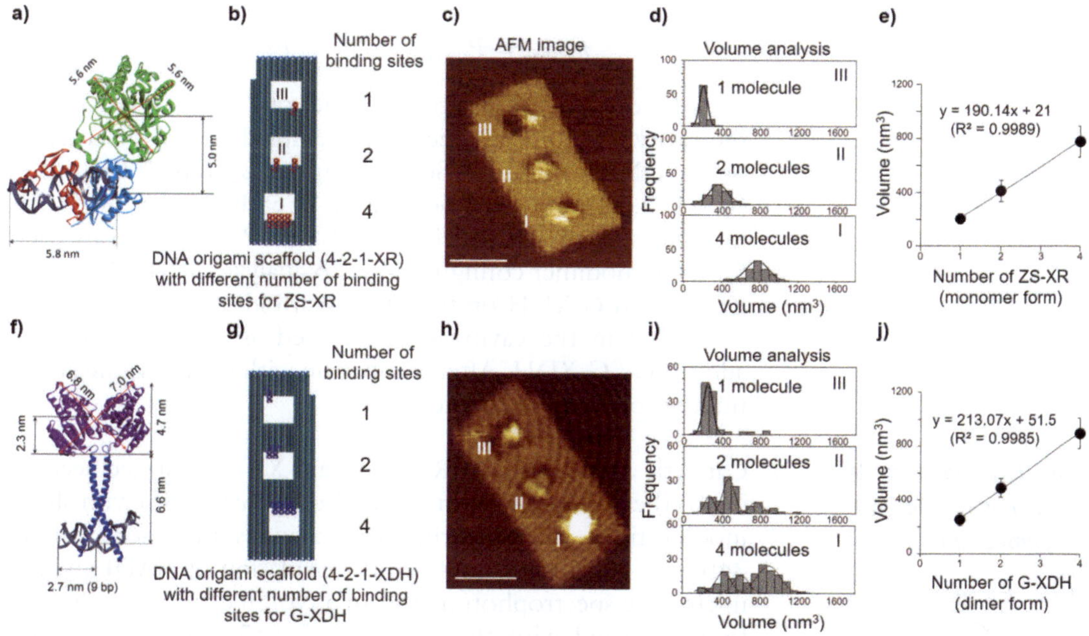

Fig. 7 (**a**) A molecular model for the complex of ZS-XR and a hairpin DNA [24]. (**b**) Illustration of the DNA origami scaffold (4-2-1-XR) with different numbers of binding sites for ZF-XR. (**c**) An AFM image of ZS-XR bound to the DNA origami scaffold (4-2-1-XR). The scale bars represent 100 nm. (**d**) Frequency distributions of molecular volumes of ZS-XR for each cavity (I, II, III) with different numbers of binding sites. (**e**) Standard curve for the number of binding sites for ZS-XR, which indicates the maximum number of bound ZS-XR molecules versus molecular volumes. (**f**) A molecular model for the complex of G-XDH and a hairpin DNA [19]. (**g**) Illustration of the DNA origami scaffold (4-2-1-XDH) with different numbers of binding sites for G-XDH. (**h**) An AFM image of G-XDH bound to the DNA origami scaffold (4-2-1-XDH). The scale bars represent 100 nm. (**i**) Frequency distributions of molecular volumes of G-XDH for each cavity (I, II, III) with different numbers of binding sites. (**j**) Standard curve for the number of binding sites for G-XDH, which indicates the maximum number of bound G-XDH molecules versus molecular volumes

(255 ± 47 nm^3). In contrast, frequency distributions covered a broader range of volumes in the case of cavities II and I, indicating a distribution from one to four G-XDH dimers in the cavity. In cavity II, the major fraction (58%) is centered at 492 ± 70 nm^3, which corresponds well to the volume of two G-XDH dimers. In cavity I, the major fraction (50%) is centered at 899 ± 113 nm^3, which corresponds well to the volume of four G-XDH dimers (*see* Fig. 7i). Standard curve for the number of dimeric G-XDH molecules versus molecular volumes is obtained as shown in Fig. 7j.

18. These standard curves (*see* Fig. 7e, j) are used to determine the actual number of ZS-XR molecules and G-XDH molecules (in the dimeric form) bound to the specific DNA sequences on the DNA scaffold.

19. The actual number of bound ZS-XR or G-XDH is determined by using the following equation:

$$n_{\text{actual}} = P_{\text{specific}} \times \sum_{k=1}^{4} F_k \times k$$

where n_{actual} is the actual number of ZS-XR or G-XDH in the cavity, P_{specific} is the yield of assembled ZS-XR or G-XDH on the DNA scaffold (*see* Subheading 3.3.2, **steps 7–9**), and F_k ($k = 1, 2, 3$, or 4) is the population of each number of ZS-XR molecules. In the case of G-XDH, F_k ($k = 2, 4, 6$, or 8) are used for its homodimer configuration. To analyze the co-assembled ZS-XR and G-XDH on the DNA scaffold, the actual number of ZS-XR in the cavity is determined as above, before the addition of G-XDH. After incubation with G-XDH, the actual number of G-XDH is determined.

3.4 Enzyme Cascade Reactions on the DNA Origami Scaffold

3.4.1 Enzyme Assay for a Single Enzyme

1. Catalytic activity of ZS-XR (or mutant XR) is analyzed according to the previously reported methods (*see* Fig. 8a) with slight modifications by measuring the changes in absorbance at 340 nm (25 °C) derived from the oxidation of NADH with a microplate spectrophotometer. In a typical experiment, a reaction is started with the addition of 0.15 mM NADH to a mixture of 25 nM ZS-XR and 200 mM xylose in enzyme assay buffer.
2. Catalytic activity of G-XDH is analyzed according to the previously reported methods (*see* Fig. 8b) with slight modifications by measuring the changes in absorbance at 340 nm (25 °C) derived from the reduction of NAD^+ with a microplate spectrophotometer. In a typical experiment, a reaction is started with the addition of 1 mM NAD^+ to a mixture of 56 nM G-XDH and 300 mM xylitol in enzyme assay buffer.

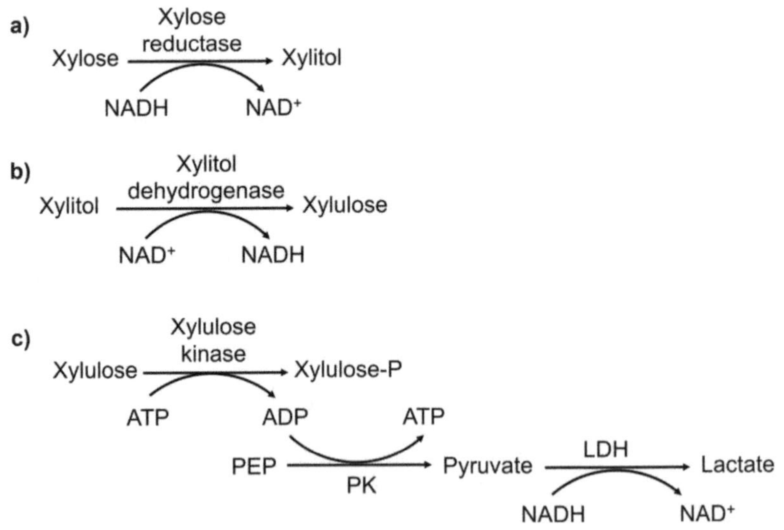

Fig. 8 Enzyme reactions of (**a**) XR derivatives or (**b**) XDH derivatives. (**c**) Coupled enzyme assay for studying the reaction of XK derivatives

3. Catalytic activity of xylulose kinase derivatives (AC-XK, and AH-XK) is analyzed by the coupled enzyme assay (*see* Fig. 8c) with pyruvate kinase (PK) and lactate dehydrogenase (LDH) as described in the previous report with slight modification [33, 34]. The XK activity is measured by monitoring NADH oxidation by LDH at 340 nm with a microplate spectrophotometer. The assays are performed in XK activity assay buffer, and the indicated concentrations of xylulose (0.03–2 mM). The reaction is started by the addition of xylulose and carried out at 25 °C.

3.4.2 Two Enzyme Cascade Reaction on DNA Origami Scaffold

1. In the bimolecular intermediates' transport system (xylitol and NAD^+), the reaction is started with the addition of 2 mM NADH to a mixture of 21 nM ZS-XR and/or G-XDH located on the DNA scaffold and 12.5 mM xylose in enzyme assay buffer, and the progress of reaction is monitored by measuring the time-course changes of absorbance at 340 nm of NADH. The reaction is also monitored by using [1-^{3}H]-labeled xylose as the substrate. The production of xylitol and xylulose from 200 nM [1-^{3}H] xylose is analyzed by HPLC (*see* Fig. 9). The production of xylitol in the reaction containing 200 nM [1-^{3}H] xylose and 200 mM non-labeled xylose is determined by HPLC. The HPLC conditions are as follows: COSMOSIL

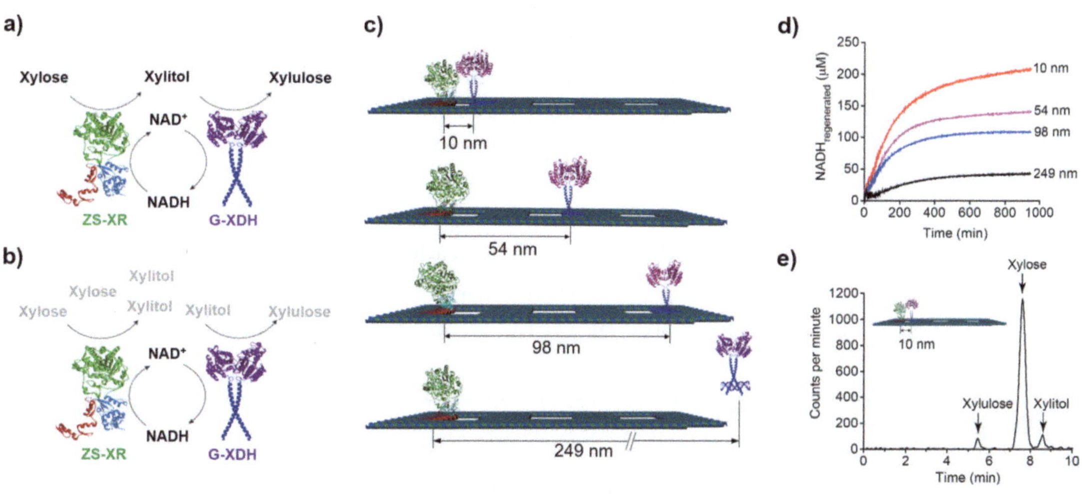

Fig. 9 Artificial enzyme cascade reactions of the xylose pathway assembled on the DNA scaffold based on the D-xylose metabolic pathway with (**a**) bimolecular intermediates' (xylitol and NAD^+) transport system and (**b**) unimolecular (NAD^+) transport system. (**c**) Illustration of assembled ZS-XR with or without G-XDH on the DNA scaffold with different inter-enzyme distances. (**d**) Time-course reaction profiles for NADH when two enzymes were co-assembled with inter-enzyme distances of 10, 54, and 98 nm and for the free diffusion system, in which ZS-XR was located on DNA scaffold, while G-XDH was free in solution with a theoretically estimated interspacing distance of 249 nm (*see* **Note 14**). (**e**) HPLC chromatogram shows the production of xylitol and xylulose in the reaction mixture upon incubation with [1-^{3}H] xylose for 16 h with the inter-enzyme distance of 10 nm (inset)

parked column sugar-D (4.6 i.d. × 250 mm) with isocratic elution solvent (acetonitrile/water = 80:20), flow rate 1 mL/min, and column temperature 30 °C detected by radiodetectors (β-RAM 5C Lablogic).

2. For the unimolecular (NAD^+) transport reaction system, the reaction conditions of the bimolecular system, except for the presence of 300 mM xylitol, are applied. The reaction progress is monitored by measuring the time-course changes in absorbance at 340 nm.
3. The processes to normalize the initial rate of NADH regeneration and the concentration of the products are described in *see* **Note 15**. The turnover frequency for each system is calculated from the normalized concentration of the products divided by the concentration of the second enzyme G-XDH on DNA scaffold in the time unit.

3.4.3 Three Enzyme Cascade Reaction on DNA Origami Scaffold

1. An enzyme cascade reaction catalyzed by three enzymes (ZS-XR, G-XDH, and AC-XK) is designed and analyzed on the three cavity DNA scaffold (*see* Fig. 10) with or without the binding sites for ZS-XR at cavity I (4 binding sites), G-XDH at cavity I or II (4 binding sites), and AC-XK at cavity I or III (1 binding site). The calculated distances between the binding site of ZS-XR and G-XDH represented as d(ZSXR/G-XDH) and G-XDH and AC-XK represented as d(G-XDH/ACXK) are determined as remarked in Fig. 10. DNA scaffold (I-4XR/II-4XDH/III-1XK): d(ZS-XR/G-XDH) = 54 nm, d(G-XDH/AC-XK) = 44 nm. DNA scaffold (I-4XR/I-4XDH/I-1XK): d(ZS-XR/G-XDH) = 10 nm, d(G-XDH/AC-XK) = 6–30 nm.
2. The reaction is started with the addition of 200 mM xylose to a mixture of ZS-XR, G-XDH, and AC-XK located on the DNA scaffold in 3-enzyme cascade reaction buffer.
3. The amounts of cofactors (ATP, ADP, NADH, and NAD^+) are monitored by HPLC at 260 nm (*see* **Note 16**).

4 Notes

1. Reversible type of DNA binding adaptor: sequence-specific DNA binding proteins are one of the well-studied class of proteins that are characterized by structural domains, such as the zinc finger, the helix-turn-helix, and the basic leucine zipper. Among them, the zinc finger protein (ZFPs) is one of the best-characterized class of sequence-specific DNA binding proteins, and the artificially designed ZFPs bind to a wide variety of DNA sequences [35]. Each zinc finger domain recognizes a tract of four base pairs in the major groove of DNA duplex. A

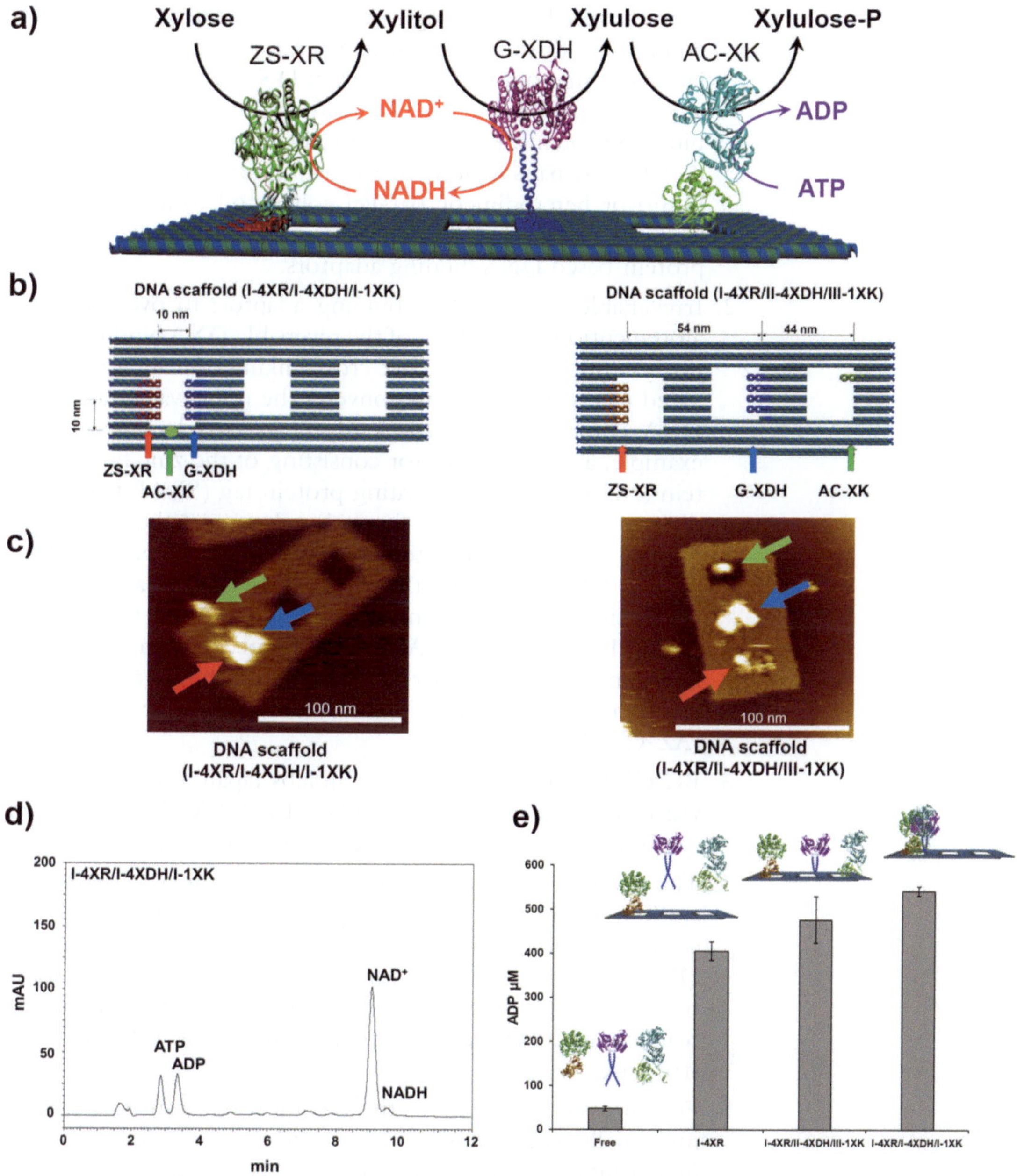

Fig. 10 (**a**) An illustration showing the three-step cascade reaction from xylose to xylurose-6-phosphate (xylulose-P) adapted from xylose pathway assembled on the DNA scaffold. (**b**) Design of DNA scaffolds with different inter-enzyme distances for co-assembled ZS-XR, G-XDH, and AC-XK. (**c**) AFM images of three enzymes bound on the DNA scaffold. ZS-XR: red arrow, G-XDH: blue arrow, and AC-XK: green arrow. (**d**) HPLC analysis of cofactors in the three-enzyme cascade reaction. (**e**) Efficiency of the three-enzyme cascade reaction on the DNA scaffold or in bulk solution

three-fingered ZFP recognizes a tract of 10 base pairs with equilibrium dissociation constants (K_d) ranged from nanomolar to sub-micromolar for the target DNA sequence. The basic leucine zipper protein (bZIP), also a well-studied class of DNA binding proteins, recognizes a specific DNA sequence around 8 or 10 base pairs with K_d in low nanomolar range through the homo or heterodimeric parallel coiled-coil formation. These well-characterized ZFP and bZIP have been utilized as the protein-based DNA binding adaptors.

2. Irreversible type of DNA binding adaptor: to overcome the aforementioned drawback of the reversible DNA binding adaptor systems, a chemoselective cross-linking domain was conjugated to the adaptor. This converts the noncovalent adaptor–DNA complex into a covalently cross-linked complex. For example, a modular adaptor consisting of the zinc finger protein (zif268) and a self-ligating protein tag (SNAP-tag) expedites a covalent linkage formation between the SNAP-tag domain and its substrate tethered to the target DNA sequence on a DNA scaffold. Furthermore, the modular adaptor was loaded in an almost quantitative yield. In order to assemble several POIs on the DNA scaffold at specific locations in the functional forms, modular adaptors with orthogonality and fast reaction kinetics under mild conditions were developed (AZ-CLIP and AZ-Halo).
3. BC-GLA-NHS (New England Biolabs Japan, cat. No. S9237) was discontinued from 2015. Thus, BC-GLA-NHS have been synthesized by us following the previous report [26].
4. The staple strands are used in 5- to 10-time molar excess to the scaffold strand to ensure their proper incorporation to DNA structure. The presence of Mg^{2+} ion is critical for the folding of DNA origami structures, especially for 3D structures. It is recommended to screen the Mg^{2+} concentration in the folding reaction when you start working with a new DNA origami structure.
5. For an easier preparation, the common mix can be arranged into smaller subgroups (or boxes) and then from subgroups to final solution. Mastermixes can be stored at −20 °C.
6. The extinction coefficient of the DNA origami nanostructure is estimated from that of the scaffold in the double strand form. Input the sequence of the appropriate scaffold strand to "DNA Calculator" function at http://www.molbiotools.com/; select "DNA," "double strand," and "circular" as input; the output will show the approximate 260 nm extinction coefficient of the scaffold strand. If you add any attached sequence, the estimated extinction coefficient of the attached sequence at 260 nm should be included for accuracy.

7. For the irreversible type of assembly, a set of three types of modular adaptor (ZF-SNAP, AZ-CLIP and AZ-Halo) are now available for orthogonal reactions [20, 21]. These modular adaptors show fast reaction kinetics to specific DNA sequence to respective substrate of them. It should be noted that their substrates (BG for SNAP-tag, BC for CLIP-tag, and CH for Halo-tag) are stable enough to stand the DNA origami preparation procedure (heat and anneal under pH 8.3). Thus, substrate-modified ODN as staple strand for DNA origami scaffold are mixed with the unmodified staple strands and heat and anneal for folding DNA origami without reducing their reactivity of protein-tags. The modular adaptors are different in molecular size (ZF-SNAP *m/z* 33,520, AZ-CLIP *m/z* 32,196, AZ-Halo *m/z* 47,768), which could show certain influence on the activity and the stability of POI fused to the adaptor.
8. POIs could be fused to the C-terminal or the N-terminal of the modular adaptor with the appropriate linker, e.g., GGSGGS [36] without disturbing the performance of the modular adaptor.
9. As the reversible assembling adaptor, monomeric ZFPs, such as zif268 and AZP4, were fused to monomeric protein of interest, and homodimeric bZIP protein GCN4 was fused to a homodimeric protein of interest. Molecular modeling of the complex of adaptor-fused POI based on the crystal structures of POIs and adaptor (PDB ID of ZFP: 1ZAA; GCN4: 1DGC; SNAP-tag and CLIP-tag: 3KZY; Halo-tag: 1CQW) by using a modeling software, such as Discovery Studio, is quite useful for the design of adaptor-fused POI. This helps to decide the position of the adaptor, the N- or C-terminal, to fuse POIs with evaluating an appropriate linker in between them. His-tag is included at the N- or C-terminal of adaptor-fused POI for easy purification by Ni-NTA agarose or related affinity column for His-tag containing recombinant proteins.
10. $ZnCl_2$ helps to maintain the native folding structure of zinc finger proteins, and at least 1 μM $ZnCl_2$ should be added in the assembling solution. The reducing environment is crucial for the activity of SNAP-tag and CLIP-tag. Therefore, addition of 1 mM DTT or 5 mM β–mercaptoethanol in stock solution or in reaction is suitable for maintaining the activity. Additionally, $ZnCl_2$ and reducing reagent should be contained during the purification process.
11. To visualize the DNA scaffolds in agarose gel, direct labeling of the DNA nanostructure with fluorophore-modified staple strands (e.g., Alexa 488) as the indicator for DNA scaffold is more convenient and reproducible than the conventional post

staining of DNA by EtBr. Upon labeling with the fluorophore, the gel should be scanned at an appropriate wavelength of the gel imager. The concentration of DNA scaffold is estimated from the standard calibration curve of the fluorophore.

12. For details, see the instruction of the RIBM (Tsukuba, Japan) high-speed AFM system [37].

13. The volume of POI assembled on DNA origami scaffold is estimated by using SPIP as follows: (1) open the .bcrf file of AFM images to SPIP interface. (2) Select "plane correction" → "global leveling" function in Modify panel to flatten the image. Select "zero background" tool to set the background z-value to zero. (3) Select the Particle and Pore analysis module in the "Analyze" tab group. The user interface will appear with three tabs (from left to right) corresponding to three steps of analysis. Detection: choose Threshold as a detection method and Auto-calculate threshold option. Post processing: choose "preserve holes in shapes" and "Suppress pixel noise" options, (optional) set up filters to discriminate unwanted shapes by selecting "Z-volume" as a parameter, and choose the minimum and maximum values for the target protein's volume. Output: choose the display z-volume value in the Parameters tab, choose to show the measurement result on Detect, "Filled" display mode, "Z-volume" label with "Value" option. Click Detect to start the analysis. (4) The detected volumes of POIs on DNA origami structures are now displayed directly in the image window; the proteins dots are in semitransparent colors with z-volume value nearby and in the same colors. Otherwise, in Shape measurements window (appears after the analysis), the results are displayed in a table where the color ID corresponds to the color of protein dots in the main window, and the z-volume value is shown in the next column. It should be noted that the threshold value is adjusted manually by increasing (or decreasing) the threshold level in the "Detection" tab of the "Particle and Pore analysis" window. The threshold values usually are set at 2 nm (corresponding to the height of DNA origami layer). (5) Collect the volumes of the target protein for analysis. Plot the value in a histogram plot as a frequency distribution of molecular volumes. Use the "Peak analysis" function in the data analysis software Origin to estimate the volume of target proteins by analyzing the plot of data.

14. Calculation of the inter-enzyme distance between ZS-XR and G-XDH in the solution. The average distance separating molecules in the solution is calculated by the following equation [38]. Briefly, in a 1 M enzyme solution, there are 6×10^{23} molecules/liter (0.6 molecules/nm^3) or inverting; the volume per molecule is $V = 1.66$ nm^3/molecule at 1 M. Therefore, for

a concentration C, the volume per molecule is $V = 1.66/C$. If we consider the cube root of the volume per molecule, the average distance (d) could be obtained from the following equation:

$$d = V^{1/3} = 1.18/C^{1/3}$$

where C is in molar and d is in nanometer.

In a solution containing 21 nM of ZS-XR assembled on the DNA scaffold and 85 nM free dimeric G-XDH, the total concentration is 106 nM. Based on the equation, the average separation of molecules is ca. 249 nm.

15. To avoid any discrepancies in the cascade reactions due to the yield of co-assembly, the amount of regenerated NADH and the V_{ini} are normalized according to the following equation:

$$\Upsilon_{obs} = P_{coassembled} \times \Upsilon_{coassembled} + P_{unassembled} \times \Upsilon_{unassembled}$$

where Υ_{obs} is the observed V_{ini} or the concentration of products (regenerated NADH or the final product xylulose) that consists of contributions from both co-assembled ($\Upsilon_{coassembled}$) and unassembled ($\Upsilon_{unassembled}$) enzymes ZS-XR and G-XDH. The $P_{coassembled}$ and $P_{unassembled}$ are the co-assembled and unassembled fractions of ZS-XR and G-XDH on the DNA scaffolds, respectively (determined by volume analysis of AFM images as described in Subheading 3.3.2, **steps 16–19**). The free diffusion reaction is used to determine $\Upsilon_{unassembled}$ in the co-assembled system for the normalization.

16. HPLC conditions: COSMOSIL packed column PBr (4.6 mm i.d. × 150 mm); eluent A, 10% methanol in 20 mM phosphate buffer (pH 7.0); eluent B, 50% methanol in 20 mM phosphate buffer (pH 7.0); gradient of eluent B increased from 0% to 20% in 1 min and to 30% in 10 min; and flow rate of 1 mL·min^{-1}.

References

1. Bonacci W, Teng PK, Afonso B, Niederholtmeyer H, Grob P, Silver PA, Savage DF (2012) Modularity of a carbon-fixing protein organelle. Proc Natl Acad Sci 109:478–483
2. Lodhi IJ, Semenkovich CF (2014) Peroxisomes: a nexus for lipid metabolism and cellular signalling. Cell Metab 19:380–392
3. Agapakis CM, Boyle PM, Silver PA (2012) Natural strategies for the spatial optimization of metabolism in synthetic biology. Nat Chem Biol 8:527–535
4. Chen AH, Silver PA (2012) Designing biological compartmentalization. Trends Cell Biol 22:662–670
5. Seeman NC (1982) Nucleic acid junctions and lattices. J Theor Biol 99:237–247
6. Rothemund PWK (2006) Folding DNA to create nanoscale shapes and patterns. Nature 440:297–302
7. Seeman NC (2007) An overview of structural DNA nanotechnology. Mol Biotechnol 37:246–257

8. Wilner OI, Willner I (2012) Functionalized DNA nanostructures. Chem Rev 112:2528–2556
9. Hong F, Zhang F, Liu Y, Yan H (2017) DNA origami: scaffolds for creating higher order structures. Chem Rev 117:12584–12640
10. Nummelin S, Kommeri J, Kostiainen MA, Linko V (2018) Evolution of structural DNA nanotechnology. Adv Mater 30:1–12
11. Wilner OI, Weizmann Y, Gill R, Lioubashevski O, Freeman R, Willner I (2009) Enzyme cascades activated on topologically programmed DNA scaffolds. Nat Nanotechnol 4:249–254
12. Fu J, Liu M, Liu Y, Woodbury NW, Yan H (2012) Interenzyme substrate diffusion for an enzyme cascade organized on spatially addressable DNA nanostructures. J Am Chem Soc 134:5516–5519
13. Linko V, Eerikäinen M, Kostiainen MA (2015) A modular DNA origami-based enzyme cascade nanoreactor. Chem Commun 51:5351–5354
14. Zhao Z, Fu J, Dhakal S, Johnson-Buck A, Liu M, Zhang T, Woodbury NW, Liu Y, Walter NG, Yan H (2016) Nanocaged enzymes with enhanced catalytic activity and increased stability against protease digestion. Nat Commun 7: 10619
15. Linko V, Nummelin S, Aarnos L, Tapio K, Toppari J, Kostiainen M (2016) DNA-based enzyme reactors and systems. Nano 6:139
16. Chandrasekaran AR, Anderson N, Kizer M, Halvorsen K, Wang X (2016) Beyond the fold: emerging biological applications of DNA origami. Chembiochem 17:1081–1089
17. Rajendran A, Nakata E, Nakano S, Morii T (2017) Nucleic-acid-templated enzyme cascades. Chembiochem 18:696–716
18. Nakata E, Liew FF, Uwatoko C, Kiyonaka S, Mori Y, Katsuda Y, Sugiyama H, Morii T (2012) Zinc-finger proteins for site-specific protein positioning on DNA-origami structures. Angew Chem Int Ed 51:2421–2424
19. Ngo TA, Nakata E, Saimura M, Kodaki T, Morii T (2014) A protein adaptor to locate a functional protein dimer on molecular switchboard. Methods 67:142–150
20. Nakata E, Dinh H, Ngo TA, Saimura M, Morii T (2015) A modular zinc finger adaptor accelerates the covalent linkage of proteins at specific locations on DNA nanoscaffolds. Chem Commun 51:1016–1019
21. Nguyen TM, Nakata E, Saimura M, Dinh H, Morii T (2017) Design of modular protein tags for orthogonal covalent bond formation at specific DNA sequences. J Am Chem Soc 139: 8487–8496
22. Yang YR, Liu Y, Yan H (2015) DNA nanostructures as programmable biomolecular scaffolds. Bioconjug Chem 26:1381–1395
23. Madsen M, Gothelf KV (2019) Chemistries for DNA nanotechnology. Chem Rev 119:6384–6458
24. Ngo TA, Nakata E, Saimura M, Morii T (2016) Spatially organized enzymes drive cofactor-coupled cascade reactions. J Am Chem Soc 138:3012–3021
25. Keppler A, Gendreizig S, Gronemeyer T, Pick H, Vogel H, Johnsson K (2003) A general method for the covalent labeling of fusion proteins with small molecules in vivo. Nat Biotechnol 21:86
26. Gautier A, Juillerat A, Heinis C, Corrêa IR Jr, Kindermann M, Beaufils F, Johnsson K (2008) An engineered protein tag for multiprotein labeling in living cells. Chem Biol 15:128–136
27. Los GV, Encell LP, McDougall MG, Hartzell DD, Karassina N, Zimprich C, Wood MG, Learish R, Ohana RF, Urh M, Simpson D, Mendez J, Zimmerman K, Otto P, Vidugiris G, Zhu J, Darzins A, Klaubert DH, Bulleit RF, Wood KV (2008) HaloTag: a novel protein labeling technology for cell imaging and protein analysis. ACS Chem Biol 3:373–382
28. Castro CE, Kilchherr F, Kim DN, Shiao EL, Wauer T, Wortmann P, Bathe M, Dietz H (2011) A primer to scaffolded DNA origami. Nat Methods 8:22
29. Douglas SM, Marblestone AH, Teerapittayanon S, Vazquez A, Church GM, Shih WM (2009) Rapid prototyping of 3D DNA-origami shapes with caDNAno. Nucleic Acids Res 37:5001–5006
30. Douglas SM, Dietz H, Liedl T, Högberg B, Graf F, Shih WM (2009) Self-assembly of DNA into nanoscale three-dimensional shapes. Nature 459:414
31. Sobczak JPJ, Martin TG, Gerling T, Dietz H (2012) Rapid folding of DNA into nanoscale shapes at constant temperature. Science 338: 1458–1461
32. Kurokawa T, Kiyonaka S, Nakata E, Endo M, Koyama S, Mori E, Tran NH, Dinh H, Suzuki Y, Hidaka K, Kawata M, Sato C, Sugiyama H, Morii T (2018) DNA origami scaffolds as templates for functional tetrameric Kir3 K^+ channels. Angew Chem Int Ed 57: 2586–2591

33. Shamanna DK, Sanderson KE (1979) Uptake and catabolism of D-xylose in Salmonella typhimurium LT2. J Bacteriol 139:64–70
34. Eliasson A, Christensson C, Wahlbom CF, Hahn-Hägerdal B (2000) Anaerobic xylose fermentation by recombinant Saccharomyces cerevisiae carrying XYL1, XYL2, and XKS1 in mineral medium chemostat cultures. Appl Environ Microbiol 66:3381–3386
35. Pavletich NP, Pabo CO (1991) Zinc finger-DNA recognition: crystal structure of a Zif268-DNA complex at 2.1 A. Science 252: 809–817
36. Chen X, Zaro JL, Shen WC (2013) Fusion protein linkers: property, design and functionality. Adv Drug Deliv Rev 65:1357–1369
37. Uchihashi T, Kodera N, Ando T (2012) Guide to video recording of structure dynamics and dynamic processes of proteins by high-speed atomic force microscopy. Nat Protoc 7:1193
38. Erickson HP (2009) Size and shape of protein molecules at the nanometer level determined by sedimentation, gel filtration, and electron microscopy. Biol Proced Online 11:32

Chapter 17

Aptamers as Functional Modules for DNA Nanostructures

Simon Chi-Chin Shiu, Andrew B. Kinghorn, Wei Guo, Liane S. Slaughter, Danyang Ji, Xiaoyong Mo, Lin Wang, Ngoc Chau Tran, Chun Kit Kwok, Anderson Ho Cheung Shum, Edmund Chun Ming Tse, and Julian A. Tanner

Abstract

Watson-Crick base-pairing of DNA allows the nanoscale fabrication of biocompatible synthetic nanostructures for diagnostic and therapeutic biomedical purposes. DNA nanostructure design elicits exquisite control of shape and conformation compared to other nanoparticles. Furthermore, nucleic acid aptamers can be coupled to DNA nanostructures to allow interaction and response to a plethora of biomolecules beyond nucleic acids. When compared to the better-known approach of using protein antibodies for molecular recognition, nucleic acid aptamers are bespoke with the underlying DNA nanostructure backbone and have various other stability, synthesis, and cost advantages. Here, we provide detailed methodologies to synthesize and characterize aptamer-enabled DNA nanostructures. The methods described can be generally applied to various designs of aptamer-enabled DNA nanostructures with a wide range of applications both within and beyond biomedical nanotechnology.

Key words DNA nanostructures, Aptamers, Atomic force microscopy, Circular dichroism, Transmission electron microscopy, Droplet microfluidic SELEX, Biophysical assays, Bioanalytical sensors

1 Introduction

The research field of DNA nanotechnology emerged from Nadrian Seeman's pioneering work in the self-assembly of synthetic nucleic acids [1–3]. Since then, fundamental Watson-Crick base-pairing principles have been applied in various ways to allow extraordinary control of nanoscale DNA design in two and three dimensions. The programmability and biocompatibility of DNA offer many advantages when compared to other nanostructures. For response to environmental molecular cues, much has been done to demonstrate DNA nanostructure response to other nucleic acids. For recognition of or response to other molecules, nucleic acid nanostructures are often decorated with protein antibodies. However, single-stranded nucleic acid aptamers can offer high-affinity discriminatory recognition to a similar extent as antibodies, yet at the same

Julián Valero (ed.), *DNA and RNA Origami: Methods and Protocols*, Methods in Molecular Biology, vol. 2639,
https://doi.org/10.1007/978-1-0716-3028-0_17,

time are elegantly built from the same material as the underlying nanostructure with clear synthesis, cost, and simplicity advantages. Aptamers are generated from an evolutionary process called Systematic Evolution of Ligands by EXponential enrichment (SELEX) [4, 5]. A random library of DNA sequences is incubated with a target molecule to fish out the best binder for next round. Counter-selection is usually performed to guarantee the specificity of selected aptamer. The latest developments in DNA nanotechnology provide a new opportunity for aptamers as ideal functional modules for response to molecular cues in biomedical applications.

As an example of integrating aptamer and DNA nanotechnology approaches, we previously demonstrated selection of an aptamer against the malaria biomarker *Plasmodium falciparum* lactate dehydrogenase (PfLDH) and subsequently appended that aptamer onto various nanostructures for biomedical sensing applications [6–9]. The binding between PfLDH and aptamer induced a conformational change, which we coupled to larger nanostructure architectural changes in both DNA nanotweezers and a DNA nanobox. For the nanotweezers, a split aptamer induced a closed nanostructure leading to the formation of a G-quadruplex that mediated colorimetrically observable peroxidase activity [7]. For the nanobox, we demonstrated that PfLDH led to box opening with subsequent change in a FRET signal [8]. The original aptamer was also appended to the vertices of DNA nanostructure polyhedra as an approach to improve sensitivity for biosensing applications [9]. These approaches exemplify how DNA nanostructures can synergize with DNA aptamer-mediated recognition for new avenues in biosensing and diagnostics.

The versatility of DNA also allows various advanced techniques for selection and characterization. In addition to the most common affinity-based SELEX [10, 11], *in vitro* compartmentalization using microfluidics has become a powerful approach to evolve functional nucleic acids for biomedical applications [12–14]. Droplet-based microfluidic technologies have been widely used in the field of high-throughput screening through encapsulation in micro-sized droplets at relatively low cost of reagents and samples [15]. For structural characterization of aptamers, circular dichroism is a spectroscopic technique that can quickly identify the presence of secondary structure or specific configuration within the aptamer [16, 17]. Circular dichroism is particularly useful for rapid characterization of G-quadruplex aptamers. The essential parameters of affinity and specificity can be efficiently characterized by combining particle display and flow cytometry for quantification [18–20]. Aptamers covalently conjugated to beads are stable at 4 °C for over 1 year indicating the consistency and reliability of analysis [19]. If the aptamer selected binds away from the active site of target like the PfLDH aptamer [6], it is possible to couple enzymatic reactions to generate colorimetric signals useful for simple applications with implications for point-of-care diagnostics [9, 21, 22]. When an aptamer is integrated to a larger DNA nanostructure, then direct observation through microscopic techniques becomes possible. The most popular methods of observation are transmission electron microscopy (TEM) and atomic force microscopy (AFM). While TEM is useful in observing rigid DNA origami [8, 23, 24], AFM allows observation in the aqueous state facilitating investigation of the dynamics of the interactions between protein and DNA nanostructures [25, 26].

In this chapter, we provide detailed protocols for DNA nanostructure assembly (3.1), aptamer evolution using microfluidic SELEX (3.2), aptamer refolding (3.3), structural characterization by circular dichroism (3.4), on-bead aptamer affinity assays (3.5), aptamer strand-displacement to determine optimal aptamer-duplex competition (3.6), G-quadruplex peroxidase assay (3.7), fluorescence resonance energy transfer (FRET) assay for observing DNA origami box opening (3.8), aptamer-tethered enzyme capture (APTEC) assay (3.9), nanostructure characterization by transmission electron microscopy (TEM) (3.10), and nanostructure characterization by atomic force microscopy (AFM) (3.11). These protocols can be integrated into research workflows on aptamer-integrated DNA nanostructures for biomedical applications.

2 Materials

Prepare all solutions using pyrogen-free, DNase-free, and RNase-free ultrapure water and analytical grade reagents. Prepare and store all reagents at room temperature (unless indicated otherwise).

2.1 Buffers

1. 0.5 M Ethylenediaminetetraacetic acid (EDTA), pH 8.0: Weigh 73.06 g EDTA and transfer to a glass beaker. Add water to a volume of 400 mL. Mix and adjust pH with NaOH (*see* **Note 1**). Make up to 500 mL with water. Store at room temperature.
2. 0.5 M $MgCl_2$: Weigh 203.3 g of $MgCl_2 \cdot 6H_2O$ and transfer to a glass beaker. Add water to a volume of ca 800 mL and dissolve. Transfer to a graduated measuring cylinder and make up to 1 L with water. Store at room temperature.
3. 10× Phosphate buffer saline (PBS): Purchase PBS tablets (Sigma-Aldrich) and follow the protocol of reconstitution to make up 1 L of 10× concentration. Store at room temperature.
4. 10× TAEM buffer: 400 mM Tris, 200 mM acetic acid, 10 mM EDTA (pH 8.0), 125 mM $MgCl_2$. Weigh 48.44 g of Tris and transfer to a glass beaker filled with ca 600 mL water (*see* **Note 2**). Add and mix 11.45 mL of acetic acid, 20 mL of 0.5 M EDTA, and 250 mL of 0.5 M $MgCl_2$. Transfer to a graduated

cylinder and make up to 1 L with water. Store at room temperature. Concentrated TAE buffer (e.g., Thermo Scientific 50× TAE buffer) is also available commercially as an alternative to mixing individual solutions.

5. 1% agarose gel: Add 1 g of agarose powder to 100 mL of TAEM buffer and microwave until all powder is dissolved. Pour the gel solution into a gel casting tray and wait until the gel is set.
6. 10% ammonium persulfate (APS): Add 1 g APS into 10 mL water. Mix thoroughly until APS is completely dissolved. Store at −20 °C. Thaw completely before use (*see* **Note 3**).
7. Native polyacrylamide gel: To prepare a 12% resolving gel, add 4 mL 30% acrylamide/bis-acrylamide solution 29:1 (Merck), 1 mL 10× TBE (Invitrogen), 4.89 mL H_2O, 0.1 mL 10% APS, and 10μL tetramethylethylenediamine (TEMED) into a 50 mL falcon tube. Gently swirl the tube to prevent bubble formation and mix completely. Add the mixture to a gel caster and wait until it is set. To prepare a 6% stacking gel on top of the resolving gel, add 2 mL 30% acrylamide/bis-acrylamide solution 29:1, 1 mL 10× TBE, 6.89 mL H_2O, 0.1 mL 10% APS, and 10μL TEMED into 50 mL falcon tube (*see* **Note 4**). Gently swirl the tube to prevent bubble formation and mix completely. Add the mixture to a gel caster and place a 10-well comb in it. Wait until it is set.
8. Staining buffer: Add 4μL of 10,000× SYBR Gold nucleic acid gel stain (Thermo Fisher Scientific) into 40 mL water (*see* **Note 5**). Swirl the mixture completely for mixing.

2.2 DNA Strands, Proteins, and Purification Columns

1. 100μM single-stranded DNA: Add correct volume of water to the freeze-dried DNA with mixing in order to make a 100μM stock solution. Single-stranded oligo DNA (desalted grade) is commercially available (e.g., from Integrated DNA Technologies). See supplementary information of published work for the sequence of staple strands for our DNA nanostructures. Store at 4 °C in dark for short term or at −20 °C in dark for long periods.
2. Scaffold DNA: M13mp18 single-stranded DNA (circular, 7249 nucleotides in length) is used and is commercially available.
3. Illustra MicroSpin S-400 HR column (GE Healthcare).
4. Purified recombinant PfLDH: Express and purify protein as reported (Full methods text available online) [6].
5. PCR purification columns.

2.3 Droplet PCR, Picoinjection, and Transcription

1. 1× Pfu DNA Polymerase Master Mix (M7745, Promega).
2. 10% Tween 20 stock solution.
3. 20% PEG 6000 stock solution.
4. Syringe pump (neMESYS low pressure module, Cetoni GmbH).
5. Polytetrafluoroethylene tubing (BB311-24, Scientific Commodities Inc.).
6. 1 mL glass syringe.
7. 0.2 mL PCR tubes.
8. FC-40 fluorinated oil.
9. FC-40 fluorinated oil with 5% (w/w) fluorosurfactant.
10. Outer oil phase: HFE-7500 fluorinated oil (3M) with 5% (w/w) fluorosurfactant (Ran biotechnologies, Inc.).
11. Picoinjection oil: HFE-7500 fluorinated oil supplemented with 1% (w/w) fluorosurfactant.
12. In vitro transcription (IVT) mixture (AS3107, Lucigen): 1× buffer, each NTPs, DTT, T7 RNA polymerase, DFHBI-1T (410-5 mg, Lucerna).
13. High voltage amplifier (5/80, Trek).
14. 1*H*,1*H*,2*H*,2*H*-perfluoro-1-octanol.

2.4 Microfluidic Devices

1. Negative photoresist (SU-8 3025, Microchem Corp.).
2. 4-inch wafer (University Wafer, Inc.).
3. UV exposure system (ABM).
4. Polydimethylsiloxane (PDMS) (Sylgard 184, Dow Corning).
5. Metal cutters.
6. 1 mm biopsy punch (33-31A, Integra Miltex).
7. Air plasma cleaner (PDC-002, Harrick Plasma).
8. Aquapel (PPG Industries).

2.5 Circular Dichroism

1. DNA aptamer strands from commercial companies. Prepare binding buffer (*see* **Note 22**) and all the other solutions at room temperature by using nuclease-free ultrapure distilled water and analytical grade reagents.
2. CD spectrometer.
3. 1-cm-pathlength quartz cuvette.
4. 2M HCl.

2.6 Synthesis and Quality Control of Aptamer Beads

1. TE buffer: 10 mM Tris-HCl, pH 8.0, 0.1 mM EDTA.
2. Forward priming (FP) reaction mixture for FP beads: 200 mM NaCl, 0.2 mM 5′-amine-modified forward primer (5′-amino-

PEG18-FP with HPLC purification), 1 mM imidazole chloride, 250 mM EDC (1-ethyl-3-(3-dimethylaminopropyl)carbodiimide hydrochloride) in water, dissolved fresh before use, 50% DMSO. Prepare just before use on Day 1.

3. Activation solution for FP beads: 250 mM EDC, 100 mM N-hydroxysuccinimide in 2-(N-morpholino) ethanesulfonic acid (100 mM MES buffer, pH 4.7). Prepare just before use on Day 2.
4. MA(PEG)12 Methyl-PEG-Amine Compound (Thermo Scientific™ 26114), warm up to RT and dissolve to 20 mM in DMSO at first use.
5. TT buffer: 250 mM Tris, 0.1% Tween 20, pH 8.0.
6. PCR master mix, e.g., GoTaq® G2 Master Mix (Promega M7832).
7. Additional polymerase, e.g., GoTaq® G2 Hot Start Taq Polymerase (Promega M7405).
8. qPCR master mix, e.g., LightCycler® 480 SYBR Green I Master (Roche 04707516001).
9. PBSMCT buffer (also for binding assay): buffer DPBS (Dulbecco's phosphate buffer) supplemented with 2.5 mM $MgCl_2$, 1 mM $CaCl_2$, 0.05% Tween 20 (*see* **Note 6** about mixing).

2.7 On-Bead Binding Assay

1. Biotinylated protein, e.g., recombinant human TNF-alpha, biotinylated protein (R&D Systems BT210).
2. Biotinylation kit (if necessary), e.g., EZ-Link NHS-PEG_4-Biotinylation Kit (Thermo Scientific™ 21455).
3. Dye-labeled streptavidin, e.g., Streptavidin, Alexa Fluor™ 488 Conjugate (Invitrogen™ S32354).
4. Streptavidin beads, e.g., Dynabeads™ MyOne™ Streptavidin C1 (Invitrogen™ 65001).
5. PBST buffer: Dulbecco's PBS, 0.1% Tween-20.
6. Flow cytometer (e.g., BD FACS Aria III (BD Biosciences)).

2.8 G-Quadruplex Peroxidase Assay

1. Hemin stock solution (0.1 mM in DMSO). Store in the dark at 4 °C.
2. 5× Assay buffer: 250 mM Tris-Cl pH 7.0, 750 mM NH_4Cl, 100 mM KCl, 0.15% Triton X-100.
3. 2,2′-azino-bis(3-ethylbenzothiazoline-6-sulphonic acid (ABTS) stock solution (50 mM in ultrapure water).
4. Hydrogen peroxide stock solution (50 mM in ultrapure water).
5. 96-well plate.
6. Plate reader (e.g., Varioskan Flash Multimode Reader (Thermo Fisher Scientific Inc.)).

2.9 Aptamer-Tethered Enzyme Capture (APTEC) Assay

1. Streptavidin-coated 96-well plates.
2. PBST washing buffer: Dulbecco's PBS, 0.1% Tween-20.
3. L-lactate/nitrotetrazolium blue chloride (NTB). Reagent was prepared immediately before use and required 12 mL L-lactate buffer (0.2 M sodium L-lactate, 100 mM Tris HCl, 0.2% Triton X-100, pH 9.1) 158μL NAD^+ solution (50 $mgmL^{-1}$ in H_2O), 48μL NTB solution (25 $mgmL^{-1}$ in H_2O), and 25μL PES solution (5 $mgmL^{-1}$ in H_2O). L-lactate buffer was stored at 4 °C and other stock solutions at −20 °C for up to 2 months.
4. Quenching buffer: PBS, 5% acetic acid.
5. Cell lysis buffer: PBS, 0.5% Triton X-100.
6. Binding buffer: 2× PBS.

2.10 Nanostructure Characterization by Transmission Electron Microscopy (TEM)

1. 2% aqueous uranyl acetate: Prepare uranyl acetate solution using fume hood and plastic blockage. Dissolve 2 g uranyl acetate in 100 mL of distilled water in a volumetric flask. Filter the solution with 0.22μm millipore filter and cover the flask with aluminum foil to avoid light exposure.
2. Carbon-formvar coated grids.
3. Sputter coater.
4. Filter paper.
5. Transmission Electron Microscope (e.g., Philips CM100 TEM).

2.11 Nanostructure Imaging by Atomic Force Microscopy (AFM)

1. Mica discs (diameter = 15 mm).
2. Steel substrates (diameter = 15 mm).
3. Roll of magnetic tape.
4. Tweezers.
5. Lintless paper cloth.
6. Cantilever (In air: FESPA-V2 with 2.8 N/m spring constant. In liquid: DNP-S10 with 0.06 N/m spring constant).
7. Atomic force microscope.
8. Vacuum grease.

3 Methods

Carry out all procedures at room temperature unless otherwise specified.

3.1 DNA Nanostructure Assembly

1. Mix 50 nM staple strands and 10 nM scaffold DNA (or 10μM of each strand for simple nanostructure) in 1× TAEM buffer (*see* **Notes** 7 and **8**).

2. In a thermal cycler, anneal the mixture by heating to 95 °C and then cool to 25 °C at a steady rate of −1 °C per minute (*see* **Note 9**).
3. Remove excess staple strands by using an Illustra MicroSpin S-400 HR column. Resuspend the resin in the column by vortexing. Spin at 750 g for 2 min to elute all buffer from the column. Carefully add 500μL TAEM buffer to the column. Spin at 750 g for 2 min. Discard the eluent. Repeat this step three times. For the last centrifugation, set at longer spinning time (e.g., 2.5 min) to remove residual TAEM. Add 50–100μL of DNA origami sample carefully to the center of the column. Avoid disturbing the beads inside. Spin at 750× *g* for 2 min. Collect flow-through as purified sample.
4. Optional: Quantify the concentration of DNA origami sample by measuring absorption at 260 nm.
5. To evaluate the assembly by gel imaging, run ca 30 fmol product of DNA nanostructures onto a 1% agarose gel in TAEM buffer. First, load 5μL of DNA sample with DNA gel loading dye into the well of the gel and operate at 80 V for 1 h at 4 °C to prevent overheating of gel. Stain the gel in staining buffer and image using a gel documentation system (*see* **Note 10**). Example of DNA nanotweezers is shown on Fig. 1 [7].

1A		✓			✓	✓		✓
1B			✓		✓		✓	✓
S3				✓		✓	✓	✓
10 bp Ladder	✓							

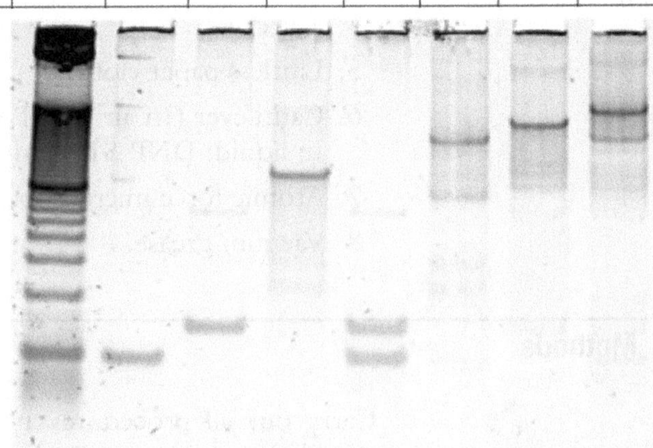

Fig. 1 Formation of aptamer-mediated DNA nanotweezers. The migration of sample became slower when the number of strands increased. Specificity of formation can be confirmed by different combinations of sequences

3.2 Aptamer Evolution Using Microfluidic SELEX

3.2.1 Aptamer Library Preparation

Eight rounds of classical binding capture SELEX are performed to enrich an aptamer library for target binding. To prepare target-labeled beads, target is conjugated to agarose beads via hexamine linker. The chemistry used depends on the target or target analogue available and may include copper(I)-catalyzed azide-alkyne cycloaddition, carbodiimide amine-carboxyl coupling, or thiol linkage. The amount of target conjugated to the agarose beads is determined by measuring the target concentration in solution before and after the conjugation reaction. Target bound beads are washed ten or more times to remove unbound target molecule. For each selection round, the amount of beads equivalent to 500 nanomoles of target is used.

Each selection cycle consists of the following:

1. Transcription of DNA into RNA (*see* **Notes 11** and **12**).
2. DNAse treatment of transcription mix for 15 min.
3. Heat deactivation of DNAse at 75 °C for 15 min.
4. Ethanol precipitation (*see* **Note 13**).
5. For later selection rounds [2–8], a negative selection is performed by incubating the library with mock beads, conjugated with hexamine linker only, for 1 h before separation of non-binding sequences for the subsequent selection.
6. The library is incubated with conjugated agarose beads for 1 h before washing 6–10 times with binding buffer and elution with free target for 4 h.
7. The eluted library is then ethanol precipitated (*see* **Note 13**).
8. The pellet is hydrated and PCR amplified (*see* **Note 14**).
9. The PCR sample is then purified using a PCR purification column with a 30μL elution volume. This purified DNA library sample seeds the next round of selection.
10. After 8 SELEX rounds, the aptamer library is ready for microfluidic selection.

3.2.2 Fabrication of Microfluidic Devices

The microfluidic devices are fabricated as described previously [27]. Briefly, the mold for making PDMS replica is fabricated by standard soft lithography techniques.

1. Use negative photoresist to pattern channels on 4-inch wafer by UV exposure.
2. Pour polydimethylsiloxane (PDMS) over the mold and cure at 65 °C in an incubator for at least 2 h.
3. Trim the cured PDMS with metal cutters and punch it with 1 mm biopsy punch to create fluidic inlets and outlets.

4. Bond the PDMS slab to a glass slide using air plasma cleaner to produce the microfluidic devices.
5. Treat the channels with Aquapel, dry with compressed air, and heat at 90 °C for 1 h to make the channels hydrophobic.

3.2.3 Droplet PCR

1. Add the oligonucleotide library (-Sequences-) at 1 pM concentration to a mixture of PCR reagents, including 1× Pfu DNA polymerase master mix, 0.4μM forward primer (-sequences-) and reverse primer (-sequences-), 2% Tween 20, and 5% PEG 6000 (*see* **Note 15**).
2. Withdraw the PCR mixture at 2000μL/hour using syringe pump (into a polytetrafluoroethylene tubing, which is connected to a 1 mL glass syringe preloaded with FC-40 fluorinated oil (*see* **Notes 16–18**).
3. Inject the PCR mixture into a flow-focusing droplet generator, carried by outer oil phase containing HFE-7500 fluorinated oil (3M) supplemented with 5% (w/w) fluorosurfactant.
4. Collect the droplets in 0.2 mL PCR tubes, where the bottom oil layer is replaced with FC-40 fluorinated oil with 5% (w/w) fluorosurfactant to improve thermal stability.
5. Place the droplet samples in a thermal cycler using the following program: incubate at 98 °C for 3 min, followed by 40 cycles with 2°C/second ramp rates of 98 °C for 10 s, 52 °C for 20 s, and 72 °C for 20 s, ended by a hold at 4 °C.

3.2.4 Picoinjection and In Vitro Transcription

1. Reinject the PCR droplets into a picoinjection device and spaced with HFE-7500 fluorinated oil supplemented with 1% (w/w) fluorosurfactant.
2. Merge the droplets at the picoinjection junction with the in vitro transcription (IVT) mixture containing 1× buffer, each NTPs, DTT, T7 RNA polymerase, and DFHBI-1T (410-5 mg, Lucerna).
3. Inject the mixture by applying a 10 kHz square wave at 400 V with a high voltage amplifier (5/80, Trek) to the electrodes on-chip (*see* **Note 19**).
4. Collect the droplets in a 0.2 mL PCR tube, and replace the bottom oil layer with FC-40 fluorinated oil with 5% (w/w) fluorosurfactant (*see* **Notes 20** and **21**).
5. Incubate the droplets at 37 °C for 1 h. RNA transcription occurs during the incubation and transforms the DFHBI fluorogenic substrate into fluorescent state.

3.2.5 Droplet Sorting

1. Reinject the incubated droplets into droplet sorter and space them with surfactant free HFE-7500 fluorinated oil. The droplets are reinjected at a frequency of ~250 Hz by adjusting the

flow rates. Droplets with 1% most green fluorescence (concentration of the fluorogenic product) are gated and sorted into the positive channel. The bright droplets are deflected into the positive channel by applying a 30 kHz square wave at 1000 V with a high voltage amplifier. The optical setup and programs are described in our previous work [27].

2. Recover the sorted droplets by mixing with 1*H*,1*H*,2*H*,2*H*-perfluoro-1-octanol followed by brief centrifugation.
3. Pipette the upper aqueous phase into a new tube for PCR to seed the next selection round or for downstream DNA sequencing.

3.3 Aptamer Refolding

1. Spin down the purchased freeze-dried DNA.
2. Reconstitute the freeze-dried DNA with nuclease-free water to 100μM.
3. Dilute the DNA into 10μM in PBS or 1× TAEM.
4. Place the mixture in a thermal cycler.
5. Increase temperature to 95 °C and hold for 5 min.
6. Reduce temperature to 20 °C (*see* **Note 9**).
7. Storage under room temperature.

3.4 Aptamer Characterization by Circular Dichroism

1. Mix aptamer and target in binding buffer (*see* **Note 22**), and then incubate the mixture for binding reactions.
2. Purge the instrument with purified nitrogen gas for 10–20 min before turning on the instrument (*see* **Note 23**). Regulate the pressure of the gas according to the manufacturer's guidelines.
3. Turn on the computer, and then turn on the instrument (*see* **Note 24**). Next, turn on Peltier temperature controller and water bath. Adjust the water bath temperature to the desired temperature.
4. Double click to open the software named "Spectra Manager" on computer's desktop, and then select "Spectrum Measurement" program. After the initialization process is finished (around 5 min), the shutter will open, and the lamp will be turned on automatically. Then, wait 20 min for the light to warm up.
5. To wash the cuvette (*see* **Note 25**), first add 2 M HCl to the cuvette and incubate for 10–20 min to dissolve the insoluble residue. Discard the HCl solution following the acid waste disposal regulations.
6. Wash the cuvette for 5–10 times with nuclease-free water and 5–10 times with 100% ethanol. Dry the curette with nitrogen gas.

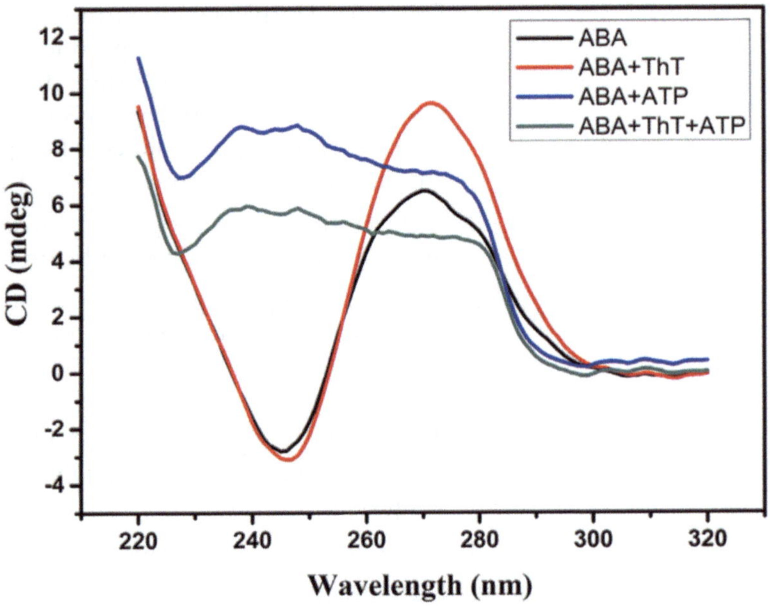

Fig. 2 CD absorption response of 10μM ATP binding aptamer (ABA) upon titration of 15 mM ThT and 5 mM ATP. 20 mM Tris-HCl (pH 8.3) was used as buffer. Spectra were taken from 220 to 320 nm at 0.5 nm intervals. ABA showed a positive peak at 270 nm and negative peak at 245 nm in CD spectrum (black curve), which demonstrated the parallel G-quadruplex conformation of ABA. After binding to ATP, the G-quadruplexes conformation of ABA was unfolded and converted to a random-coil conformation (blue curve) (Reproduced with permission from Ref. [28])

7. To measure the spectra, first click on the "Measurement" menu and select "Parameter." Input appropriate parameters according to the needs of sample measurement.
8. Select Channels 1 and 2 under "Data Mode" menu, which are for CD signal and HT (Voltage) monitoring, respectively (*see* **Note 26**).
9. Name the spectra in "Data File" under the "Parameter" menu and save it in a folder under your name.
10. Record the CD and HT signal of buffer and save it as blank (*see* **Note 27**, Fig. 2).
11. Clean the cuvette and use the same cuvette to detect the CD and HT signal of samples under the same instrumental condition, and cuvette should be put in the same orientation as that for blank detection.
12. After the scanning is finished, a new window showing a spectrum will appear. Subtract the blank from the sample spectrum and save the corrected spectrum. Then select "File" and

"Export." Export the spectra in formats (e.g., *.csv, *.txt) that can be transferred to data analysis software (e.g., Excel, Origin) for further analysis.

13. After finishing all the samples detection and data export, close the software first. Then, turn off the instrument and nitrogen gas. Sign out of the computer.

3.5 On-Bead Aptamer Affinity Assay

3.5.1 Synthesis and Quality Control of Aptamer Beads

1. Forward priming beads (FP beads) are constructed by conjugating forward primers (*see* Subheading 2.6, **item 2** – 5′-amine-modified forward primer with PEG18 spacer) and PEG spacer molecules to Dynabeads MyOne™ Carboxylic Acid beads (*see* Subheading 3.5.2) following Wang et al.
2. Aptamers are attached to beads through polymerase chain reaction (bPCR); *see* Subheading 3.5.4.

3.5.2 Synthesis of Forward Priming Beads

1. Day 1: Wash 500μL of Dynabeads MyOne™ Carboxylic Acid beads once with 500μL 100 mM NaOH, and then thrice with 1 mL water.
2. Resuspend the washed beads in the reaction mixture. Vortex and sonicate briefly (*see* **Note 28**). Rotate overnight.
3. Day 2: Separate the beads from the supernatant and resuspend them in 150μL of the activation solution. Vortex and rotate for 30 min to activate remaining carboxylic acid groups on the beads.
4. Prepare 150μL of 20 mM amino-PEG12 in PBS buffer and mix well.
5. Separate the beads, remove the supernatant, and resuspend them in 150μL of amino-PEG12 solution. Rotate for 1 h.
6. Separate the beads, then wash twice with 500μL of TT buffer, and resuspend in 500μL of TE buffer. Beads can be stored at 4 °C.

3.5.3 Concentration Measurement

The concentrations of beads are measured with UV-Vis spectroscopy on a plate reader. Prepare triplicate wells for each standard and sample.

1. Prepare a series of standards using MyOne beads (5000, 10,000, 15,000, 20,000 beads/μL) in TE buffer.
2. Dilute the FP beads 1:400 in TE buffer (*see* **Note 28**).
3. Distribute 200μL of diluted solutions into triplicate wells.
4. Measure the absorption at 595 nm (or a convenient wavelength within 595–700 nm).
5. Calculate the sample concentrations using a linear fit to the standard curve.

3.5.4 On-Bead Polymerase Chain Reaction (bPCR) to Generate Aptamer Particles

1. Each bPCR contains 1X GoTaq PCR Master Mix, 25 mM $MgCl_2$, 0.025 U/μL G2 Hot-Start polymerase, 3×10^5 FP beads (*see* **Note 28**), 2μM reverse primer, and 10 nM of template.
2. After mixing the reactions, vortex thoroughly then spin quickly.
3. Perform bPCR using the following program:

	Temperature	Time
	95 °C	3 min
28 cycles	95 °C 65 °C (*see* **Note 29**) 72 °C	20 s 30 s 1 min
	72 °C	5 min

4. For efficient and even amplification over the beads, briefly vortex (and quick spin) the samples every 3–4 cycles.
5. After PCR, wash the beads thrice with TE buffer.
6. Quantify bead concentration by UV-Vis spectroscopy as described above. 1:200 dilutions give a reasonable concentration for measurement.
7. Measure the number of sequences per bead by quantitative PCR (qPCR) for quality control.
8. Each 10μL reaction contains 1× PCR Master Mix, 0.25μM forward primer, 0.25μM reverse primer, and, if necessary, 1× SYBR Green dye (*see* **Note 30**).
9. Standard reactions contain 1000 FP beads and 10^9, 10^8, 10^7, 10^6, and 10^5 molecules of the starting template. Sample reactions contain 1000 beads from bPCR.
10. If using the same polymerase as that used for bPCR, the PCR program can be the same except without the final extension during qPCR. You may need to optimize the PCR program for different combinations of primers and polymerases.
11. Data analysis: Plot standard curve and use it to calculate the average number of DNA per bead (*see* **Note 30**). The correct sequence can be inferred by comparing the melting curves between bPCR products and the standard reactions.

3.5.5 Aptamer Bead Quality Control by Flow Cytometry

The presence of complete products and relative coverage on the beads can be checked using flow cytometry. The on-bead PCR products are first dehybridized with sodium hydroxide and then hybridized with reverse primers labeled with a dye at the 5′ end. We selectively provide details relevant for analyzing aptamer beads and suggest some resources for flow cytometry.

1. For dehybridization, first separate beads from the supernatant and resuspend in 0.1 M NaOH. After incubating at 50 °C for 5 min, discard the supernatant. Repeat this step once.
2. Wash the beads thrice with TE buffer. Beads can be stored at 4 °C.
3. For hybridization, incubate ~1000 beads/μL with 0.2μM 5′ dye–reverse primer in 50μL, at 60 °C for 5 min in PBSMCT (*see* **Note 31**).
4. Cool the beads to RT.
5. Wash the beads twice with PBSMCT, and then resuspend in 200μL TE buffer.
6. Perform the flow cytometry analysis of the aptamer functionalized beads (e.g., on a BD FACS Aria III (BD Biosciences)) using the 70-micron nozzle and no neutral density filter during analysis for all measurements described in this section. If using a fluorescein dye, we use the "FITC" setting, which combines the 488 nm laser for excitation, a 530/30 nm emission/bandpass filter, and 502 nm longpass filter. The voltage settings for the photomultiplier tube (PMT) detectors are set to 300, 360, and 500–600 V for the forward scattering, side scattering, and FITC, respectively, adjusted according to day-to-day variations of the instrument (*see* **Note 32**).
7. Use the negative control beads to gate for singlets beads in the scattering signal [19].

3.5.6 On-Bead Aptamer Binding Assay

This fluorescence bead-based binding assay is adapted from Wang et al. and is similar to other protocols [10, 20, 29]. Aptamer beads of one known concentration are incubated with varying concentrations of biotinylated protein in triplicate, labeled with Alexa488-streptavidin (Alexa488-SA) conjugate, and analyzed by flow cytometry. The procedure below uses an aptamer for tumor necrosis factor alpha (TNFα) as an example [10] (*see* **Note 33**).

1. To perform binding reactions, we recommend the following strategy for setting up the reactions with 10^2 beads/μL and the desired protein concentration.

Ingredient	Vol (μL)
Buffer	160
10× beads	20
10× proteins	20
Total	200

2. Wash SA beads in PBST thrice. Resuspend to 10^6 beads/μL.
3. Prepare 10× bead solutions by diluting them to 10^3 beads/μL in PBSMCT buffer (*see* **Note 28** and **32**). Sonicate ~20 s immediately before adding to the reactions.
4. Prepare 10× peptide series in PBSMCT buffer. For this example, the peptide is TNFα.
5. Distribute to each reaction tube in this order: buffer, 10× beads, 10× peptide.
6. Rotate for 2 h. This can be done a Ferris wheel style rotator or vertical rotating mixer at "low" to "medium" speed, perhaps 0.5–1 rotation per second. The important part is that it is continuously and gently mixed by inversion.
7. Dilute Alexa488-SA solution by 1:1000 in PBSMCT, and spin at 1000 *g* for 3 min to remove aggregates.
8. "Rinse" the beads by separating them on a magnetic rack, removing the supernatant, adding 200μL of PBSMCT, and then removing it while keeping the tubes on the rack.
9. Resuspend the beads in 200μL SA-dye solution. Do NOT vortex. Rotate for at least 15 min.
10. Rinse again.
11. Resuspend beads gently in 200μL PBSMCT.
12. Immediately before measurement, mix each sample by gently pipetting ~10 times. Do NOT vortex or sonicate.
13. For the flow cytometry measurements, first set up the appropriate settings on the flow cytometer (*see* **Note 34**).
14. Choose a flow rate and measurement time that yields a sufficient number of data points while limiting measurement to a reasonable amount of time. We measure each sample using a flow rate of ~18μL per minute for 3 min or until 2000 singlet beads are counted, whichever comes first.
15. Following data acquisition, calculate the median fluorescence intensities (MFI) of triplicate samples. Subtract the baseline signals, and plot with standard deviations as the error bars. A relative dissociation constant can be calculated by fitting the averaged MFI and concentrations to the Hill equation:

$$I - b = \frac{\mathrm{I}_{\mathrm{max}} - b}{1 + \left(\frac{K_{\mathrm{D}}}{C}\right)^{n}}$$

where I is the measured MFI, C is the known concentration of the peptide or protein, and b is the MFI at measured for $C = 0$. The maximum MFI at saturation, I_{max}, and relative dissociation constant, K_{D}, are calculated by the fitting algorithm. Scripts can be written (e.g., Matlab) to analyze the data efficiently and consistently.

3.6 Determine Optimal Aptamer-Duplex Competition

1. Mix 50μL of 100 nM PfLDH aptamer with 50μL of 400 nM complementary DNA (12bp1) in TAEM buffer (*see* **Note 35**) [8].
2. In a thermal cycler, heat the mixture to 95 °C and then cool to 20 °C at a steady rate of 1 °C per minute. Keep the product at 20 °C (*see* **Note 9**).
3. Evaluate assembly by running native-PAGE. Load 10μL of product at 100 nM on a native polyacrylamide gel, and run it at 100 V for 1 h at 4 °C to prevent overheating of samples (*see* **Note 36**). Stain the gel in staining buffer and image using a gel documentation system (*see* **Notes 10** and **37**).
4. Dilute assembled aptamer lock module with TAEM buffer to 50 nM.
5. Mix 50 nM aptamer lock with 2μM PfLDH in 1:1 ratio (*see* **Note 35**).
6. Incubate the mixture for 1 h at 25 °C.
7. Evaluate the strand displacement reaction efficiency by native-PAGE as **step 3**.

Example of aptamer lock in DNA nanobox is shown on Fig. 3 [8].

3.7 G-Quadruplex Peroxidase Assay [7]

1. Mix 500 nM of DNA nanostructure with quadruplex and 250 nM of hemin.
2. Incubate 30 min with constant shaking in the presence of 1× assay buffer.
3. Add 2 mM ABTS and 2 mM hydrogen peroxide.
4. Transfer mixture in a 96-well plate.
5. Measure absorbance at 420 nm using a plate reader.

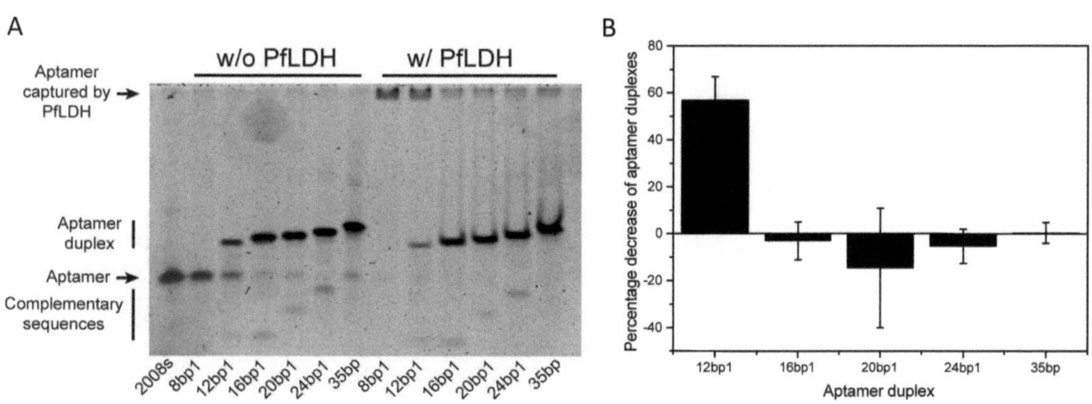

Fig. 3 Analysis of strand displacement. (**a**) Electrophoretic mobility shift assay (EMSA) to observe the extent of strand displacement in the absence and presence of PfLDH in the presence of each of the duplex pairs. (**b**) Comparison of intensity to look at the percentage reduction of aptamer duplex

3.8 Quantification of Conformational Change using Fluorescence Resonance Energy Transfer (FRET)

3.8.1 Standard Curve Preparation

1. Mix 50μL of 40 nM Cy3-PolyA with 50μL of 40 nM Cy5-PolyT in PBS.
2. In a thermal cycler, heat the mixture to 95 °C and reduce to 20 °C at a steady rate of 1 °C per minute.
3. Dilute the product with PBS into different concentrations such as 2.5 nM, 5 nM, 10 nM, 15 nM. and 20 nM.
4. Gently transfer 20μL of sample into a 384-well plate (*see* **Note 38**).
5. Insert the 384-well plate into the plate reader. Measure the fluorescence signal with excitation at 558 nm and emission at 680 nm (*see* **Note 39**).
6. Plot the results on graph as concentration (x-axis) against emission (y-axis) using appropriate software (e.g., Origin). Fit the data with linear regression ($y = mx + c$).
7. Compare signal response from 20μL of 10 nM DNA nanobox to the regression line to find the experimental concentration (Fig. 4) [8].

3.8.2 PfLDH-Mediated Opening of DNA Origami Box Assessed by FRET Assay

1. Mix 20 nM DNA nanobox with 1μM PfLDH in 1:1 ratio in PBS (*see* **Note 35**).
2. Prepare the plate reader by setting the temperature to at 25 °C and wait until the temperature is stable.

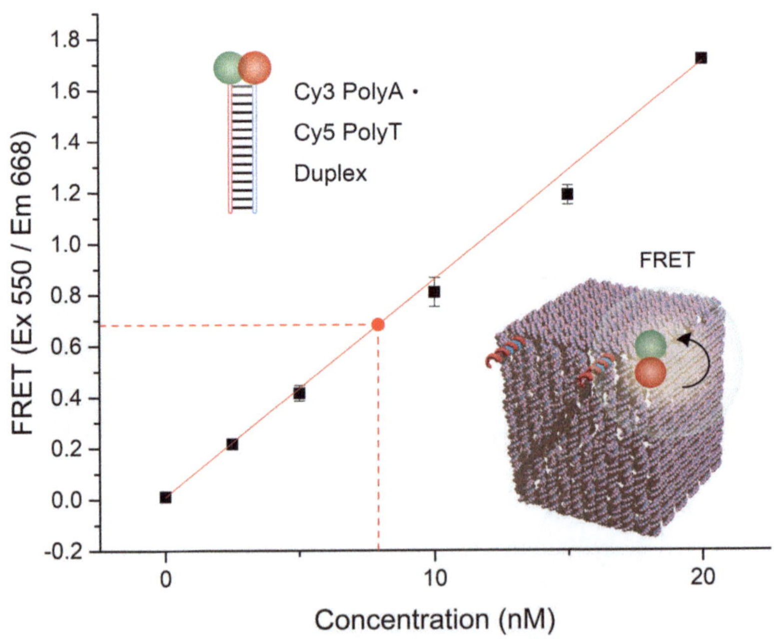

Fig. 4 Comparison of FRET signal between a duplex and DNA nanobox. The yield of the nanobox showed here was 79.1%

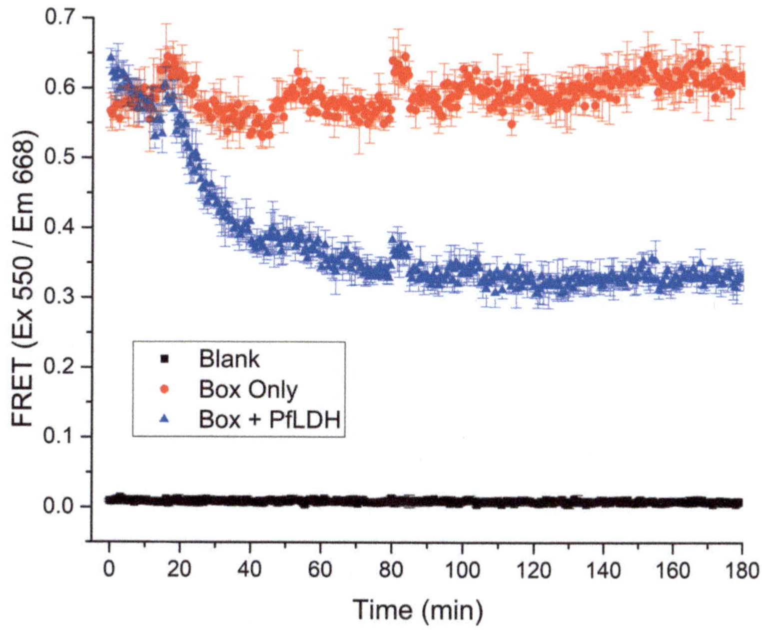

Fig. 5 Measurement of FRET signal from DNA nanobox in the absence and presence of PfLDH

3. Gently transfer 20μL of sample into a 384-well plate (*see* **Note 38**).
4. Insert the 384-well plate into the plate reader. Incubate for 3 h with shaking at 60 rpm.
5. Measure the fluorescence signal with excitation at 558 nm and emission at 680 nm. Make the measurement at 30 s intervals (*see* **Note 38**, Fig. 5) [8].

3.9 Aptamer-Tethered Enzyme Capture (APTEC) Assay [9, 21]

3.9.1 Functionalization of 96-Well Plates

1. Wash the streptavidin-coated 96-well plate with PBS, 0.1% Tween-20 (PBST). Repeat three times.
2. Add 100μL biotinylated PfLDH aptamer or aptamer functionalized nanostructure in PBS into the well and incubate for 2 h.
3. Wash the plate with PBST. Repeat three times.
4. Use the plate directly for the assay or store at 4 °C.

3.9.2 APTEC Assay

1. Add 100μL of sample to the plate in triplicate and incubate for 1 h.
2. Wash the plate three times with PBST.
3. Add 120μL L-lactate/nitrotetrazolium blue chloride (NTB) solution.
4. Incubate 45 min with mild shaking for color development (*see* **Note 40**).

5. Add 100μL PBS with 5% acetic acid to quench the reaction.
6. Measure absorbance at 570 nm.

3.9.3 APTEC Assay in Whole Rat Blood

1. Use 2× PBS as the binding buffer for incubation.
2. Mix 50μL whole blood and 50μL PBS, 0.5% Triton X-100 to allow red blood cell lysis. Perform APTEC assay as indicated in 3.9.2.

3.10 Nanostructure Imaging by Transmission Electron Microscopy (TEM) [8]

*3.10.1 Negative Staining (see **Note 42**)*

1. Prepare of 2% aqueous uranyl acetate (*see* **Note 41**)
2. Gently agitate the solution with preassembled nanostructure.
3. Aliquot 6μL preassembled nanostructure.
4. Discharge carbon-formvar coated grid with sputter coater for 200 s.
5. Pipette 5μL of nanostructure onto the grid.
6. Place the grid on the stain droplet with sample face down using a metal tweezer (*see* **Note 43**).
7. Incubate for 1 min.
8. Transfer the grid to a water droplet for washing. Repeat twice.
9. Dry the grid using a filter paper with the grid positioned perpendicularly to the filter paper.
10. Put the grid in the specimen container for at least 1 day of drying.

3.10.2 Observation Under an Electron Microscope

1. Place the grid on specimen holder and insert into the electron microscope (e.g., Philips CM100 TEM).
2. Use 100 kV accelerating voltage for observation.
3. Select optimum magnification (28,500 times for DNA nanobox) (*see* **Note 44**).
4. Adjust the electron beam intensity for maximized brightness (depends on the mean percentage brightness).
5. Adjust the electron beam coordinates to place the beam at the center.
6. Adjust focus step (higher magnification usually needs smaller focus step).
7. Adjust focus until a clear image can be observed.
8. Use the joystick to move to another area if necessary.

1. For the dry sample preparation, first cut wide magnetic tapes into thin strips (width = 4 mm).
2. Attach thin strips of magnetic tapes onto flat surfaces (*see* **Note 45**, Fig. 6).

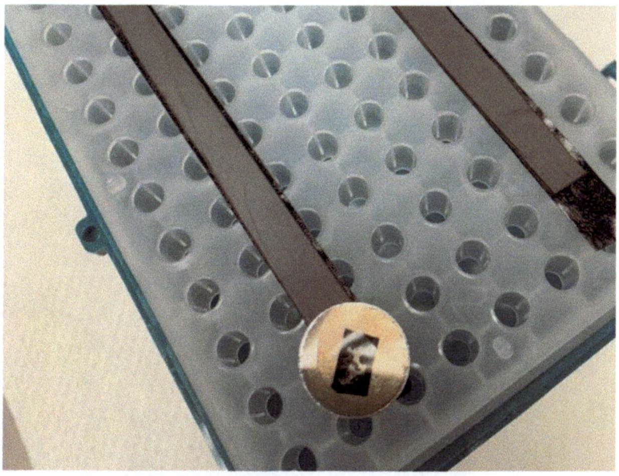

Fig. 6 AFM: Preparation of holder for mica. Picture demonstrating a plastic pipette box with a magnetic stripe on a double-sided carbon tape as a sample holder to bind a mica piece on a steel plate via magnetic interactions

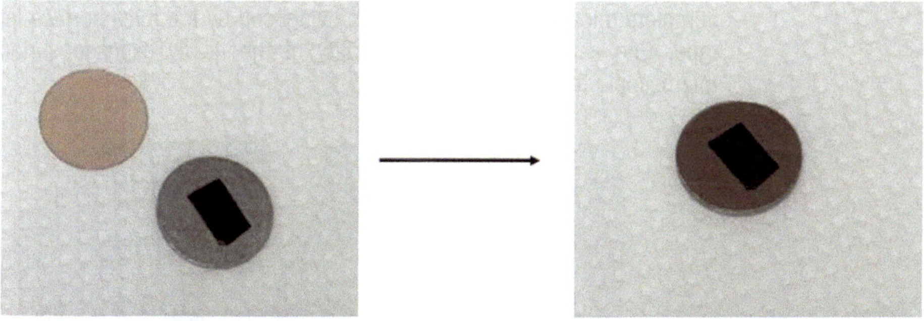

Fig. 7 AFM: Attachment of mica on steel plate

3.11 Nanostructure imaging by Atomic Force Microscopy (AFM)

3.11.1 AFM Air Mode Protocol [26, 30]

3. Adhere a mica disc onto a steel substrate using a piece of double-sided tape (4 mm). The steel substrate provides the necessary magnetic interaction with the magnetic tape to hold the mica disc in place during storage and transport between lab and AFM instrument.
4. Place a piece of double-sided tape on top of the mica surface of the mica/steel construct (*see* **Note 46**, Fig. 7).
5. Carefully slide a pair of tweezers between the mica and the tape, the topmost layer of the mica disc is removed, and remain on the tape (*see* **Note 47**, Fig. 8).
6. Repeat the above mica cleavage (a.k.a. peel-off) procedure at least three times to ensure the complete removal of the topmost layer of mica.
7. Clean the cleaved mica surface using 1 mL ultrapure water at least three times, respectively (*see* **Note 48**, Fig. 9), and then

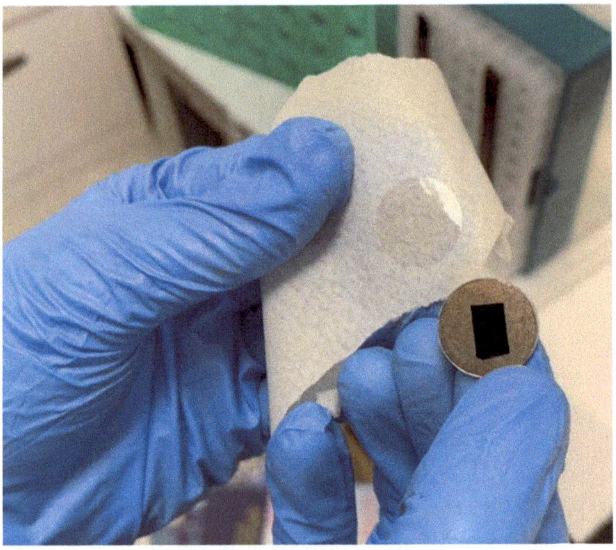

Fig. 8 AFM: Preparation of freshly cleaved mica. Cleaving (or peeling off) the top layer of a piece of mica to expose a fresh mica surface using tape. Ideally a complete mica surface is cleaved as evidenced by the mica surface left on the tape. One can also check the quality of the newly exposed surface using microscopic techniques

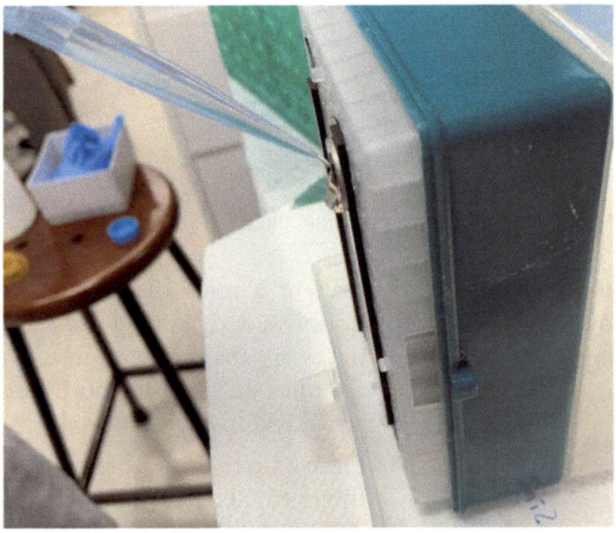

Fig. 9 AFM in air: Rinsing of freshly cleaved mica. Aim the pipette tip at the top of the mica surface without touch during the rinse process. Angle the sample holder in a vertical position. Having a secondary container in the bottom helps collect the rinse liquid for waste collection and disposal later

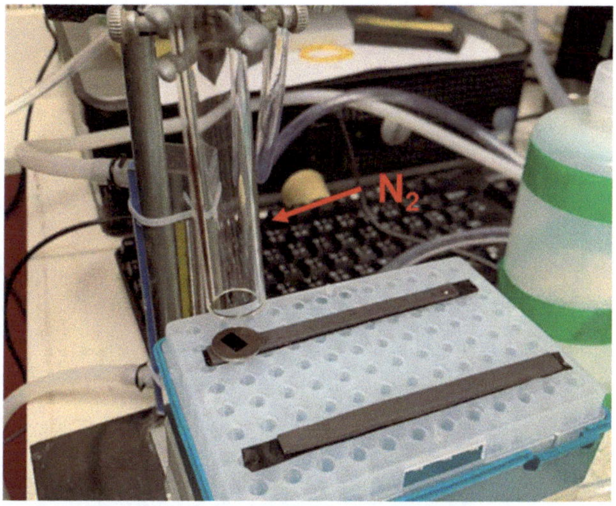

Fig. 10 AFM in air: Removal of excess fluid on mica. First blow N_2 on the sample while the AFM sample is still in the vertical position to collect the excess fluid at the bottom and then remove the excess fluid using lint-free cloth. Then change the AFM sample holder to a horizontal position and dry the AFM sample under a N_2 flow

dry under ultrapure or glass-wool-filtered N_2 at room temperature (*see* **Note 49**, Fig. 10).

8. Maintain the freshly-cleaved mica/steel construct in a horizontal position on the magnetic tape/pipette box construct.
9. For depositing DNA origami/structures on the mica discs, pipette 10μL of solution onto the cleaned and dried mica surface, ensuring that the pipette tip does not touch the substrate (or else the freshly cleaved mica surface might be scratched).
10. Wait for 5 min to ensure complete DNA adsorption on the surface.
11. Rotate the mica/steel/magnetic tape/pipette box construct into a vertical position, and then immediately rinse the mica surface using 2 mL ultrapure water by aiming the pipette at (but not touching) the top of the mica surface and with the water flowing in a vertical direction downwards. This step serves to untangle the overlapped DNA strands and align the DNA strands in one direction. This step also serves to remove weakly-bound (or physisorbed) DNA strands and buffer salts (*see* **Note 48**).
12. Dry the mica surface under ultrapure N_2 (or Ar) in room temperature, and make sure to remove excess water trapped at the bottom rim (or edge) or the mica surface using lintless paper cloth (*see* **Note 49**).

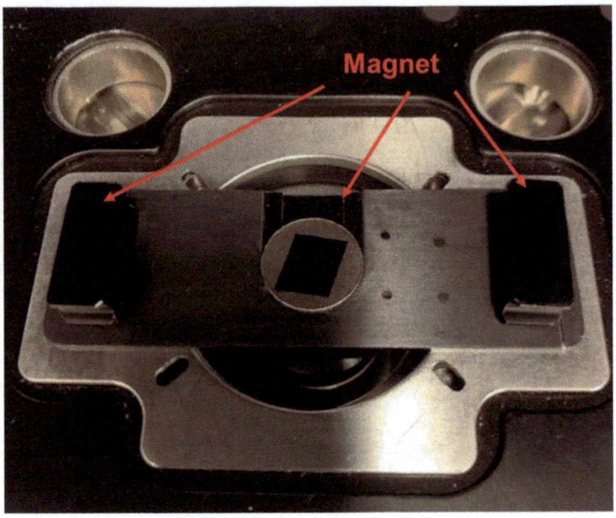

Fig. 11 AFM: Affix of sample holder on the AFM stage using magnets

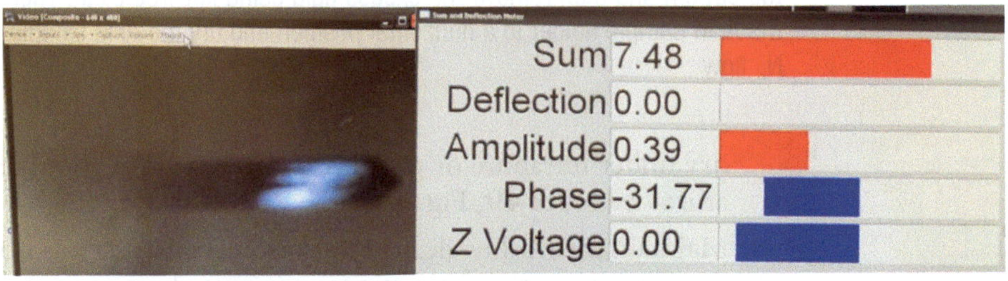

Fig. 12 AFM: Optimization of laser deflection. Optimal laser position on the cantilever to give a maximized "Sum" (different values for different cantilevers) and "Deflection" value of 0

13. Affix the mica/steel construct firmly onto the sample stage, and then affix the sample stage onto the AFM instrument using two removable magnets with one on each side of the mica/steel construct (*see* **Note 50**, Fig. 11).
14. Place the cantilever holder on the cantilever loading stage, load the cantilever using a pair of tweezers, and then secure with the screw provided (*see* **Note 51**, Fig. 12).
15. Attach the cantilever holder onto the AFM scan-head, and flip the AFM scan-head over, and place on top of the mica sample.
16. Place the bubble leveler on the top of the AFM scan-head. Make sure the AFM scan-head is leveled at all times if possible.
17. Open the AFM control software, and choose standard air topography mode (choose tapping mode for soft-landing of the AFM cantilever).
18. Turn on the top-side camera (if available), and adjust the camera focus to find the cantilever.

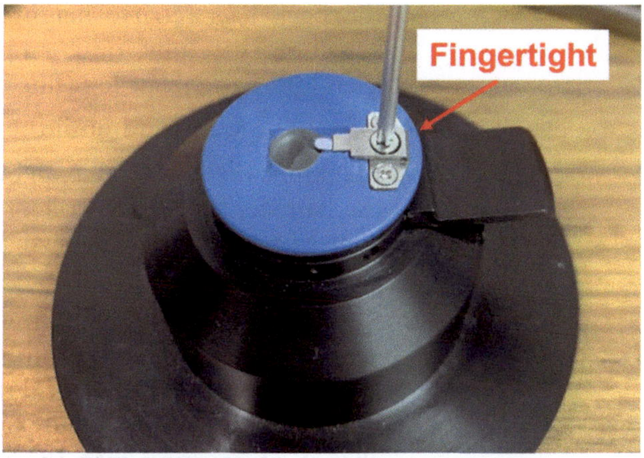

Fig. 13 AFM: Positioning of cantilever. Tighten the screw by hand to secure the cantilever in place

19. Adjust the position of laser beam with two X-Y knobs. The laser should be positioned where the value of the "Sum" signal is at its maximum (*see* **Note 52**, Fig. 13).
20. Adjust the deflection to 0. This value will fluctuate during cantilever approaching and measurement. Try to maintain the value between ±0.01.
21. Set the initial scan parameters in the Main Control Panel as follows:

Scan area: depends on the sample to be investigated; scan rate: 1 or 0.75 Hz (depending on scan area); point and line: 512 (for lower resolution but faster scans, use 256); and initial set point: 950 mV

22. Change the image storage path.
23. Conduct background thermal noise scans for force-distance calibration. Input the spring constant value stated by the manufacturer on the cantilever box, and then capture thermal data.
24. Wait for about 30 cycles, then initialize fit, and then fit thermal data. The data will automatically be transferred to the Main Control Panel.
25. Set the low and high frequency to 50 kHz and 100 kHz, respectively, in the Tune Panel. Set target percentage value to −5%. Set the amplitude to 1 V. Auto-tune the AFM system.
26. Click engage on the Main Control Panel to initiate the cantilever approaching procedure. The AFM computer system can now control the cantilever.
27. Lower the cantilever unit with the knobs on the side of the AFM scan-head (automatic procedure is available for certain AFM brands and models). Make sure the whole AFM scan-head unit is leveled by monitoring the bubble leveler. Lower

the back legs (coarse height adjustments); adjust the AFM scan-head carefully, once amplitude value reaches 0.95; and then lower the front leg (fine height adjustment).

28. Lower the scan-head until the Z voltage begins to drop.
29. Maximize the Z voltage (can be done by activating and turning the hamster wheel controller) until the Z voltage bar in red is approximately equal in size (or area) to that of the phase parameter bar in blue.
30. Repeat the adjustment of the scan-head until the Z voltage cannot be maximized though turning the hamster wheel (during this procedure, be careful that the Z voltage bar should not be turn from red to blue; otherwise, the tip might have a chance of crashing into the sample, thereby ruining the tip and a new cantilever is needed and the whole AFM procedure above needs to be repeated).
31. Close the AFM chamber doors carefully. Do scan and save images.
32. Adjust the set point, drive amplitude, and integral gain to optimize the image quality.
33. Right click on the real-time imaging window, choose "fix all scale." On the signal graph, two overlapping trace and retrace lines will appear if the cantilever approaching procedure is successful. If the trace and retrace lines are not overlapping or the image is fuzzy, then further adjust the Z voltage, drive amplitude, gain, and other related AFM scan parameters. (For the drive amplitude, only increase its value by 20 mV at max every time. Increasing too much at single change may ruin the tip).
34. After collecting enough AFM images, withdraw the cantilever or stop the AFM scan.

3.11.2 AFM Liquid Mode Protocol (Should Be Conducted Swiftly to Minimize Evaporation) [31, 32]

1. For liquid sample preparation, repeat all Air Mode Sample Preparation steps.
2. Fill a syringe with a small amount of vacuum grease (*see* **Note 53**, Fig. 14).
3. Build a sample well as a solvent reservoir by using vacuum grease to prevent the liquid from spreading too thin which would otherwise facilitate evaporation (*see* **Note 54**, Fig. 15).
4. Pipette 20μL solution into the reservoir on the mica's surface (*see* **Note 55**, Fig. 16).
5. Affix the mica/steel construct firmly onto the sample stage, and then affix the sample stage onto the AFM instrument using two removable magnets with one on each side of the mica/steel construct.

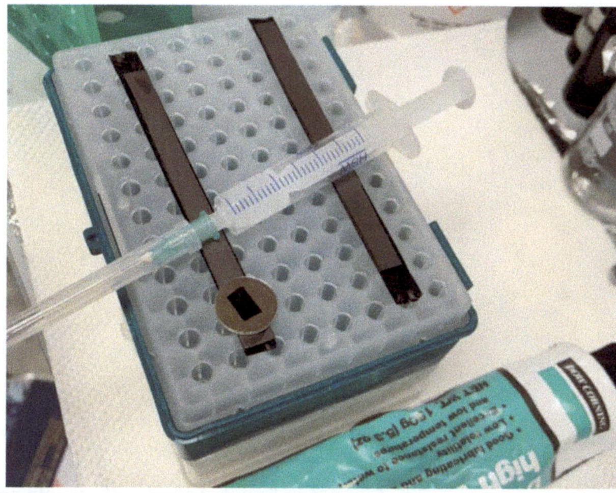

Fig. 14 AFM in liquid: Preparation of grease-filled syringe. Fill a syringe with high-vacuum grease. If the grease is too viscous, using a higher gauge needle or a syringe without a needle is an alternative method

Fig. 15 AFM in liquid: Addition of grease on mica. Apply grease on a mica surface using a syringe to form a solution well

6. Place the cantilever holder on the cantilever loading stage, and load the cantilever using a pair of tweezers, and then secure with the screw provided (finger-tight is adequate, do not overtighten).
7. Wet the cantilever using 5μL of sample solution by dispensing the imaging fluid at the side of cantilever (*see* **Note 56**, Fig. 17).

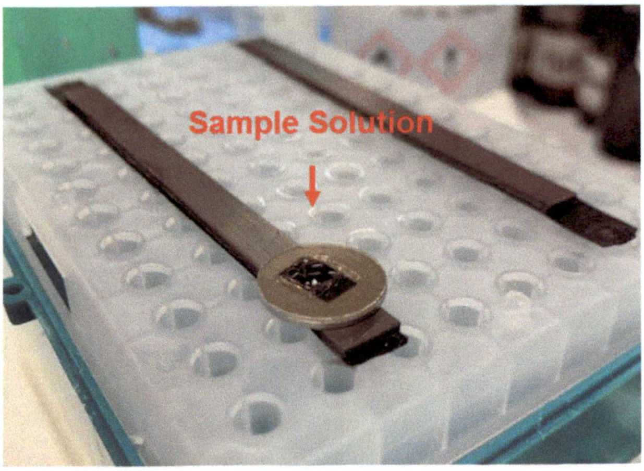

Fig. 16 AFM in liquid: Addition of sample solution on mica. Apply solution into the sample well to check if solution leaks out

Fig. 17 AFM in liquid: Removal of excess liquid from the mica. Rotate the scan-head sideways to let solution slide away from the cantilever

8. Attach the cantilever holder onto the AFM scan-head, and flip the AFM scan-head over, and place on top of the mica sample.
9. Place the bubble leveler on the top of the AFM scan-head. Make sure the AFM scan-head is leveled at all times if possible.
10. Open the AFM control software, and choose standard *liquid* topography mode (choose tapping mode for soft-landing of the AFM cantilever).

11. Turn on the top-side camera (if available), and adjust the camera focus to find the cantilever.
12. Adjust the position of the laser beam with two X-Y knobs. The laser should be positioned where the value of the "Sum" signal is at its maximum (this position usually is found at near 1/3 of the cantilever measuring from the tip of the cantilever).
13. Adjust the deflection to 0. This value will fluctuate during cantilever approaching and measurement. Try to maintain the value between ±0.01.

Set the initial scan parameters in the Main Control Panel as follows: Scan area: depends on the sample to be investigated; scan rate: <0.5 Hz; point and line: 512 (for lower resolution but faster scans, use 256); and initial set point: 950 mV.

14. Change the image storage path.
15. Conduct background thermal noise scans for force-distance calibration. Input the spring constant value stated by the manufacturer on the cantilever box, and capture thermal data.
16. Wait for about 30 cycles, then initialize fit, and then fit thermal data. The data will automatically be transferred to the Main Control Panel.
17. Set the low and high frequency to 50 kHz and 100 kHz, respectively, in the Tune Panel. Set target percentage value to −5%. Set the amplitude to 1 V. Manual-tune of the AFM system. First, append thermal data by transferring manually the drive amplitude frequency from thermal data to the Tune Panel using the one-tune function. Aim the computer mouse cursor at the most pronounced peak, and then right click at the center of that pronounced peak to set the drive frequency. Update the drive frequency in the Main Control Panel using the one-tune value manually obtained from the Tune Panel.
18. Click engage on the Main Control Panel to initiate the cantilever approaching procedure. The AFM computer system can now control the cantilever.
19. Lower the cantilever unit with the knobs on the side of the AFM scan-head (automatic procedure is available for certain AFM brands and models). Make sure the whole AFM scan-head unit is leveled by monitoring the bubble leveler. Lower the back legs (coarse height adjustments); adjust the AFM scan-head carefully, once amplitude value reaches 0.95; and then lower the front leg (fine height adjustment). Do not allow the AFM scan-head to touch the sample well (a.k.a. solvent reservoir) made of grease (*see* **Note 55**, Fig. 16).
20. Lower the scan-head until the Z voltage begins to drop.

21. Maximize the Z voltage (can be done by activating and turning the hamster wheel controller) until the Z voltage bar in red is approximately equal in size (or area) to that of the phase parameter bar in blue.
22. Repeat the adjustment of the scan-head until the Z voltage cannot be maximized though turning the hamster wheel (during this procedure, be careful that the Z voltage bar should not be turn from red to blue; otherwise, the tip might have a chance of crashing into the sample, thereby ruining the tip and a new cantilever is needed as well as the whole AFM procedure above needs to be repeated).
23. Close the AFM chamber doors carefully. Do scan and save images.
24. Adjust the set point, drive amplitude, and integral gain to optimize the image quality.
25. Right click on the real-time imaging window, choose "fix all scale." On the signal graph, two overlapping trace and retrace lines will appear if the cantilever approaching procedure is successful. If the trace and retrace lines are not overlapping or the image is fuzzy, then further adjust the Z voltage, drive amplitude, gain, and other related AFM scan parameters. (For the drive amplitude, only increase its value by 20 mV at max every time. Increasing too much at single change may ruin the tip).
26. After collecting enough AFM images, withdraw the cantilever or stop the AFM scan (*see* **Note 56**, Fig. 17).

4 Notes

1. Concentrated NaOH (10 N) can be used initially to equilibrate the pH close to 8. Then, it is suggested to use a lower concentration of NaOH to avoid a sudden increase in pH above the required pH.
2. Use a magnetic stirrer for mixing.
3. Aliquot the 10% APS solution to avoid repetitive freeze-thaw cycles.
4. The addition of reagents has to follow the sequence as mentioned. The addition of TEMED has to be performed in a fume hood with good ventilation because of the toxicity.
5. Containers of SYBR gold stain should be covered in foil to protect from light and kept at −20 °C for long-term storage and to avoid bleaching.
6. Distribute to each reaction tube in this order: buffer, 10× beads, 10× peptide. When preparing PBSMCT buffer, add

$CaCl_2$ solution FIRST and mix well. Then add 10× PBS slowly while swirling, followed by $MgCl_2$ and Tween-20. A precipitate forms when concentrated $CaCl_2$ is added to a phosphate buffer, which is very difficult to dissolve. Filter the buffer before use.

7. Pipette up and down (without bubbles) to ensure complete mixing of all components in the solution.
8. Scaffold DNA is assembled to form a box shape [33] through the use of a molar excess of staple DNA strands (fivefold excess in our typical method). Other groups employ from fourfold [34] to 100-fold excess [23].
9. Set the ramp rate of the thermal cycler to 0.1 °C per 6 s. For precise temperature control, keep sample volume at no more than 50μL avoiding wells around the edge of heat block.
10. The gel staining process needs to be carried out in the dark to avoid unnecessary bleaching of SYBR Gold stain.
11. For the first SELEX cycle, 1 nmole of DNA library is used in a 100μL transcription reaction. In the subsequent SELEX rounds, 5.3μL of the purified DNA library sample is used.
12. Transcription mix (Table 1) is incubated at 37 °C for 1 h.
13. Steps for purification using ethanol precipitation:
    - Add 1μL RNA grade glycogen.
    - Add 1/10th volume sodium acetate.
    - Add 3× volume ice cold 100% ethanol.
    - Leave in −20 °C overnight.
    - Centrifuge at 18K G for 30 min at −4 °C.

Table 1
Transcription reaction mix

Reagent	Volume (μL)
10× buffer	2
DNA template	6.3
ATP	1.8
CTP	1.8
GTP	1.8
UTP	1.8
RNAse inhibitor	0.5
DTT	2
T7 RNA polymerase	2
Total	20

Table 2
PCR reaction mix

Reagent	Volume (μL)
10× buffer	10
H_2O	68
FP	8
RP	8
NTP	3
Template	2
PFU polymerase	1
Total	100

- Pour off supernatant, add 800μL of 75% ethanol.
- Centrifuge at 18K G for 10 min at −4 °C.
- Pour off supernatant, dry pellet on heat block at 65 °C.
- Dry cool pellet on ice or in −20 °C for 1 min.

14. PCR is carried out by performing thermal cycling on the PCR mixture (Table 2) and the following thermal cycling sequence.

    Thermal cycling protocol:

    1× 1 min at 96 °C

    20× 15 s at 96 °C

    30 s at 50 °C

    30 s at 72 °C

    1× 15 s at 72 °C

    Hold at 4 °C

15. It is important to vortex and centrifuge the PCR and transcription mixtures for a few seconds each to ensure uniformity of the mixtures and to remove any bubbles before using them. If there are still some bubbles present, the centrifuge tube should be gently tapped to remove them.
16. It is important to ensure that the tubing used for the syringes is long enough to reach the PDMS device from the syringe pump and long enough to allow that a sufficient volume of mixture can be taken into it without flowing into the syringe.
17. There should be a few centimeters air gap between the oil and the mixtures/emulsion when introducing them into the tubing connected to the syringe.
18. After the experimental setup is completed, it is good to allow the continuous oil phase to pass through the channels first for a

minute or two before introducing the PCR/transcription mixtures/droplets into the channels to ensure there are no blocks in the channel.

19. After the experiment is started, it is important to check all the junctions of the channels to ensure that the experiment is running smoothly.
20. When collecting the droplets in the centrifuge tube, it is very important to empty the remaining droplets inside the outlet tubing into the centrifuge tube as a lot of droplets remain in the tubing.
21. It is important to slowly spread the oil with surfactant in the tube with droplets by pipetting it along the wall of the tube. Since the pipette touches the walls, it is important to change the pipette tip after every turn.
22. Refer to the related literature for the composition of binding buffer. For example, the binding buffer of ATP and ATP-binding aptamer in Fig. 1 was 300 mM NaCl, 5 mM $MgCl_2$, and 20 mM Tris-HCl (pH 8.3).
23. It is critical to purge the instrument by using nitrogen gas before turning on the instrument. Running with oxygen can ruin the optics of the instrument.
24. The computer should be turned on before turning on the instrument, which otherwise can cause disconnection.
25. Quartz cuvette should be used in CD detection. Glass cuvette cannot be used owing to its absorption of ultraviolet light.
26. HT (voltage) signal should be monitored while recording CD signal. For Jasco J-810 spectropolarimeter, if the voltage is above 700 V, the CD signal will be unreliable, and if the voltage is above 800 V, the signal will become invalid.
27. Strong absorption of some buffer components can cause high HT signal. If the voltage is above 700 V, dilute the buffer accordingly.
28. The beads must be as singular as possible for efficient conjugation and accurate measurements. Sonicate the beads immediately prior to use in all steps. Vortexing does not sufficiently separate aggregated beads.
29. Optimize the annealing temperature for your primers and polymerase. The settings given here work well for the TNFα aptamers [29].
30. Some templates do not amplify well or give different products for different polymerases. Check that your combination of polymerase, primers, and templates gives the desired products in both bPCR and qPCR. If necessary, use the same polymerase and add SYBR Green dye (1X final concentration) for fluorescence readout.

31. Include the FP complement during hybridization to block the FPs.
32. Filter the buffer before use.
33. Both double-stranded or single-stranded aptamer beads can be analyzed by qPCR. If the DNA on beads is double-stranded but the template is single-stranded, account for this by subtracting 1 from the C_t value reported for the beads. We recommend performing qPCR on double-stranded products to avoid uncertainty caused by incomplete dehybridization.
34. The bead concentration in the reactions (10^2 beads/μL) is too low to see by eye. One can minimize bead loss by timing the separations and not touching the side of the tube next to the magnet when moving solutions. Keep the bead concentration low so as to keep the concentration of aptamers below the K_D [29].
35. Pipette up and down (without bubbles) to ensure complete mixing of all components in the solution.
36. Make sure the running buffer in the gel tank is enough to cover the entire gel and pre-run the gel at 4 °C for 30 min before loading samples.
37. To optimize the competitive interaction between PfLDH and complementary ssDNA, we first examined the length of the complementary sequence from 8 to 24 bp in 4 bp steps and determined 12 bp as the optimal length. 8 bp was too short to hybridize to aptamer while strands longer than 16 bp hybridized too tightly and were not able to be displaced by PfLDH interaction. We next scanned for the optimal position on the aptamer for the complementary sequence in 6 bp steps and concluded that 12 bp1 was the most optimal sequence for the aptamer key module.
38. Avoid the introduction of air bubbles which may affect the fluorescence readout.
39. Measure the fluorescence signal with 500 ms measurement time and 12 nm bandwidth. FRET signal intensity will gradually decrease, reflecting PfLDH-mediated opening of the lid of the DNA box structure.
40. Perform this procedure in dark for maximized color development.
41. Prepare uranyl acetate solution using fume hood and plastic blockage.
42. Perform staining using spill tray to avoid contamination on bench surface.
43. Avoid sudden vibration that leads to damage of grid and coating of sample.

44. Adjustment of all parameters has to be performed again when the magnification is changed.
45. For cost effectiveness and environmental protection purposes, empty pipette boxes that are reusable and durable can also be used as the AFM sample holders (Fig. 6).
46. The mica surface can adhere firmly to the double-sided carbon tape on the steel plate by pressing gently with the operator's fingers (Fig. 7).
47. Instead of using tweezers, one can also use fingers with nitrile gloves worn to remove the tape (Fig. 8).
48. Depending on the experimental condition, washing with ethanol, acetone, and/or methanol is a common practice (Fig. 9).
49. Ar can be used in place of N_2 (Fig. 10).
50. Hold the mica/steel construct firmly in place using magnets on the AFM stage (Fig. 11).
51. The optimal laser position usually is found at near 1/3 of the cantilever measuring from the tip of the cantilever (Fig. 12).
52. Do not overtighten the screw that secure the cantilever; finger-tight is adequate (Fig. 13).
53. Fast-setting/curing glue can be used in place of high-vacuum grease (Fig. 14).
54. Make sure the barrier height is short enough to allow the cantilever to approach the mica surface without having the scan-head touching the high-vacuum grease (Fig. 15).
55. Make sure the solution does not overspill after the scan-head with the cantilever is lowered into the solution (Fig. 16).
56. Place the scan-head on its side after liquid AFM scans to remove sample solution from the cantilever by gravity (Fig. 17).

Acknowledgments

J.A.T. acknowledges funding provided by the Hong Kong University Grants Council General Research Fund Grants (17127515, 17163416 and 17102318). L.S.S. acknowledges Professors Angela R. Wu and H. Tom Soh, Dr. Jianpeng Wang, and Kaitlin H.Y. Chan for their support and assistance. L.S.S. is funded by the Hong Kong Innovation and Technology Commission (ITS/350/16) and a Junior Fellowship from the Hong Kong Jockey Club Institute for Advanced Study at the Hong Kong University of Science and Technology. C.K.K. acknowledges support from Shenzhen Basic Research Project [JCYJ20180507181642811]; Research Grants Council of the Hong Kong SAR, China Projects [CityU

11100421, CityU 11101519, CityU 11100218, N_CityU110/17, CityU 21302317]; Croucher Foundation Project [9500030, 9509003]; the State Key Laboratory of Marine Pollution Director Discretionary Fund; and City University of Hong Kong projects [6000711, 7005503, 9667222, 9680261]. E.C.M.T. acknowledges the Croucher Foundation, the HK RGC (HKU JLFS/P-704/18; E-HKU704/19; ECS 27301120), the EU (Horizon 2020: SABYDOMA – 862296), the PRC NSFC (22002132), and the HK ITC (Health@InnoHK Program: Laboratory for Synthetic Chemistry and Chemical Biology).

References

1. Seeman NC (1982) Nucleic acid junctions and lattices. J Theor Biol 99(2):237–247
2. Kallenbach NR, Ma R-I, Seeman NC (1983) An immobile nucleic acid junction constructed from oligonucleotides. Nature 305(5937): 829–831
3. Seeman NC (2003) DNA in a material world. Nature 421(6921):427–431
4. Ellington AD, Szostak JW (1990) In vitro selection of RNA molecules that bind specific ligands. Nature 346(6287):818–822
5. Tuerk C, Gold L (1990) Systematic evolution of ligands by exponential enrichment: RNA ligands to bacteriophage T4 DNA polymerase. Science 249(4968):505
6. Cheung YW, Kwok J, Law AW, Watt RM, Kotaka M, Tanner JA (2013) Structural basis for discriminatory recognition of *Plasmodium* lactate dehydrogenase by a DNA aptamer. Proc Natl Acad Sci U S A 110(40):15967–15972
7. Shiu SC-C, Cheung Y-W, Dirkzwager RM, Liang S, Kinghorn AB, Fraser LA, Tang MSL, Tanner JA (2017) Aptamer-mediated protein molecular recognition driving a DNA tweezer nanomachine. Adv Biosyst 1(1–2):1600006
8. Tang MSL, Shiu SC-C, Godonoga M, Cheung Y-W, Liang S, Dirkzwager RM, Kinghorn AB, Fraser LA, Heddle JG, Tanner JA (2018) An aptamer-enabled DNA nanobox for protein sensing. Nanomed Nanotechnol 14(4): 1161–1168
9. Shiu CS, Fraser AL, Ding Y, Tanner AJ (2018) Aptamer display on diverse DNA polyhedron supports. Molecules 23(7)
10. Fraser AL, Kinghorn BA, Tang SLM, Cheung Y-W, Lim B, Liang S, Dirkzwager MR, Tanner AJ (2015) Oligonucleotide functionalised microbeads: indispensable tools for high-throughput aptamer selection. Molecules 20(12)
11. Li Y, Lee J-S (2019) Recent developments in affinity-based selection of aptamers for binding disease-related protein targets. Chem Pap 73(11):2637–2653
12. Chao Y, Shum HC (2020) Emerging aqueous two-phase systems: from fundamentals of interfaces to biomedical applications. Chem Soc Rev 49:114
13. Tawfik DS, Griffiths AD (1998) Man-made cell-like compartments for molecular evolution. Nat Biotechnol 16(7):652–656
14. Ryckelynck M, Baudrey S, Rick C, Marin A, Coldren F, Westhof E, Griffiths AD (2015) Using droplet-based microfluidics to improve the catalytic properties of RNA under multiple-turnover conditions. RNA 21(3):458–469
15. Teh S-Y, Lin R, Hung L-H, Lee AP (2008) Droplet microfluidics. Lab Chip 8(2):198–220
16. Lyu K, Chen S-B, Chan C-Y, Tan J-H, Kwok CK (2019) Structural analysis and cellular visualization of APP RNA G-quadruplex. Chem Sci 10(48):11095–11102
17. del Villar-Guerra R, Trent JO, Chaires JB (2018) G-quadruplex secondary structure obtained from circular dichroism spectroscopy. Angew Chem Int Edit 57(24):7171–7175
18. Pollard TD (2010) A guide to simple and informative binding assays. Mol Biol Cell 21(23): 4061–4067
19. Siu RHP, Chan KHY, Ridzewski C, Slaugther LS, Wu AR (2019) Single particle analysis of on-bead emulsion polymerase chain reaction (Submitted)
20. Wang J, Gong Q, Maheshwari N, Eisenstein M, Arcila ML, Kosik KS, Soh HT (2014) Particle display: a quantitative screening method for generating high-affinity aptamers. Angew Chem Int Edit 53(19): 4796–4801

21. Dirkzwager RM, Kinghorn AB, Richards JS, Tanner JA (2015) APTEC: aptamer-tethered enzyme capture as a novel rapid diagnostic test for malaria. Chem Commun 51(22): 4697–4700
22. Fraser LA, Kinghorn AB, Dirkzwager RM, Liang S, Cheung Y-W, Lim B, Shiu SC-C, Tang MSL, Andrew D, Manitta J, Richards JS, Tanner JA (2018) A portable microfluidic Aptamer-Tethered Enzyme Capture (APTEC) biosensor for malaria diagnosis. Biosens Bioelectron 100:591–596
23. Rothemund PW (2006) Folding DNA to create nanoscale shapes and patterns. Nature 440(7082):297–302
24. Douglas SM, Bachelet I, Church GM (2012) A logic-gated nanorobot for targeted transport of molecular payloads. Science 335(6070):831
25. Saccà B, Meyer R, Feldkamp U, Schroeder H, Niemeyer CM (2008) High-throughput, real-time monitoring of the self-assembly of DNA nanostructures by FRET spectroscopy. Angew Chem Int Edit 47(11):2135–2137
26. Tse ECM, Zwang TJ, Bedoya S, Barton JK (2019) Effective distance for DNA-mediated charge transport between repair proteins. ACS Central Sci 5(1):65–72
27. Tang MY, Shum HC (2016) One-step immunoassay of C-reactive protein using droplet microfluidics. Lab Chip 16(22):4359–4365
28. Ji D, Wang H, Ge J, Zhang L, Li J, Bai D, Chen J, Li Z (2017) Label-free and rapid detection of ATP based on structure switching of aptamers. Anal Biochem 526:22–28
29. Wang J, Yu J, Yang Q, McDermott J, Scott A, Vukovich M, Lagrois R, Gong Q, Greenleaf W, Eisenstein M, Ferguson BS, Soh HT (2017) Multiparameter particle display (MPPD): a quantitative screening method for the discovery of highly specific aptamers. Angew Chem Int Edit 56(3):744–747
30. Tse ECM, Zwang TJ, Barton JK (2017) The oxidation state of [$4Fe_4S$] clusters modulates the DNA-binding affinity of DNA repair proteins. J Am Chem Soc 139(36):12784–12792
31. Kielar C, Xin Y, Xu X, Zhu S, Gorin N, Grundmeier G, Möser C, Smith DM, Keller A (2019) Effect of staple age on DNA origami nanostructure assembly and stability. Molecules 24(14):2577
32. Li L, Tian X, Dong Z, Liu L, Tabata O, Li WJ (2013) Manipulation of DNA origami nanotubes in liquid using programmable tapping-mode atomic force microscopy. Micro Nano Lett 8(10):641–645
33. Andersen ES, Dong M, Nielsen MM, Jahn K, Subramani R, Mamdouh W, Golas MM, Sander B, Stark H, Oliveira CL, Pedersen JS, Birkedal V, Besenbacher F, Gothelf KV, Kjems J (2009) Self-assembly of a nanoscale DNA box with a controllable lid. Nature 459(7243): 73–76
34. Wagenbauer KF, Engelhardt FAS, Stahl E, Hechtl VK, Stommer P, Seebacher F, Meregalli L, Ketterer P, Gerling T, Dietz H (2017) How we make DNA origami. Chembiochem 18(19):1873–1885

Chapter 18

Production and Testing of RNA Origami Anticoagulants

Abhichart Krissanaprasit, Carson Key, Kristen Froehlich, and Thomas H. LaBean

Abstract

Nucleic acid nanotechnology provides the ability to create unprecedented nanostructures with diverse architectures and functions that can be utilized in myriad fields. A set of self-folding, single-stranded RNA origami structures bearing thrombin RNA aptamers have been demonstrated to act as anticoagulants. Here, we describe the detailed methods of producing and testing of such RNA origami anticoagulants. This method highlights the potential of RNA origami for biomedical applications.

Key words RNA origami, Anticoagulants, In vitro transcription, Self-assembly, RNA aptamers, Nucleic acid nanotechnology, Coagulation, Direct thrombin inhibitor

1 Introduction

The coagulation cascade is a physiological system involving a series of enzymatic reactions that mediate and ultimately produce cross-linked fibrin clots in the blood [1, 2]. Monitoring the coagulation cascade is essential to prevent unwanted blood clot formation that can potentially occlude blood flow to vital organs. Chemical-based anticoagulants, such as warfarin and heparin, are commonly used in clinical settings; however, these chemical anticoagulants have several associated risks, including narrow therapeutic windows, undesired side effects, and difficulties with reversing their activity [3–5].

In the general field of nucleic acid aptamers, many molecules have been developed that bind specifically to targets such as small molecules, proteins, and whole cells [6, 7]. DNA and RNA aptamers have been developed to selectively bind with exosite-1 and exosite-2 of thrombin, a pivotal molecule in the coagulation cascade [8–13]. When bound, the aptamers can inhibit thrombin's various activities and can therefore act as anticoagulants, preventing or at least slowing cross-linked fibrin clot formation. Such nucleic

Julián Valero (ed.), *DNA and RNA Origami: Methods and Protocols*, Methods in Molecular Biology, vol. 2639,
https://doi.org/10.1007/978-1-0716-3028-0_18,

acid-based anticoagulants provide many advantages over chemical-based anticoagulants including low immunogenicity, wider therapeutic windows, no known side effects, higher biocompatibility, and the availability of an inhibitory complementary strand to act as an antidote [13–15]. RNA thrombin aptamers have progressed to clinical trials; however, they were quickly removed by the kidneys due to their small size. Modifications of RNA aptamers that were required to enlarge their size and enhance stability in a nuclease environment also created multiple extra steps for production, increased the cost of their production, and introduced immunogenicity problems that ultimately derailed their further commercial development [7, 16].

To improve the pharmacokinetics of nucleic acid-based therapeutics, DNA weave tiles have been constructed and decorated with thrombin-binding DNA aptamers. The anticoagulation activity of these structures was greater than for free aptamers or for single-stranded DNA-linked aptamers [17–19]. DNA nanostructures provide the benefits of controlling the relative positions and distances between aptamers, increasing local concentration of aptamers, and enlarging the molecule to sufficient sizes to prevent renal filtration. Moreover, the inhibited thrombin activity is potentially reversed by the addition of complementary nucleic acid antidotes. Although there are many advantages to DNA weave tiles, the formation of weave tiles typically entails the assembly of two strands of DNA; this requires stoichiometry matching and thus increases the difficulty of achieving high yields of purified product.

Recently, a single-stranded RNA origami has been developed and shown to form a honeycomb-like nanoscale pattern by intermolecular interactions between single RNA origami units [20]. The self-assembly of RNA origami enables us to design and incorporate RNA aptamers into its structure to obtain a functional RNA origami made of a self-folding, single-strand RNA [21, 22]. Recently, we reported a family of RNA origami molecules bearing two thrombin RNA aptamers that were shown to be effective anticoagulants [22]. The anticoagulation activity of RNA origami decorated with two RNA aptamers was higher than both free RNA aptamers and the DNA weave tiles decorated with DNA aptamers. In this method, we focus on the details of production of RNA origami and demonstrate its anticoagulation activity by activated partial thromboplastin time (aPTT) assays. Standard methods for design of RNA origami are not described here, since they have been previously and eloquently described by Steffen et al. [23] (Fig. 1).

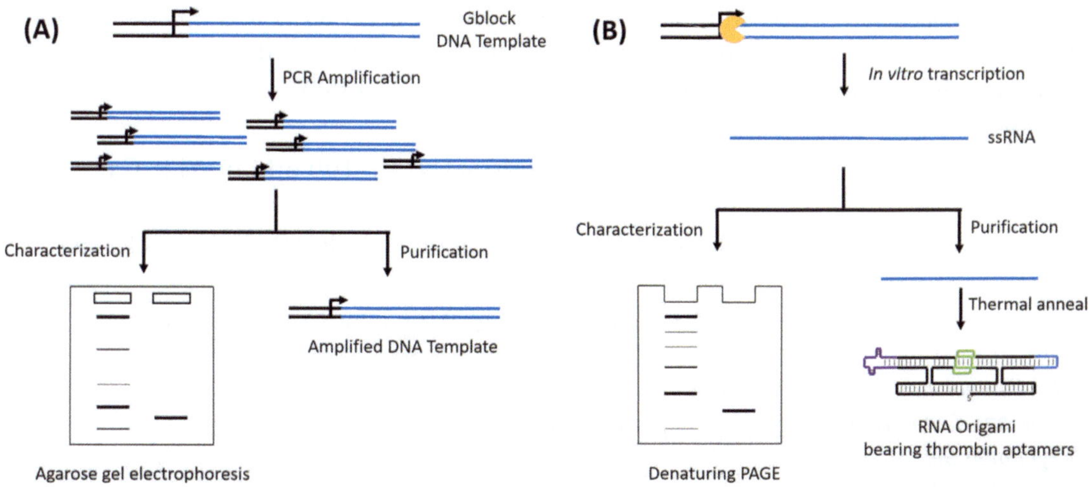

Fig. 1 Flow chart of RNA origami anticoagulant production. (**a**) Amplification of DNA template using polymerase chain reaction (PCR). (**b**) In vitro transcription and folding of RNA origami bearing two thrombin RNA aptamers

2 Materials

2.1 DNA Template Amplification and Purification

1. Double-stranded DNA fragments, Gblock (IDTDNA). The Gblock samples are received in tubes containing lyophilized powder from IDTDNA. Prior to use, centrifuge the Gblock tubes at 3000 g for 1 min (*see* **Note 1**). Dilute the Gblock DNA to 100 nM in nuclease-free water. Store the Gblock solutions at −20 °C.
2. DNA primers (e.g., IDTDNA). Dissolve primers in 1X TE buffer, pH 8.0. The stock concentrations of primers are 10 μM. Store at −20 °C.
3. dNTP mix: dATP, dCTP, dGTP, and dTTP (10 mM of each).
4. Phusion High-Fidelity DNA polymerase and 1X HF reaction buffer (ThermoFisher Scientific). HF reaction buffer is provided by ThermoFisher Scientific.
5. DNA purification kit (e.g., GFX DNA purification kit).
6. Thermocycler.
7. Spectrophotometer (e.g., Nanodrop 3000c).

2.2 Agarose Gel Electrophoresis

1. Ultrapure agarose.
2. 10x Tris-borate-EDTA buffer, 10X TBE buffer.
3. Ethidium bromide.
4. 6x loading dye: 300 μL glycerol, 2.5 mg bromophenol blue, 2.5 mg xylene cyanol. Add water to adjust to final volume of 1 mL. Store at 4 °C.

2.3 RNA Origami Production Using In Vitro Transcription and Purification

1. Mutant T7 RNA polymerase (Y639F) and 5× transcription buffer (Lucigen). The transcription buffer is provided with commercialized mutant T7 RNA polymerase.
2. T7 RNA polymerase.
3. Nucleotide triphosphate (NTP): 100 mM of ATP, CTP, UTP, and GTP.
4. 100 mM of 2′fluoro-modified dCTP (2′F-dCTP) and 2′ fluoro-modifed dUTP (2′F-dUTP).
5. RNA Clean-up kit (e.g., Monarch RNA Clean-up kit).

2.4 Components for Native and Denaturing Acrylamide Gel Electrophoresis

1. Urea.
2. 40% acrylamide solution: 19:1 acrylamide: bis-acrylamide.
3. 30% w/v ammonium persulfate (APS): 3 g of APS and add water to adjust to final volume of 10 mL (*see* **Note 2**).
4. Tetramethylethylenediamine (TEMED).
5. 2x formamide loading dye: 8 mL formamide, 2 mL 10X TBE buffer, 4 mg bromophenol blue. Add water to final volume of 10 mL. Store at 4 °C.
6. 10X Tris-borate-EDTA buffer, 10X TBE.

2.5 Folding RNA Origami by Heat-Annealing Procedure

1. 5x RNA folding buffer: 100 mM HEPES pH 7.4, 750 mM NaCl, and 10 mM $CaCl_2$.

2.6 Anticoagulation Test: aPTT Assay

1. ST4 coagulometer (Diagnostica Stago).
2. aPTT reagent (Diagnostica Stago). Dissolved in Milli-Q water: Dissolve aPPT reagent in 5 mL of Milli-Q water, and let it sit at room temperature for 30 min prior to use.
3. $CaCl_2$ solution (Diagnostica Stago). Store at 4 °C.
4. Plastic cuvettes for coagulometer (Diagnostica Stago).
5. Magnetic metal balls for coagulometer (Diagnostica Stago).
6. Pooled normal human plasma, 1 mL per vial (George King Bio-medical, Inc.). Store at −80 °C. Thaw human plasma at 37 °C in water bath (*see* **Note 3**) (Fig. 2).

3 Methods

3.1 DNA Template Production

1. Dissolve Gblock DNA template in Milli-Q water to a final concentration of 100 nM (*see* **Note 4**). Vortex and spin down.
2. Dilute forward and reverse DNA primer in Milli-Q water to a concentration of 10 μM.

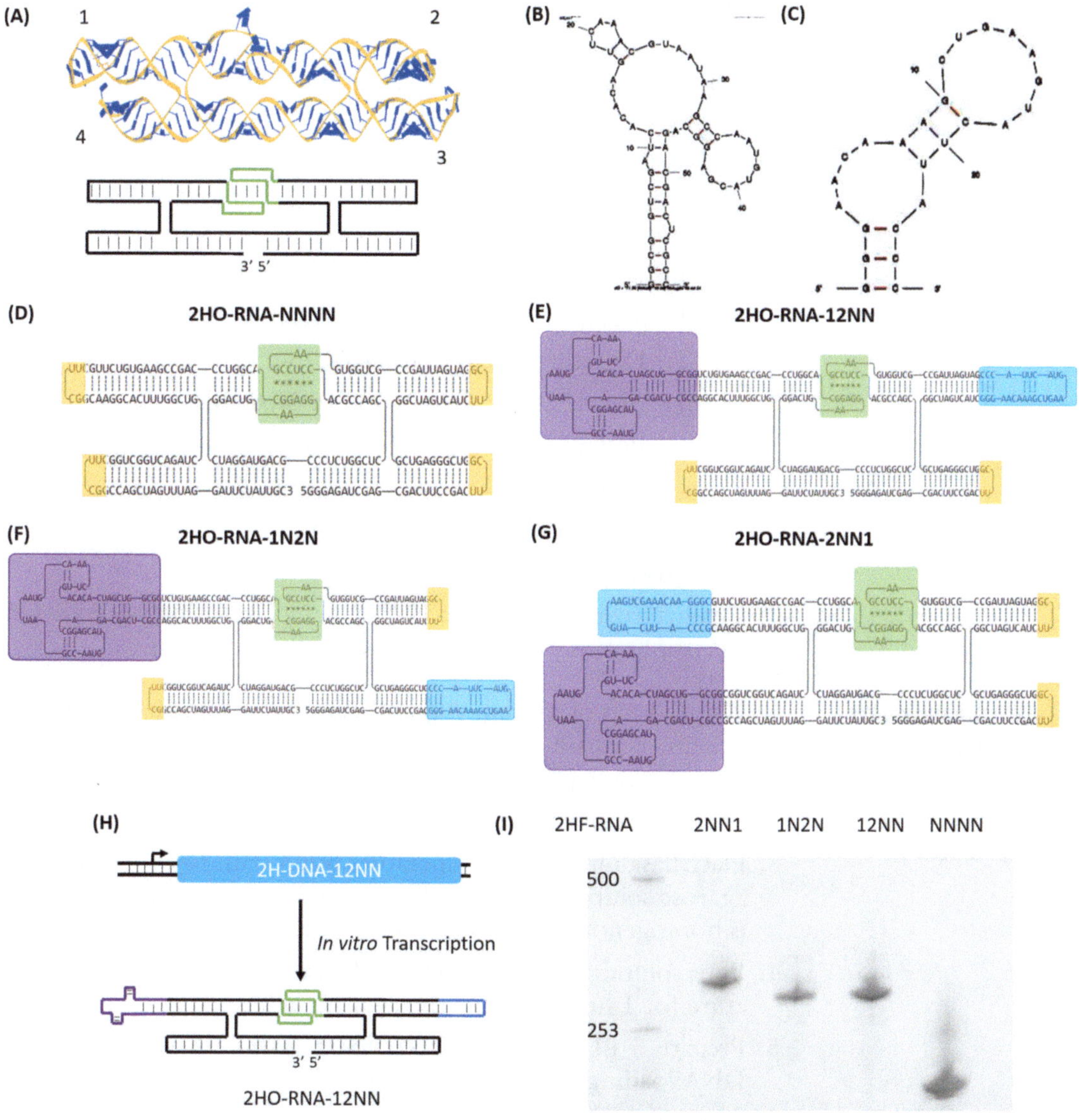

Fig. 2 RNA origami anticoagulants. (**a**) 3D model and 2D ribbon of two-helix RNA origami without aptamers. (**b**) Exosite 1-binding RNA aptamer ***(8)***. (**c**) Exosite 2-binding RNA aptamer ***(9)***. (**d**) 2D ribbon structure of RNA origami. (**e–g**) 2D ribbons of RNA origami bearing two thrombin aptamers. Green rectangles denote kissing loops. Yellow rectangles represent tetraloops. Purple and blue rectangles indicate exosite 1- and 2-binding aptamers, respectively. (**h**) Schematic drawing of production of RNA origami using in vitro transcription. (**i**) Characterization of RNA origami with denaturing polyacrylamide gel electrophoresis. Image reproduced with permission from ref. [22]. Copyright 2019 Wiley

3. Mix 2 μL 100 nM DNA template and 2.5 μL 10 μM forward, and reverse DNA Primer and 1 μL 10 mM dNTP mixtures (10 mM of dATP, dGTP, dCTP, and dTTP), 10 μL 5x Phusion HF buffer. Add Milli-Q water to adjust volume to 49.5 μL. Place sample on ice.

4. Store Phusion DNA polymerase at −20 °C. Take Phusion DNA polymerase from −20 °C storage and place on ice. Add 0.5 μL 2 unit/μl Phusion DNA polymerase to the sample mixture.
5. Mix by pipette. Do not vortex.
6. Place the sample into a thermocycler. PCR procedure as follows:
   (a) Initial denature at 95 °C for 30 s.
   (b) PCR cycle as follows:
      (i) Denature at 95 °C for 10 s.
      (ii) Anneal at 63 °C for 15 s (*see* **Note 5**).
      (iii) Polymerize at 72 °C for seconds.
      (iv) Repeat for 30 cycles.
   (c) Final extension at 72 °C for 2 min.
   (d) Store at 4 °C.
7. Characterize the amplified DNA template using agarose gel electrophoresis and purify with GFX PCR clean up kit.

3.2 Characterization of the Amplified DNA Template Using Agarose Gel Electrophoresis

1. Weigh 0.5 g of ultrapure agarose and pour agarose in 100 mL Erlenmeyer flask. Add 50 mL 0.5× TBE buffer into the flask.
2. Place the flask in the microwave and heat the solution just until boiling in order to dissolve agarose (*see* **Note 6**).
3. Place flask into room temperature water bath to cool down the agarose solution. Periodically swirl the flask to ensure the liquid is homogenous. Do not allow the solution to solidify.
4. Pour solution in gel casting mold and let it sit at room temperature for 1 hr. to solidify the agarose gel.
5. Prepare 5 μL amplified DNA template mixed with 1 μL 6X DNA loading buffer. Mix by pipette.
6. Load DNA samples and appropriate DNA maker into the wells. Run at 150 V for 30 min.
7. Post-stain the gel with ethidium bromide solution at room temperature for 15 min. (*see* **Note** 7).
8. Image the gel with UV imager.

3.3 In Vitro Transcription of RNA Origami

3.3.1 Transcription of Native RNA Origami

1. Thaw materials, excluding T7 RNA polymerase, on ice. Keep polymerase in freezer until immediately before use (*see* **Note 8**).
2. Combine ATP, UTP, CTP, and GTP in equivalent ratios to create the NTP mix. The concentration of each NTP is 25 mM. Mix by pipetting.

3. Mix transcription reaction on ice in nuclease-free tubes. Add 5X Lucigen reaction buffer (final concentration of 1x), DNA template (final concentration 5 ng/μL), NTP mix (final concentration 2.5 mM), fresh DTT (final concentration 10 mM), and nuclease free water. Pipette to mix on ice (*see* **Note 9**).
4. Take T7 polymerase from freezer and put on ice. Add 1.25 units/μL T7 polymerase to solution and pipette thoroughly to mix. Do not vortex.
5. Place into a thermocycler at 37 °C for 18 h and hold at 4 °C (*see* **Note 10**).
6. After transcription, purify using Monarch RNA Clean-up Kit. The purification procedure follows the RNA Clean-up kit's instruction.

3.3.2 Transcription of 2′-Fluoro Modified RNA Origami

1. Thaw all materials, excluding mutant RNA T7 polymerase (Y639F), on ice. Keep polymerase in freezer until immediately before use.
2. Combine equivalent ratios of ATP, GTP, 2′fluoro-modified CTP, and 2′fluoro-modified UTP to create the modified NTP mix. The concentration of each NTP is 25 mM. Mix by pipetting.
3. Mix transcription reaction on ice in nuclease-free tubes. 5X Lucigen reaction buffer (final concentration 1X), DNA template (final concentration 5 ng/μL), modified NTP mix (final concentration 2.5 mM each), fresh DTT (final concentration 10 mM), and nuclease free water. Pipette to mix on ice (see **Note 9**).
4. Remove mutant T7 polymerase from freezer and put on ice. Add 1.25 units/μL T7 polymerase to solution and pipette thoroughly to mix. Do not vortex.
5. Place in a thermocycler at 37 °C for 18 h (*see* **Note 10**).
6. After transcription, purify using Monarch RNA Clean-up Kit.

3.4 Characterization of RNA Origami

1. Cast denaturing PAGE gel. Determine the correct percentage of acrylamide necessary for the length of RNA being tested. In our case, 6% denaturing PAGE was used to characterize RNA origami.
2. Weigh 24 g of urea in the 50 mL tube. Then, add 7.5 mL of 40% acrylamide solution and 5 mL of 10x TBE, and fill with water up to 50 mL (*see* **Note 11**).
3. Microwave the mixture solution until urea dissolves (*see* **Note 12**).
4. Cool the solution by placing in room temperature water bath.

5. Add 166 μL of 30% APS and shake the mixture solution vigorously.
6. Add 20 μL of TEMED and shake the mixture solution vigorously.
7. Pour the solution between the glass plates for casting and let solidify for 1 h.
8. Once gel is polymerized, add 1x TBE running buffer in the gel chamber.
9. Using a syringe, rinse each well to clean the wells and pre-run for 30 min at 20 W in order to heat the gel up for denaturing condition.
10. Aliquot 5 μL of RNA origami samples to be used for denaturing PAGE gel.
11. Add 2x formamide loading dye to each sample as well as a nucleic acid ladder (*see* **Note 13**).
12. Denature RNA samples at 95 °C for 5 min in a thermocycler. Remove from the thermocycler and immediately place samples on ice.
13. Prior to loading the denatured samples, use a syringe to rinse the desired well and immediately load the sample with a pipette. Alternate between rinsing and loading until all samples are loaded.
14. Run the gel at 20 W. The running time will vary depending on the length of the sample.
15. Remove gel and stain with EtBr for 15 min.
16. View under UV imager.

3.5 Folding RNA Origami

1. Measure concentration of the purified RNA origami by nanodrop.
2. Determine the desired final volume. Combine RNA (final concentration of 5 uM), 5X buffer (final concentration of 1x), and nuclease-free water (to desired volume).
3. Heat to 90 °C for 5 min in thermocycler.
4. Let cool at room temperature for 30 min.
5. The heat-annealed RNA origami is ready to use or be stored at 4 °C for up to 3 months (*see* **Note 14**) (Fig. 3).

3.6 Anticoagulation Activity Test by aPTT Assay

1. Aliquot aPTT and $CaCl_2$ into 1.5 mL tube for later use. Keep aliquoted aPTT on ice and $CaCl_2$ heated to 37 °C (*see* **Note 15**).
2. Set coagulometer to aPTT test settings and wait for temperature to reach 37 °C.

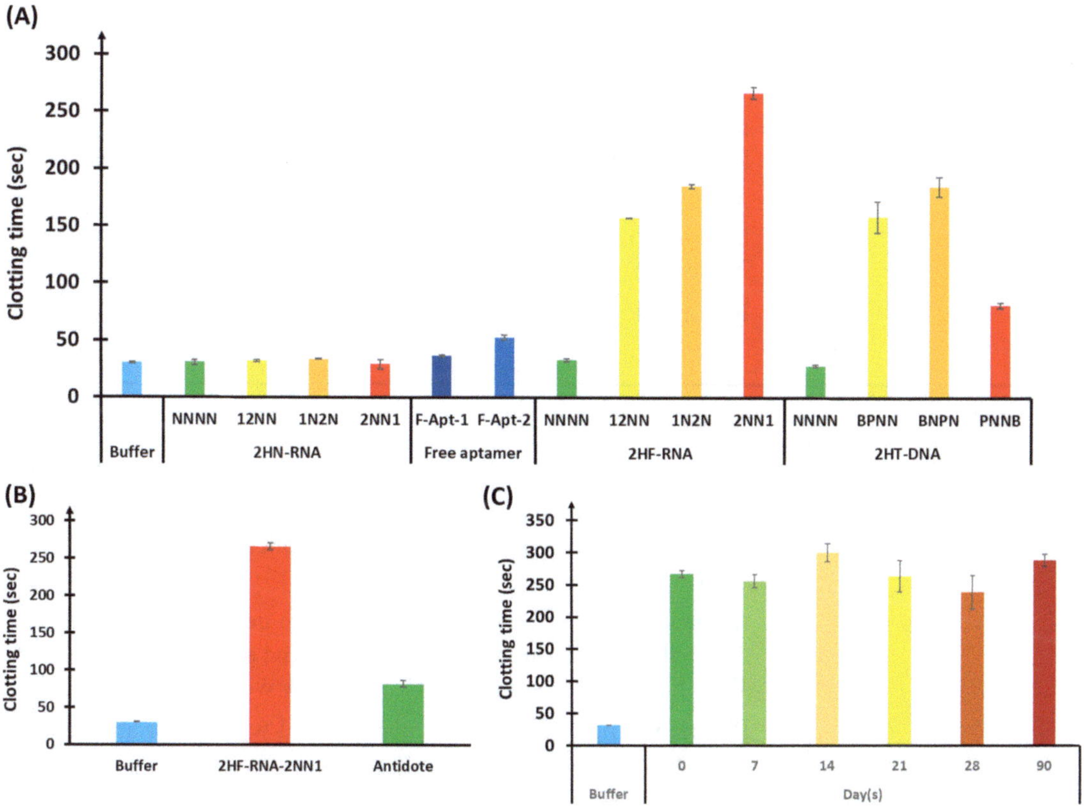

Fig. 3 Anticoagulation activities of RNA origami bearing two thrombin RNA aptamers. The anticoagulation activities were tested by activated partial thromboplastin time (aPTT) assay. (**a**) RNA folding buffer represented a negative control. 2HR-RNA refers to non-modified RNA origami that has no anticoagulation activity due to instability in plasma. 2HF-RNA refers to 2′fluoro-modified RNA origami. 2HT-DNA refers to DNA weave tiles appended with thrombin DNA aptamers. (**b**) Reversal of anticoagulation activity using ssDNA antidotes. (**c**) Stability of RNA origami anticoagulant stored at 4 °C and tested with aPTT assay. RNA origami stored at 4 °C up to 3 months. (Image reproduced with permission from ref. [22]. Copyright 2019 Wiley)

3. Warm water bath to 37 °C.
4. Using gloves, lab coat, and eye protection, retrieve normal pooled human blood plasma from −80 °C storage and keep on ice (*see* **Note 16**).
5. Thaw plasma in a 37 °C water bath. Plasma must be used within 2 h after thawing.
6. Place a cuvette into position in the coagulometer.
7. Add one magnetic metal ball to each cuvette well.
8. Add 50 μL of blood plasma followed by 50 μL of aPTT and let incubate at 37 °C for 300 s.
9. Add 16.67 μL of 5 μM RNA origami anticoagulant. 1x RNA folding buffer can be added as a negative control. Let it incubate at 37 °C for 300 s. For reversal of thrombin activity, add

13.67 μL of 6.09 μM RNA origami and incubate at 37 °C for 300 s. Then, add 3 μL of 500 μM DNA antidote strands and let it incubate for 300 s (*see* **Note 17**).

10. Move the cuvette to the rightmost slot and press the pipette key to begin the movement of the magnetic metal balls.
11. Add 50 μL $CaCl_2$ to the first cuvette and press the pipette handle to start the clot timer. Do this for each well (*see* **Note 18**).
12. Record clot times for analysis.

4 Notes

1. As the dried DNA products are spread around the entire tube wall, we highly recommend centrifuging at 3000 g for at least 1 min in order to let all products collect at the bottom of the tube.
2. Aliquot 500 μL of 30% w/v ammonium persulfate into Eppendorf tubes and store at −20 °C. Thaw the solution prior to use. The thawed APS solution is good for 1 week.
3. Plasma is single use and should be used within 2 h after thawing. Do not use after multiple freeze-thaw cycles.
4. For example, add 12.50 μL Milli-Q water to 1250 fmol of lyophilized Gblock DNA template to get 100 nM DNA template. The mole value of DNA template was provided from IDTDNA.
5. Annealing temperature depends on the melting temperature (T_m) of DNA primers, type of DNA polymerase, and PCR buffer. We calculated annealing temperature for PCR by using online tools provided by New England Biolabs (https://tmcalculator.neb.com/).
6. Caution: flask will be very hot. Wear heat protectant gloves.
7. Ethidium bromide solution: add 5 μL ethidium bromide in 50 mL Milli-Q water.
8. Always wear gloves during experiments in order to avoid DNase and RNase contamination.
9. DTT solution should be made from DTT powder and stored at −20 °C until use.
10. To ensure that the solution is heated thoroughly, we aliquot transcription mixture into tubes of 50 μL before incubation at 37 °C.
11. Wear gloves while working with acrylamide. Unpolymerized acrylamide solution is a neurotoxin.

12. Keep the cap loose while heating solution in microwave. Heat the solution for 15 s and mix solution by swirling or shaking the tube. Repeat until urea dissolved. Do not heat solution for more than 1 min; it can cause boiling of the acrylamide solution.
13. Use single-strand RNA ladders for denaturing PAGE.
14. Do not freeze the folded RNA origami.
15. aPTT must be kept at 4 °C and used within 1 week after dissolving. $CaCl_2$ can be kept at room temperature during storage, but needs to be warmed to 37 °C for usage in coagulation assay.
16. Plasma should not be refrozen. We suggest ordering 1 mL vials and disposing after a single use.
17. Due to the sensitivity of coagulation tests, care should be taken to prevent components from sticking to the sides of the cuvette. Place the pipette tip to the bottom of the cuvette before expelling liquid.
18. For the most accurate clot time, start the timer for each cuvette well as quickly as possible after adding $CaCl_2$.

Acknowledgments

The authors would like to thank Prof. Ebbe S. Andersen and Dr. Steffen M. Sparvath for generous support during RNA origami design and Prof. Jørgen Kjems and Dr. Daniel Miotto Dupont for supplying mutant T7 RNA polymerase and Dr. Rasmus P. Thomsen for assistance in the lab. This project was supported by NSF grants: IRES-1559077, BMAT-1709010, and BME-1603179.

References

1. Crawley JT, Zanardelli S, Chion CK, Lane DA (2007) The central role of thrombin in hemostasis. J Thromb Haemost 1:95–101
2. Huntington JA (2005) Molecular recognition mechanisms of thrombin. J Thromb Haemost 3:1861–1872
3. Thaler J, Pabinger I, Ay C (2015) Anticoagulant treatment of deep vein thrombosis and pulmonary embolism: the present state of the art. Front Cardiovasc Med 2:30
4. Harter K, Levine M, Henderson SO (2015) Anticoagulation drug therapy: a review. West J Emerg Med 16:11–17
5. Wolberg AS, Rosendaal FR, Weitz JI, Jaffer IH, Agnelli G, Baglin T, Mackman N (2015) Venous thrombosis. Nat Rev Dis Primers 1: 15006
6. Keefe AD, Pai S, Ellington A (2010) Aptamers as therapeutics. Nat Rev Drug Discov 9:537–550
7. Zhou J, Rossi J (2016) Aptamers as targeted therapeutics: current potential and challenges. Nat Rev Drug Discov 16:181–202
8. Bompiani KM, Monroe DM, Church FC, Sullenger BA (2012) A high affinity, antidote-controllable prothrombin and thrombin-binding RNA aptamer inhibits thrombin generation and thrombin activity. J Thromb Haemost 10:870–880

9. Long SB, Long MB, White RR, Sullenger BA (2008) Crystal structure of an RNA aptamer bound to thrombin. RNA 14:2504–2512
10. White R, Rusconi C, Scardino E, Wolberg A, Lawson J, Hoffman M, Sullenger B (2001) Generation of species cross-reactive aptamers using "toggle" SELEX. Mol Ther 4:567–573
11. Müller J, Wulffen B, Pötzsch B, Mayer G (2007) Multidomain targeting generates a high-affinity thrombin-inhibiting bivalent aptamer. Chembiochem 8:2223–2226
12. Bock LC, Griffin LC, Latham JA, Vermaas EH, Toole JJ (1992) Selection of single-stranded DNA molecules that bind and inhibit human thrombin. Nature 355:564–566
13. Rusconi CP, Roberts JD, Pitoc GA, Nimjee SM, White RR, Quick G Jr, Scardino E, Fay WP, Sullenger BA (2004) Antidote-mediated control of an anticoagulant aptamer in vivo. Nat Biotechnol 22:1423–1428
14. Rusconi CP, Scardino E, Layzer J, Pitoc GA, Ortel TL, Monroe D, Sullenger BA (2002) RNA aptamers as reversible antagonists of coagulation factor IXa. Nature 419:90–94
15. Chan MY, Rusconi CP, Alexander JH, Tonkens RM, Harrington RA, Becker RC (2008) A randomized, repeat-dose, pharmacodynamic and safety study of an antidote-controlled factor IXa inhibitor. J Thromb Haemost 6:789–796
16. Chan MY, Cohen MG, Dyke CK, Myles SK, Aberle LG, Lin M, Walder J, Steinhubl SR, Gilchrist IC, Kleiman NS, Vorchheimer DA, Chronos N, Melloni C, Alexander JH, Harrington RA, Tonkens RM, Becker RC, Rusconi CP (2008) Phase 1b randomized study of antidote-controlled modulation of factor IXa activity in patients with stable coronary artery disease. Circulation 117:2865–2874
17. Hansen MN, Zhang AM, Rangnekar A, Bompiani KM, Carter JD, Gothelf KV, LaBean TH (2010) Weave tile architecture construction strategy for DNA nanotechnology. J Am Chem Soc 132:14481–14486
18. Rangnekar A, Zhang AM, Li SS, Bompiani KM, Hansen MN, Gothelf KV, Sullenger BA, LaBean TH (2012) Increased anticoagulant activity of thrombin-binding DNA aptamers by nanoscale organization on DNA nanostructures. Nanomedicine 8:673–681
19. Rangnekar A, Nash JA, Goodfred B, Yingling YG, LaBean TH (2016) Design of potent and controllable anticoagulants using DNA aptamers and nanostructures. Molecules 21:1–13
20. Geary C, Rothemund PW, Andersen ES (2014) A single-stranded architecture for cotranscriptional folding of RNA nanostructures. Science 345:799–804
21. Jepsen MDE, Sparvath SM, Nielsen TB, Langvad AH, Grossi G, Gothelf KV, Andersen ES (2018) Development of a genetically encodable FRET system using fluorescent RNA aptamers. Nat Commun 9:1–18
22. Krissanaprasit A, Key C, Fergione M, Froehlich K, Pontula S, Hart M, Carriel P, Kjems J, Andersen ES, LaBean TH (2019) Genetically encoded, functional single-Strand RNA origami: anticoagulant. Adv Mater 31: 1808262
23. Sparvath SL, Geary CW, Andersen ES (2017) Computer-aided design of RNA origami structures. In: 3D DNA nanostructure. Methods in molecular biology, vol 1500. Springer, New York, pp 51–80

Index

Julián Valero (ed.), *DNA and RNA Origami: Methods and Protocols*, Methods in Molecular Biology, vol. 2639, https://doi.org/10.1007/978-1-0716-3028-0,

MIX
Papier aus verantwortungsvollen Quellen
Paper from responsible sources
FSC® C105338

If you have any concerns about our products,
you can contact us on
ProductSafety@springernature.com

In case Publisher is established outside the EU,
the EU authorized representative is:
Springer Nature Customer Service Center GmbH
Europaplatz 3, 69115 Heidelberg, Germany

Printed by Libri Plureos GmbH
in Hamburg, Germany